VOLUME 6

of Special Supplements to
Journal of Atmospheric and Terrestrial Physics

SOLAR ECLIPSES AND THE IONOSPHERE

1. Mr. G. Millington
2. Mr. M. Ryle
3. Dr. G. Elwert
4. Prof. K. R. Ramanathan
5. Mr. H. Hvatum
6. Mr. S. Olving
7. Mr. G. M. Brown
8. Mr. A. E. Lilley
9. Mr. J. A. Ratcliffe
10. Mr. D. M. Parkyn
11. Mr. F. Delobeau
12. Mr. A. F. Wilkins
13. Mr. T. W. Bennington
14. Dr. M. Waldmeier
15. Mr. J. Mawdsley
16.
17. Dr. R. M. MacRae
18. Dr. K. Davies
19. Dr. J. P. Wild
20. Mr. D. Lepechinsky
21. Dr. K. Rawer
22. Mr. P. L. M. van Berkel
23. Miss A. R. Robbins
24.
25. Dr. H. Daene
26. Mr. W. W. Kellogg
27. Mr. F. T. Haddock
28. Prof. C. W. Allen
29. Prof. S. Chapman
30. Dr. R. Lindquist
31. Dr. W. Becker
32. Mr. A. J. Lyon
33. Dr. N. Fukushima
34. Dr. E. A. Lauter
35. Dr. R. W. Wright
36. Mr. N. J. Skinner
37.
38. Mr. J. Egedal
39. Prof. J. Kaplan
40. Prof. J. Sayers
41. Prof. T. Hatanaka
42. Dr. R. L. Smith-Rose
43. Prof. A. Ehmert
44. Father P. Lejay
45. Mr. B. Landmark
46. Mr. M. H. Davenport
47. Dr. L. V. Berkner
48. Dr. M. A. Ellison
49. Mr. C. M. Minnis
50. Dr. C. de Jager
51. Mr. J. C. de Munck
52. Mr. A. H. Shapley
53. Mr. A. H. de Voogt
54. Dr. J. Houtgast
55.
56. Mr. W. R. Piggott
57. Mr. S. Skribeland
58. Dr. J. Bartels
59. Prof. L. Vegard
60. Dr. D. H. Menzel
61. Dr. S. D. Gossner
62. Prof. D. R. Bates
63. Dr. Y. Aono
64. Dr. R. Rivault
65. Dr. O. Burkard
66. Dr. M. Laffineur
67. Mr. W. Stoffregen
68. Dr. G. Righini
69. Mr. A. Rohdin
70. Dr. B. H. Briggs
71. Dr. W. J. G. Beynon
72. Dr. J. P. Hagen
73. Prof. T. Nagata

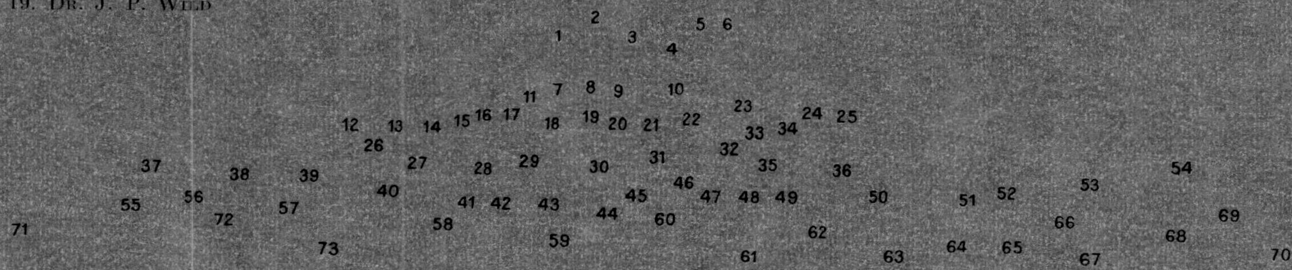

Group photograph of delegates attending the conference

SOLAR ECLIPSES AND THE IONOSPHERE

A SYMPOSIUM HELD UNDER THE AUSPICES OF
THE INTERNATIONAL COUNCIL OF SCIENTIFIC UNIONS
MIXED COMMISSION ON THE IONOSPHERE
IN LONDON IN AUGUST 1955

Editors

W. J. G. BEYNON and G. M. BROWN

Department of Physics
University College of Swansea

PERGAMON PRESS
London & New York

First published 1956

Published by Pergamon Press, 4 & 5 Fitzroy Square, London W.1
and 122 E. 55th Street, New York 22, N.Y.

Printed in Northern Ireland at The Universities Press, Belfast
by Sir Isaac Pitman & Sons, Ltd.

CONTENTS

I. INTRODUCTION

II. THEORETICAL CONSIDERATIONS

III. IONOSPHERIC ECLIPSE RESULTS

IV. SOLAR RADIATION AND LAYER THEORY

V. RECOMBINATION

VI. GEOMAGNETIC ECLIPSE PHENOMENA

VII. SOLAR RADIO NOISE

VIII. MISCELLANEOUS

IX. CONCLUSION

PREFACE

ECLIPSE phenomena in the ionosphere were discussed at a symposium held in the rooms of the Royal Society, London, on 22–24 August 1955. The meeting was organized by the Mixed Commission on the Ionosphere and sponsored by the International Council of Scientific Unions. Some eighty delegates from eighteen countries participated, and fifty-three papers were received. In addition to the presentation of formal papers there were several valuable discussion sessions, and summaries of these are included in this report.

In presenting the proceedings of the symposium we have grouped the papers and discussions into sections corresponding, as far as possible, to separate aspects of the subject. An introductory survey is first given and this is followed by a series of papers describing recent ionospheric eclipse results. Interpretation of the observations necessitates a knowledge of the solar ionizing radiation, and a section is devoted to this. Consideration is also given to the present position with respect to theories of formation of the ionized layers. A further section is devoted to measurements on solar radio noise radiations. Determinations of recombination coefficients from eclipse measurements and also by laboratory methods are considered. Papers describing recent work on the influence of eclipses on the terrestrial magnetic field are also included. Finally, we have added an exhaustive bibliography of literature on solar eclipses and the ionosphere. This is essentially based on the Netherlands Postal and Telecommunications Services Publication No. 768a, together with additions provided by several delegates to the symposium. The Netherlands P.T.T. very graciously provided copies of their document for distribution to all delegates at the symposium, and we wish to record our thanks for permission to make use of it and to all who helped in the compilation of the comprehensive bibliography here presented.

Our thanks are due to Sir EDWARD APPLETON (Chairman of the Mixed Commission on the Ionosphere) and to Dr. L. V. BERKNER for their invaluable assistance in organizing the symposium. Grateful acknowledgement is also made to the Executive Board of the International Council of Scientific Unions for a grant towards the expenses of delegates, to the Council of the Royal Society for the excellent facilities provided during the course of the symposium, and to the publishers for their co-operation in producing this volume.

W. J. G. BEYNON
G. M. BROWN

Department of Physics
University College of Swansea

Solar Eclipses and the Ionosphere (Special Suppl. Vol. 6 to J. Atmosph. Terr. Phys.), Pergamon Press, London, 1956

SECTION I

INTRODUCTION

A Survey of Solar Eclipses and the Ionosphere

J. A. RATCLIFFE
Cavendish Laboratory, Cambridge

1. INTRODUCTION

IT is the purpose of this survey to discuss the results of experiments made on the ionosphere at times of solar eclipse, and to consider how far they have contributed, or can contribute, to our understanding of the physics of the upper atmosphere and of the sun. The information obtained during past eclipses is summarized in Table 1, which should be referred to in connection with the following discussion.

The earliest object of ionospheric experiments made during solar eclipses was to establish whether the ionizing radiation emitted by the sun consisted of photons (ultra-violet) or particles. It was clearly established that the main ionizing agent in the *E* and *F1* layers was photons (summary in APPLETON and NAISMITH, 1938); there still remains some doubt about what causes eclipse effects in the *F2* layer, and about whether any ionization is produced by particles emitted from the sun.

In what follows, the different ionospheric layers will be considered in turn with special reference to specific problems. First, however, the results to be expected on the simplest theory will be considered.

2. THE EFFECT OF AN ECLIPSE ON A SIMPLE (CHAPMAN) LAYER

A first attempt at explaining the effect of an eclipse has usually been based on the theory of a simple (Chapman) layer, as given, for example, by CHAPMAN and BARTELS (1951). The rate $I(\chi, h)$ of production of electrons at any level h is given by their equation (32) in terms of the solar zenith angle χ, the constant flux S of ionizing radiation, and other quantities denoted by standard symbols defined in the book. During an eclipse S is a function $S(t)$ of time, and if $\chi(t)$ is also written as a function of time, their equation (32) may be written

$$I(t, h) = \beta A \rho_0 S(t) \exp\left\{-\frac{h}{H} - A\rho_0 H \varepsilon^{-h/H} \sec \chi(t)\right\} \tag{1}$$

If the electrons disappeared at a rate given by αN^2 (corresponding to a recombination-like process) or γN (attachment-like process) then the number $N(t, h)$ present at time t and height h would be given by one or other of the equations

$$\mathrm{d}N(t, h)/\mathrm{d}t = I(t, h) - \alpha\{N(t, h)\}^2 \tag{2}$$

or

$$\mathrm{d}N(t, h)/\mathrm{d}t = I(t, h) - \gamma N(t, h). \tag{3}$$

From these equations, given the initial conditions, an assumed value of α or γ, and an assumed form of $S(t)$, $N(t, h)$ can be calculated for comparison with the results of experiment.

It is not usual from experiments, however, to determine N at a fixed height; it is easier, and therefore more usual, to determine the value $N_p(t)$ of N at the peak of the layer. Now the peak of electron *production* occurs at a height $h_p(\chi)$ given by

$$h_p(\chi) = H \log (A\rho_0 H \sec \chi) \tag{4}$$

and has a magnitude $I_p(\chi)$ given by

$$I_p(\chi) = \{\beta S(t) \cos \chi\}/\{H\varepsilon\}. \tag{5}$$

It moves up and down as χ varies (but not as $S(t)$ varies), but, because of the finite recombination (or attachment) time, the peak of N occurs at a different height from that of I. Equations (2) and (3), which relate to a fixed height, do not apply to values $I_p(t)$ at the peak of I; they apply even less to values at the peak of N, which is at a different level.

Most analyses of eclipse results have, however, been made on the supposition that these differences are small and that equations (2) and (3) refer to the value $N_p(t)$ at the peak of N, and that this peak is in the same place as the peak of I, so that $I(h, t) = I_p(t)$. Substitution from equation (5) into equations (2) and (3) then leads to

$$\mathrm{d}N_p/\mathrm{d}t = (\beta/H\varepsilon)S(t) \cos \chi - \alpha N_p^2 \tag{6}$$

and

$$\mathrm{d}N_p/\mathrm{d}t = (\beta/H\varepsilon)S(t) \cos \chi - \gamma N_p. \tag{7}$$

The errors introduced by these assumptions have been considered, for the case when $S(t)$ is constant, by APPLETON and LYON (1955). It would be interesting to evaluate them for the cases of eclipses when $S(t)$ is not constant.

RYDBECK and WILHELMSSON (1954) have considered the general solution of equation (6) and have shown how to obtain simple approximate solutions, appropriate to values of α which are likely to occur in the E and $F1$ regions. They also show how to compute $S(t)$ from simple parameters describing the eclipse, total or partial (but not annular). In particular, they give curves which show how the minimum value of N, and the time lag (Δt_{min}) of that minimum, depend on the value of α for different circumstances of the eclipse (their Figs. 4 and 5). Several eclipse observers have computed similar curves appropriate to the circumstances of their own particular eclipses, and attempts have been made to deduce values of α from the results. It is therefore of interest to consider, quite simply, what information could, in principle, be derived from observations of this kind.

Suppose that, during an eclipse, the curve representing $N(t)$ has a form such as that shown by the continuous lines of Figs. 1(a) or (b), so that two times t_0 and t_1 can be selected when $\mathrm{d}N/\mathrm{d}t = 0$.

Equation (6) then shows that

$$\frac{N_1^2}{N_0^2} = \frac{\alpha_0 S_1 \cos \chi_1}{\alpha_1 S_0 \cos \chi_0} \tag{8}$$

where the values of α at times t_1 and t_0 are written α_1 and α_0. On simple theories $\alpha_1 = \alpha_0 = \alpha$, and equation (8) may be written

$$N_1^2/N_0^2 = I_1/I_0 \tag{9}$$

where $I(\equiv S \cos \chi)$ represents the rate of production of electrons. Equation (9) shows that N_1^2/N_0^2 does not depend fundamentally on α. In most papers an

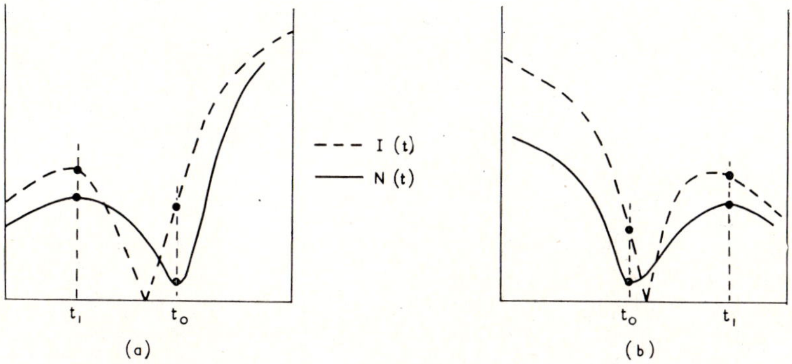

Fig. 1. To illustrate variations in the rate (I) of production, and the number density (N), of electrons during an eclipse.

attempt has, nevertheless, been made to estimate the value of α from the observed ratio N_1^2/N_0^2, and it is interesting to ask how this can be possible. The answer is found in the fact that the time t_0 depends on α, and it is this time which determines the magnitude of I_0, as may be seen by considering the (dotted) curves of I in Fig. 1. *Thus* α *determines the time at which the minimum of* N *occurs, but the minimum magnitude of* N *is then determined by equation* (9).

3. The *E* Layer

An eclipse always has a marked effect on the *E* layer. The results have usually been expressed in terms of the critical frequency and, although the occurrence of sporadic *E* has sometimes interfered with the measurements, they are, on the whole, reliable and accurate. The results on control days have also usually been consistent and reliable.

To a very rough first approximation the results agree with those calculated from equation (6) if $S(t)$ is made proportional to the area of the exposed visible disk, and if $\alpha \doteqdot 10^{-8}$ cm^3 sec^{-1}. Equation (7) is much less satisfactory, and it seems safe to suppose that a recombination type of electron-loss process is operative. It does not, however, need any great accuracy of experiment to show that this simple result is not at all satisfactory. In the first place, equation (6) does not represent the facts even on control days, when there is no eclipse. With the large value (10^{-8}) for α, equation (6) shows that we should expect to a close approximation to find

$$N_p{}^2 \propto \cos \chi \tag{10}$$

$$N_p \propto (\cos \chi)^m \tag{11}$$

with $m = 0{\cdot}5$. The most accurate experiments have shown, however, that values of m near 0·67 more nearly represent the facts. This complication cannot be attributed to the numerical value assumed for α, and there has been much discussion about what it means. It has been supposed that the recombination coefficient might vary with height, or with time; that the distribution of the ionizable gas might not be that postulated on the simple theory; or that movements of the layer might be important. There is, as yet, no generally accepted explanation. Our lack of knowledge of the normal E layer is illustrated by the fact that the most detailed experiments, and discussions of results, are in papers where the primary interest is in eclipses (MINNIS, 1955; ESTRABAUD, 1953, 1954).

If equation (6) does not apply to control days, how can it be applied to eclipse results? Not all workers have asked themselves this question, but even when they have, it has been considered best to take it as it stands and to write $m = 0{\cdot}5$ in equation (11). Most workers have then used equation (6) to calculate N_p on the assumption that $S(t)$ was proportional to the area of the unobscured visible disk (as it would appear at a height of 100 km), and have found, as stated previously, that it was impossible to assign a value of α which fitted the results well. The difficulty is that, if methods such as those of RYDBECK and WILHELMSSON (1954) are used to calculate α, first from the magnitude of N_1/N_0, and then from the time delay (Δt) between the optical centre of the eclipse and the occurrence of the minimum critical frequency, quite different values are obtained.

It has been usual to deduce the value of α from the ratio N_1/N_0 rather than from Δt. The values shown in column 8 of Table 1 were mostly calculated in this way.

Attempts have been made, in two different ways, to make the theory approximate more closely to the experimental results. In one it is supposed that the recombination coefficient α alters during the eclipse, so that equation (9) is replaced by equation (8). For example, on one occasion (McLEISH, 1948), α appeared to decrease from 4×10^{-8} cm^3 sec^{-1} at first contact to $0{\cdot}8 \times 10^{-8}$ at totality. In support of this kind of explanation it has been suggested (PIERCE, 1948; McLEISH, 1948) that, if α were merely an "effective" value of the recombination coefficient, determined in reality by a complicated process of attachment and detachment of electrons to and from neutral atoms (see BATES and MASSEY, 1948), then the balance would be upset when the detaching radiation was removed by the eclipse, and the magnitude of α_{eff} would change. Explanation along these lines appears at present, however, to require an unacceptably large value of λ, the ratio of ion density to electron density.

In the other type of explanation it is supposed that $S(t)$ is not simply proportional to the area of the unobscured visible disk, but that an appreciable proportion of the ionizing radiation comes from "active areas" on the sun, and possibly from coronal regions beyond the visible disk. In the most complete analysis of this kind (MINNIS, 1955) observed values of N_p and dN_p/dt were inserted into equation (6) to compute a curve to show how $S(t)$ varied with time, and this was compared with what was known, from visual observation, about the sun. In the analysis, the recombination coefficient was found from values of N_p and dN_p/dt observed at second contact, together with the assumption that then $S(t) = 0$. This value of α was then used in the rest of the calculation. The curve finally deduced for $S(t)$ corresponded to radiation of which 64 per cent was emitted uniformly over the visible disk, 25 per cent from a source on the west limb, 3 per cent from a spot, and 8 per cent from near the east limb. MINNIS also states that similar results were obtained from observations made at a second place where the eclipse was only partial. The "active areas" deduced in this way from the ionosphere measurements correspond reasonably closely to those indicated by enhancement of the coronal green line.

Other workers, using the same type of explanation, but less complete methods of analysis, have reached some conclusions which are less easy to accept. Thus HIGGS (1942) found himself forced to conclude that *all* the radiation came from three active areas and none from the uniform solar disk, while DELOBEAU (1953) found it necessary to postulate that a visible active area was responsible for 54 per cent of the total ionizing radiation; but when, after a few days, the area disappeared behind the sun, there was no noticeable change in the density of the E layer.

Before this type of explanation is accepted as completely satisfactory it is well to notice one or two points which are indicated by Table 1. Equation (9) relates the values (N_1 and N_0) of the electron density at times such as t_1 and t_0 in Fig. 1, to the values of I_1 and I_0 at those times. Observed values of N_1^2/N_0^2, and values of the ratio I_1/I_0 estimated on the supposition that S is proportional to the area of the unobscured visible disk, are shown in columns 9 and 10 of Table 1. It is seen that, in all cases, $N_1^2/N_0^2 < I_1/I_0$, and this was particularly noticeable when t_0 occurred during the time of totality, so that I_1/I_0 was infinite. These results point to the conclusion that, at the time of minimum N_p, i.e., at or slightly after totality, the value of I_0 was greater than was expected. On any hypothesis of "active regions" which have a chance distribution over the sun this is difficult to explain; it would be more reasonable to expect that I_0 would sometimes be too small and sometimes too great.

A value of I_0 which is always greater than the value expected on the simple "visible-area" basis implies that some ionizing radiation comes from beyond the visible disk, so that it is not removed at, or near, totality. As PIDDINGTON (1951) has emphasized, it would seem, on this argument, that about 10 to 15 per cent of the total ionizing radiation must come from beyond the edge of the visible disk. If that were so, the discrepancy between N_1^2/N_0^2 and I_1/I_0 would not be so large in the case of a partial eclipse, and indeed reference to Table 1 seems to show that it is not. If appreciable ionizing radiation were present even when the visible disk was obscured, the calculated value of α would be increased.

It should be noticed that, in the results of MINNIS (1955), which appear to be the most satisfactory so far, it has to be supposed that about 30 per cent of the total radiation comes from near the E and W edges of the solar disk. Moreover, in the calculation it had been *assumed* that all the radiation was cut off during totality; it was this assumption which led to the value of α listed in the Table.

If explanations in terms of areas of intense radiation from the sun are to be considered to be the most satisfactory, there are still several points to be settled. When there is no eclipse, does the E layer density vary, from day to day, by the amount which would be expected if intensely emitting regions of the type suggested are sometimes present and sometimes absent? Has the effect of the disappearance of these regions behind the sun ever been noticed? How is the normal-day variation, with $m \risingdotseq 0{\cdot}67$, to be explained, and what would be the effect of including the explanation in the analysis of eclipse results?

If the explanation in terms of active regions is accepted, it is usually possible to select, by trial and error, the value of α which will give the most "reasonable" curve for $S(t)$, i.e., the curve which makes the background contribution from the undisturbed disk as uniform as possible. Values of α estimated in this way are not far from 10^{-8} cm^3 sec^{-1}.

4. The *F1* Layer

The $F1$ layer does not always show itself on $h'f$ records made on ordinary days, and when it does it is most pronounced at midday, and rapidly disappears, or merges into the $F2$ layer, near sunset. The frequency which is denoted by f_0F1 usually corresponds to a point of inflexion on the N,h curve and only seldom to the "peak" of a layer. There is not much precise evidence about the variation of f_0F1 with the solar zenith distance χ, but what there is seems to show that $f_0F1 \propto (\cos\chi)^{1/4}$ fairly accurately, and probably more accurately than for the E layer, though Meriau and Rawer (1954) have stated that $(\cos\chi)^{1/5}$ fits the results better.

Eclipses always have an effect on f_0F1. The most remarkable effect is that the decrease of solar radiation during an eclipse makes the difference between the $F1$ and $F2$ layers more marked, whereas on ordinary days the opposite happens and a decrease of radiation makes the $F1$ layer become less obvious. Sometimes an eclipse can even cause the $F1$ layer to show itself when, on the control days, it had not been noticeable. There seems to have been little discussion of this remarkable fact.

The height of the $F1$ layer has sometimes been found to increase markedly during an eclipse (Meriau and Rawer, 1954) and on other occasions it did not change appreciably (Minnis, 1955).

Several workers have used equations (6) and (7) to consider the variation of f_0F1. Their use in this way is even less justifiable than for the E layer, partly because f_0F1 does not refer to the peak of a layer, and partly because there are often considerable changes of height of the layer. It would be useful to investigate the changes of N at a fixed level in the $F1$ layer during an eclipse.

The results agree with equation (6) better than with equation (7), and it is concluded that electrons are lost in the $F1$ layer by a process of recombination. The agreement with equation (6) is no better, in detail, than it is for the E layer results, and there are the same difficulties in estimating α. Again it has been usual to estimate α from the values of the minimum electron density, and the results are shown in column 13 of Table 1. The unsatisfactory nature of this procedure is seen from columns 14 and 15, where magnitudes of N_1^2/N_0^2 and I_1/I_0 are compared.

On one or two occasions attempts have been made to improve the agreement by making allowance for the presence of areas of intense radiation on the sun. In the most satisfactory of these (MINNIS, 1955) the distribution of radiating sources which had been deduced from the E-layer results was found to agree well with the $F1$-layer results if α-$F1$ was taken to be 8×10^{-9} cm^3 sec^{-1}. This value, which is greater than those deduced by other workers, depends on the assumption that the ionizing radiation was completely removed at optical totality. It is particularly noticeable that the values of α deduced for the E and the $F1$ layers are nearly the same. This fact needs to be remembered in assessing the value of theories of the E layer which require α to change markedly with height.

It is interesting to notice that there is considerable consistency in the observed values of Δt listed in columns 11 and 16 of Table 1 for the E and $F1$ layers. Thus all those who recorded Δt found approximately the same values for both layers for the same eclipse, as follows: 1940, $\Delta t = 0$; 1945, $\Delta t = 5$ or 10 min; 1952, $\Delta t = 0$, or was negative. If it could be shown that Δt depends on how the source of radiation is distributed over the solar disk, these results would appear to indicate that the radiations which ionize the two layers come from the same regions on the sun.

5. THE *F2* LAYER

The effects of eclipses on the $F2$ layer have been very varied. On most occasions some noticeable changes have occurred near the time of the visible eclipse, but there has been little regularity in the phenomenon. In particular, the eclipse has seldom resulted in a simple decrease of f_0F2 followed by an increase to its normal value. Before discussing any details, it will be well to consider some facts about the normal $F2$ layer.

It is well known that the normal variation of f_0F2 is not at all like that expected on a simple theory, and that it is much altered on days of magnetic disturbance. Moreover, even on the magnetically quietest days there are variations which do not repeat from day to day. It is therefore very difficult to decide what the "average control day" would be like, and it has often been found that changes which were apparently associated with an eclipse were of the same order of magnitude as those which occurred when there was no eclipse.

There is much evidence that the shape of the "normal" $F2$ layer, and even the existence in it of a peak of electron density, is primarily controlled by large-scale movements of the electrons, and that these are caused by diffusion under gravity and by electromagnetic forces, the required electric field originating in polarization charges built up elsewhere in the ionosphere (see, for example, MARTYN, 1955).

It has often been suggested, following BRADBURY (1938), that one and the same ionization process forms both the $F1$ and $F2$ layers, and that the $F2$ owes its separate existence to the fact that the electron-loss rate becomes considerably smaller higher up. Detailed analysis of the height distribution of electrons in the $F2$ layer has recently (RATCLIFFE *et al.*, 1956) given some precision to these ideas, and it has been possible to suggest a tentative model in which the loss is by an attachment-like process, with the loss coefficient varying with height as given approximately by

$$\gamma = 10^{-4} \exp \left\{\frac{300 - h(\text{km})}{50}\right\} \sec^{-1} \tag{12}$$

On this model the time constants of the electrons, at different heights, would be as shown in Table 2.

Table 2. Estimated time constants for the life of electrons at different heights.

Height, km	250	300	350	400
Time constant, hours	1	3	7·5	20

If these time constants are of the right order, it does not seem likely that an eclipse lasting about two hours would have much direct effect on the electron density at heights above about 300 km, where the peak of the $F2$ layer is often found. This circumstance, together with the observed fact that the effect of an eclipse on the $F2$ layer is not always a simple decrease of critical frequency, led PIERCE (1948) to suggest that a large part of the observed effect might be a secondary one associated with movements of the electrons. These movements could arise if the eclipse were to produce changes in electrostatic polarization in the E layer, and if these in turn affected the $F2$ layer in virtue of their electric field.

Since all modern theories suggest that the height of the peak of the $F2$ layer and the magnitude of f_0F2 are considerably influenced by movements, it is important that, during eclipses, attention should not be concentrated on these two quantities alone. It is desirable that the distribution $N(h)$ of electron density with height should be deduced from the $h'f$ curves, and that then the temporal variation $N(t)$ of N at a series of fixed heights should be studied. This was done by SAVITT (1950) for some results from the 1947 eclipse, but he took no account of possible movements, although the results for control days indicated that they must have been important. The whole of the difference between the control and the eclipse days was assigned to a variation of photon intensity proportional to the unobscured area of the visible disk, and the results were used to assign values of α or γ to the different heights. It was not possible to match the experimental results at all accurately from the theory, but the best agreement was obtained on the supposition that the electrons were lost by attachment with a coefficient γ which varied roughly linearly with height between 7×10^{-4} sec^{-1} at 200 km and 3×10^{-4} sec^{-1} at 320 km.

Recently, some of my colleagues have re-examined these results, together with results from the 1952 and 1954 eclipses, on the assumption that the model $F2$ region of RATCLIFFE *et al.* (1956) is correct. They have then deduced the movements which would have to be postulated to explain the observations. These results are described in the paper by THOMAS and ROBBINS (pp. 94–101, this volume).

Although, in general, the long time-constant at the greater heights will prevent an eclipse having much direct effect on the electron density in the $F2$ layer, there might be an important direct effect near sunrise. In equation (3) let us add a term $N \,.\, M(t)$ proportional to N, and to a function $M(t)$ of time, to represent, at a given height h, the effect of movements. It then becomes

$$\mathrm{d}N/\mathrm{d}t = I(t) - \gamma N + N \,.\, M(t) \,. \tag{13}$$

Near sunrise when N is small the terms γN and $N \,.\, M(t)$ are probably small compared with $\mathrm{d}N/\mathrm{d}t$, so that the value of N is largely determined by the magnitude of $I(t)$. We have shown (RATCLIFFE *et al.*, 1956) that analysis of this kind leads to interesting conclusions on normal days. If an eclipse occurred near sunrise, the direct effect of changes in $I(t)$ should be noticeable, and useful results could be obtained. WELLS (1952) observed an eclipse at sunrise, and it would be most interesting to examine his $h'f$ records in detail.

If a full analysis of the $h'f$ records cannot be made, something may be deduced from observations of the height of the layer, and some observers have attempted to record this. The *absolute* height h_mF2 of the electron peak, calculated in the usual way on the hypothesis of a parabolic layer, has been given, and is useful. Some workers have also listed values of $h'_{\mathrm{min}}F2$, the lowest *equivalent* height of the $F2$ region, but this is a much less satisfactory quantity, since it is so much influenced by retardation in the $F1$ layer, which is known to change markedly during an eclipse.

Some of those who have observed only f_0F2 have tried to deduce something about the distribution of radiation over the solar disk. DENISSE *et al.* (1947) suggested that a late appearance of an eclipse effect might have been related to the covering of an area in which the coronal green line was known to be strong. If the time-constants mentioned above are of the right order, this does not seem to be likely.

The effect of a corpuscular eclipse on the $F2$ layer has often been looked for, but the previously mentioned variability on the control days has rendered it difficult to establish any such effect with certainty. The only serious suggestion that effects had been observed at times corresponding to a corpuscular eclipse was made in connection with the 1945 eclipse by PAPALEXI (1946), and by AL'PERT and GOROZHANKIN (1946).

6. THE $F1\frac{1}{2}$ LAYER

A most interesting phenomenon is the occasional occurrence, during an eclipse, of a new critical frequency between f_0F1 and f_0F2. This was mentioned by MINNIS (1955), LEDIG *et al.* (1946), DELOBEAU (1953), ESTRABAUD (1953, 1954), and possibly by NAKATA (1950). The intermediate critical frequency has been indicated by the symbol $f_0F1\frac{1}{2}$. The phenomenon is well shown in the $h'f$ curves of MINNIS (1955), and in the associated plots of f_0F1, $f_0F1\frac{1}{2}$, and f_0F2.

It is important to know under what conditions the new critical frequency occurs. In the earlier eclipses it was not looked for, and the absence of a mention of it does not necessarily imply its absence, but, if it is supposed to have been present only when mentioned, we note from Table 1 that six out of the seven* occurrences are reported from stations near the geomagnetic equator, with dip angles less than 20 degrees. The other occurrence reported by NAKATA (1950) was not described in the same terms as those of the other three, and the $h'f$ curves which were published in connexion with it do not show it at all clearly marked. The observations of DENISSE *et al.* (1947) are the only ones made where the dip was less than 20 degrees and the phenomenon was *not* noticed.

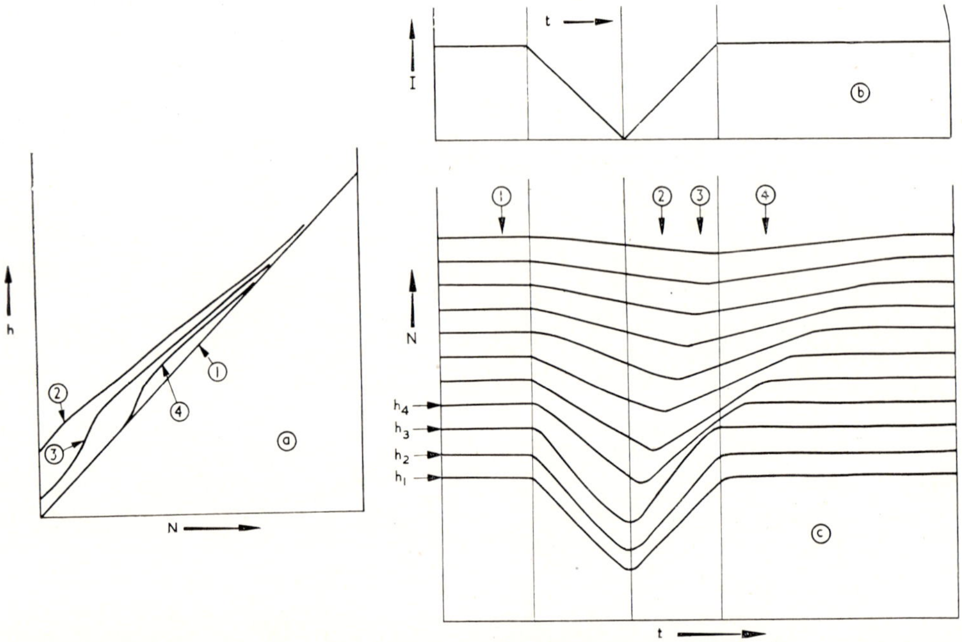

Fig. 2. To illustrate a possible way in which the extra critical frequency $f_0F1\frac{1}{2}$ could arise.

Only tentative suggestions have been made to explain the phenomenon. It has sometimes been suggested that the critical frequency $f_0F1\frac{1}{2}$ corresponds to the existence of a separate layer $F1\frac{1}{2}$ which is caused by a special process of electron production at this level. It should, however, be realized that, if the electron-loss coefficient is smaller at greater heights, it is only to be expected that the critical frequency $f_0F1\frac{1}{2}$ would be observed, for the following reasons. Suppose, before an eclipse starts, there is a linear distribution of N with height as shown at (1) in Fig. 2(a), and that during the eclipse the rate $\{I(t, h)\}$ of production of electrons is the same at all heights and is as shown at (b). Then at different heights $h_1, h_2, h_3, \ldots,$ the variations $N(t)$ of electron density will be as shown at (c) and will depend on the electron relaxation times, in such a way that when this time is greater (high up) the minimum of N will be later and its magnitude will be greater. The curves shown at (c) are "contours" of constant height (h) on an $N - t$ map, and cross-sections at constant t are easily visualized

* Examination of the curves published by SAVITT (1950) shows that the new critical frequency was in evidence.

as on an ordinary topographical map. $N(h)$ curves appropriate to the times marked 1, 2, 3 and 4 are shown at (a). Those marked 2, 3, and 4, which occur after totality, have an inflexion which moves upwards, and which would produce $h'f$ curves in which there would be an additional critical frequency like the observed $f_0F1\frac{1}{2}$. It would be interesting to examine in more detail what conditions would lead to the formation of an inflexion on the $N(h)$ curve.

7. The *D* Region

Since there is not yet any satisfactory evidence of the separate existence of a *D* layer, we here consider the *D* region, defined according to standard usage as "that region of the ionosphere situated below 90 km." Very few direct observations have been made of echoes reflected at vertical incidence from these heights during eclipses. Bracewell (1952) suggested that an effect was produced, by a small partial eclipse, on the phase-height of reflexion of waves of frequency 16 kc/s returned from a height of about 80 km.

There have been several measurements of the intensities of medium, long, and very long waves, and of atmospherics, reflected at oblique incidence during eclipse. Although it is probable that most of these refer to waves reflected in the *D* region, the results are difficult to interpret and do not seem to have added much to our knowledge.

It is generally assumed that a considerable part of the absorption of short waves, reflected at vertical incidence, occurs in the *D* region. Observations of their intensity during an eclipse should therefore be of use. Apart from demonstrating that there is a decrease of absorption, roughly centred on the visible eclipse, measurements of this kind do not seem to have yielded much of value.

8. The *Es* Region

Some solar control of the *Es* region is well established, and it is therefore of great importance to try to decide how an eclipse affects this region. Unfortunately, the occurrence of *Es* is so sporadic (hence its name) that it has proved impossible to draw any conclusions which have statistical significance. It is desirable that, in order to settle this important point, the fullest possible information about the occurrence of *Es* in future eclipses, and associated control days, should be collected, in the hope that, by considering all the evidence together, some conclusion, however tentative, can be reached.

9. Conclusions and Suggestions for Future Work

9.1. *The* E *Layer*

The considerable lack of detailed agreement between the simplest theory and observation can probably be removed most satisfactorily by supposing that electron-loss is by a recombination-like process with $\alpha = 10^{-8}\ cm^3\ sec^{-1}$, and that frequently the emission of ionizing radiation is not uniform over the solar disk, but is greater where the coronal green line is observed to be stronger.

There is considerable evidence that an appreciable fraction (perhaps 10 per cent) of the ionizing radiation comes from beyond the visible disk, and, as stressed by

PIDDINGTON (1951), it may well be that α is greater than has usually been calculated by neglecting this fraction.

There are still several important difficulties in an explanation of this kind. It does not explain the disagreement between experiment and the simple theory on normal days. There are examples of eclipses where the postulated distribution of radiation over the solar disk was unreasonable. The day-to-day variation of E-layer electron density does not seem to be as great as might be expected from the intensely active areas suggested by eclipse experiments.

It appears that useful advances could be made by considering in detail the limitations of equation (6), even when applied to a simple layer. Some calculations could profitably be made on the basis of equation (6) on the assumption that 10 per cent of the radiation comes from beyond the solar disk. It might be that the results would be at least as consistent as the present ones, and they might indicate a different value of α.

9.2. *The* F1 *Layer*

The outstanding problem is why an eclipse always makes the $F1$ layer appear more prominent.

Calculation of α, and conclusions about the distribution of the ionizing radiation over the solar disk, are at least as uncertain as for the E layer. It is usually supposed that $\alpha F1 \sim 5\text{–}9 \times 10^{-9}$ cm^3 sec^{-1}, and there is some evidence that enhanced radiation comes from areas which contribute enhanced ionization to the E layer.

It would be profitable to use $h'f$ curves to determine the variation of N at a fixed height and to analyse the results in terms of equation (6), and to try making an allowance for radiation from beyond the visible disk.

9.3. *The* F2 *Layer*

The eclipse effect is almost certainly partly caused by movements, and the only satisfactory method of analysis seems to be one in which the variations of electron density at a given level are analysed in terms of some model of the F layer to determine from experiment what the movements are. It might prove possible to calculate the movements expected theoretically and to compare them with the deductions from experiment. It would be interesting to know whether an eclipse effect, caused by movements, is observable just outside the area of the visible eclipse.

9.4. *The* F1½ *Layer*

Observations are needed at places with different dip angles, to see whether the occurrence of the $F1\frac{1}{2}$ layer is confined to places with small dip.

On the Bradbury (1938) theory of the $F2$ layer, with no movements included, it seems that a critical frequency $f_0F1\frac{1}{2}$ should be observed during the second half of an eclipse.

9.5. *The* D *Region*

Detailed observations of phase- or group-height of reflexion on long and very long waves are needed.

9.6. *Sporadic* E

It is desirable that full statistics about sporadic E reflexions both on eclipse and on control days should be collected and examined.

References

Al'pert, Y. L., and Gorozhankin, B. N. 1946 *Bull. Acad. Sci. U.R.S.S. ser. phys.*, **10**, 245.

Appleton, E. V., and Lyon, A. J. 1955 *Rep. Phys. Soc. Ionosp. Conf., Cambridge*, p. 20.

Appleton, E. V., and Naismith, R. 1938 *Gen. Ass. U.R.S.I.*

Bracewell, R. N. 1952 *J. Atmosph. Terr. Phys.*, **2**, 226.

Bradbury, N. E. 1938 *Terr. Mag. Atmos. Electr.*, **43**, 55.

Chapman, S. 1933 *Terr. Mag. Atmos. Electr.*, **38**, 175.

Chapman, S., and Bartels, J. 1951 *Geomagnetism* (Oxford Univ. Press), p. 504.

Delobeau, F. 1953 *Ann. Géophys.*, **9**, 317

1954 *Ann. Géophys.*, **10**, 148.

Denisse, J. F., Seligmann, P., and Gallet, R. 1947 *C.R. Acad. Sci. (Paris)*, **225**, 1169.

Estrabaud, S. 1953 *C.R. Acad. Sci. (Paris)*, **236**, 833.

1954 *Note prélim. de Lab. Nat. de Radio électr.*, No. 174.

Geger, S., and Akerlund, P. 1947 *Terr. Mag. Atmos. Electr.*, **52**, 479.

Gilliland, T. R. 1942 *Nat. Geogr. Soc. Solar Eclipse Series* No. 2, p. 88.

Harang, L. 1945 *Terr. Mag. Atmos. Electr.*, **50**, 307.

Higgs, A. J. 1942 *Monthly Not. Roy. Astron. Soc.*, **102**, 24.

Kirby, S. S., Berkner, L. V., Gilliland, T. R., and Norton, K. A. 1934 *Proc. Inst. Rad. Eng.*, **22**, 247.

Ledig, P. G., Jones, M. W., Giesecke, A. A. and Chernosky, E. J. 1946 *Terr. Mag. Atmos. Electr.*, **51**, 411.

Martyn, D. F. 1955 *Rep. Phys. Soc. Ionosp. Conf., Cambridge*, pp. 212, 254.

McLeish, C. W. 1948 *Canad. J. Res.*, **26**, 137.

Meriau, P., and Rawer, K. 1954 *Ann. Géophys.*, **10**, 9.

Minnis, C. M. 1955 *J. Atmosph. Terr. Phys.*, **6**, 91.

Nakata, Y. 1950 *Rep. Ionosp. Res., Japan.*, **4**, 21.

Papalexi, N. D. 1946 *Bull. Acad. Sci. U.R.S.S. ser. phys.*, **10**, 237.

Perers, O. F. 1952 *Mixed Comm. Ionosp. 3rd Mtg., Canberra*, p. 50.

Piddington, J. H. 1951 *J. Geophys. Res.*, **56**, 409.

Pierce, J. A. 1948 *Proc. Inst. Rad. Eng.*, **36**, 8.

Ratcliffe, J. A., Schmerling, E. R., Setty, C. S. G. K., and Thomas, J. O. 1956 In publication, in *Phil. Trans. Roy. Soc.* A. **248**, 609.

Rydbeck, O. E. H. 1946 *Trans. Chalmers Univ. Tech.*, No. 53.

Rydbeck, O. E. H., and Wilhelmsson, H. 1954 *Trans. Chalmers, Univ. Tech.*, No. 149.

Savitt, J. 1950 *J. Geophys. Res.*, **65**, 385.

Smith-Rose, R. L. 1946 *J. Brit. Inst. Rad. Eng.*, **6**, 82.

Stoffregen, W. 1946 *Terr. Mag. Atmos. Electr.*, **51**, 495.

Wells, H. W. 1952 *J. Geophys. Res.*, **57**, 291.

Solar Eclipses and the Ionosphere (Special Suppl. Vol. 6 to J. Atmosph. Terr. Phys.), Pergamon Press, London, 1956

SECTION II

THEORETICAL CONSIDERATIONS

A Theoretical Study of *E*-layer Behaviour during a Solar Eclipse

O. E. H. Rydbeck

Chalmers University of Technology, Gothenburg

Abstract—Expressions are developed for the theoretical reduction in electron density N of a plane-stratified isothermal E layer during total and partial eclipses. Consideration is given to the delay of the minimum value of N on the instant of totality of an eclipse, and the changes in layer shape are computed for forenoon and afternoon eclipses, allowing for the variation of solar zenith angle during eclipse.

1.

For a plane-stratified isothermal atmosphere the relation between rate of electron production q, height h, and solar zenith distance χ is given by

$$q(z) = q_0 \exp(1 - z - e^{-z} \sec \chi) \tag{1}$$

where q_0 is the maximum rate of electron production, occurring at height h_0, when the sun's rays are vertical ($\chi = 0$). One further has

$$z = (h - h_0)/H \tag{2}$$

where H is the scale height of the atmospheric constituent ionized.

The spatial maximum of q is

$$q_m = q_0 \cos \chi \tag{3}$$

occurring at the height

$$h(q_m) = h_m = h_0 + \ln(\sec \chi) \,.\, H \tag{4}$$

It is important to note in this connexion that h_m is independent of the intensity of the primary ionizing rays falling upon the atmosphere. *This means that, neglecting the change in χ during the eclipse, the "electron production layer" $q(z)$ maintains its Chapman shape.** Making use of this fact, Rydbeck (1946) showed that for every E- and $F1$-layer level (except those of very low electron density) the ratio between the minimum electron density $N_{\min}$ and the electron density N_0 at first contact for a total eclipse is given very accurately by

$$\frac{N_{\min}(z)}{N_0(z)} \simeq \frac{0{\cdot}68}{\{\alpha N_0(z)t_0\}^{1/3}} \quad (\alpha N_0 t_0 > \text{about } 3) \tag{5}$$

where t_0 is the time between second and first contacts. Introducing the relaxation

* The fact that the first contact between sun and moon is noticed at different times at E and F level (if $\chi \neq 0$) is of small importance in this connexion. As the moon shadow travels with a speed of the order of 1 km/sec, the delay between the E and F levels is of the order of 200 sec, which is less than the E layer relaxation time $1/2\alpha N \approx 500$ sec (normal order).

time $\tau = 1/2\alpha N_0(z)$, (5) can be written

$$\frac{N_{\min}(z)}{N_0(z)} \simeq 0{\cdot}86 \left(\frac{\tau}{t_0}\right)^{1/3} \tag{5a}$$

Fig. 1 demonstrates the actual variation of the electron density for various values of $\alpha N_0 t_0$ ($= z$, say), i.e., for various αN_0 or h levels. For a study of the exact mathematical expression for $N(z)/N_0(z)$, reference is made to RYDBECK (1946).

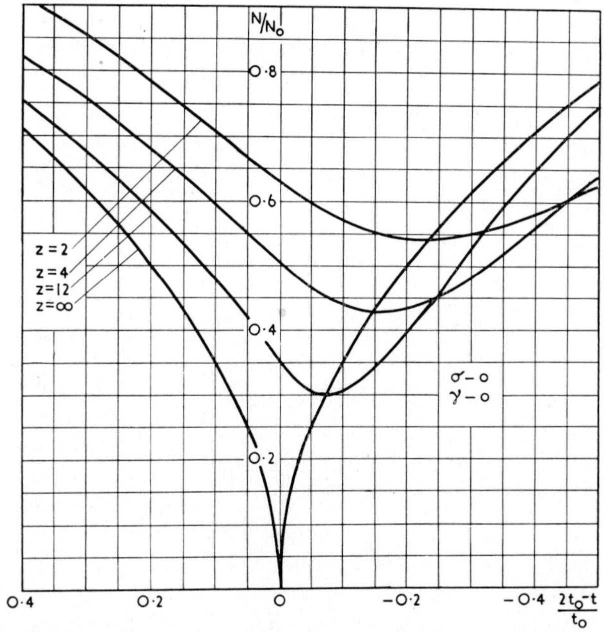

Fig. 1. The variation in electron density during a total solar eclipse for various values of $z = \alpha N_0 t_0$. (Note: This is not the z of equations 2 and 5.)

The electron-density minima are reached somewhat after maximum phase, the delay being greater the smaller is N_0. This will distort the electron-density distribution during the eclipse, as will subsequently be shown graphically.

It was shown by RYDBECK (1946) that the time-lag σt between maximum phase (second contact) and the electron-density minimum is given quite accurately by

$$\frac{\sigma t(z)}{t_0} \simeq \frac{0{\cdot}36}{(\alpha N_0(z) t_0)^{2/3}} \quad (\sigma N_0 t_0 > \text{about } 3) \tag{6}$$

It is further interesting to note that

$$\sigma t(z)\,.\,N_{\min}(z) \simeq \frac{0{\cdot}245}{\alpha}, \tag{7}$$

or

$$\sigma t(z) \simeq 0{\cdot}49(\tau)_{N=N_{\min}}, \tag{7a}$$

i.e., the time-lag is about half the actual relaxation time at electron-density minimum.

We next apply (5) and (6) to typical E-layer data. Assuming $N_0 = 1{\cdot}23 \times 10^5$ electrons/cm^3 (corresponding to a critical frequency of 3·3 Mc/s), and $\alpha = 10^{-8}$, we find, if the time difference between first and fourth contacts is 2·2 h (duration of the eclipse),

$$N_{\min} \simeq 4{\cdot}9 \times 10^4,$$

i.e., a critical frequency of 2·0 Mc/s, and $\sigma t \simeq 4{\cdot}1$ min.

Relation (5) is valid for total eclipses of short duration. For longer total eclipses, such as the 1955 example at the Philippines, corrections must be applied

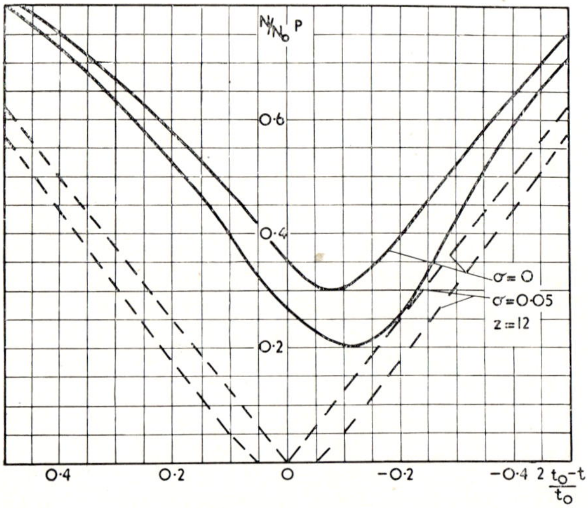

Fig. 2. The change in electron-density curve around maximum phase when the relative difference between the apparent lunar and solar diameters is not neglected.

to (6). RYDBECK and WILHELMSSON (1954) extended the original expressions to include also total eclipses of long duration.

Fig. 2 depicts the situation for the longest total eclipses, where the totality is of the order of 6 min. If the duration of totality is Δt, then $\Delta t = 2\sigma t_0$, i.e., about six or seven min for the lowest curve in Fig. 2. The dashed curves represent the corresponding exposure coefficient $p(t) = 1 - \{\text{eclipsed area/total area}\}$ (the radiation is assumed uniform over the solar disk). For the detailed relations of $N(z)/N_0(z)$, etc., reference is made to RYDBECK and WILHELMSSON (1954).

It was shown by RYDBECK (1946) that at second contact (beginning of totality)

$$\left(\frac{N}{N_0}\right)_{\text{2nd cont.}} = \frac{N(0, z)}{N_0} \simeq \frac{0{\cdot}81}{(\alpha N_0 t_0)^{1/3}} \quad (\alpha N_0 t_0 > \text{about } 3) \tag{8}$$

This means that $N_{\min}/N(0) \simeq 0{\cdot}84$, constant and independent of σt.

According to (8) and (1) at second contact

$$N(z) = N(0, z) \simeq \frac{0{\cdot}81}{(\alpha t_0)^{1/3}} \left(\frac{q(z)}{\alpha}\right)^{1/3},$$

if we assume quasistationary conditions at first contact.

If the maximum electron density at first contact is $N_{0_{max}}$, occurring at height h_m, according to (8) the maximum electron density at second contact (which occurs at the same level) is $N(0)_{max} \simeq N_{0_{max}} . 0{\cdot}81/(\alpha N_{0_{max}} t_0)^{1/3}$. Therefore*

$$N(0,z) \simeq N(0)_{max} e^{\frac{1}{3}(1-\Delta z - e^{-\Delta z})}, \tag{9}$$

where

$$\Delta z = (h - h_m)/H \tag{9a}$$

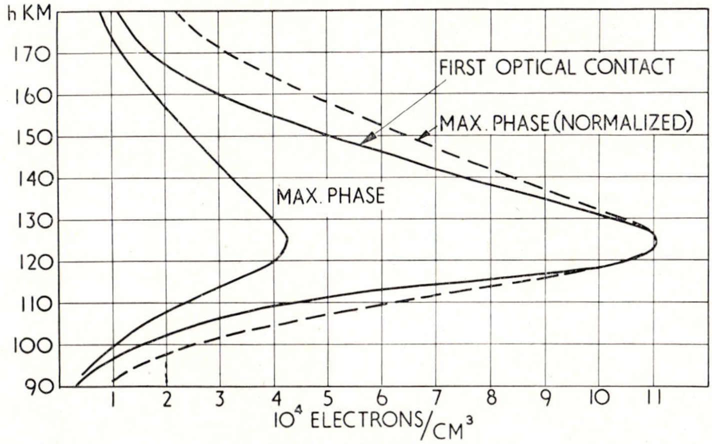

Fig. 3. E-layer shape (and relative deviation from Chapman form) at the beginning of totality (maximum phase), for an eclipse solar zenith distance of about 38°. ($H = 10$ km.)

Fig. 3 demonstrates the "distortion" of the E layer at the beginning of totality. For the undistorted layer at first contact, the electron-density variation *near* maximum level has the approximate shape

$$N_0(z) \simeq N_{0_{max}} (1 - \Delta z^2/4),$$

whereas, according to (9), the corresponding shape at second contact is

$$N(0, z) \simeq N_{0_{max}}(1 - \Delta z^2/6).$$

The apex of the layer thus appears like a Chapman layer with an effective scale height $\sqrt{3/2}H$.

2.

So far we have not considered the effect of the variation (with time) of the solar zenith distance during the eclipse. This problem is solved relatively easily using the type of first- and second-order approximations used by RYDBECK (1946) and RYDBECK and WILHELMSSON (1954). The continuity equation now has the form (neglecting drift and diffusion)

$$\frac{\partial N(z, t)}{\partial t} = p(t)q(z, t) - \alpha N^2(z, t). \tag{10}$$

* Compare: $N_0(z) \simeq N_{0max} e^{\frac{1}{2}(1-\Delta z - e^{-\Delta z})}$

At first contact $q(z, t) = q_1(z)$, and the corresponding zenith distance is χ_1. According to §1, the solution of (10) for $q = q_1(z)$ is known through the eclipse. The corresponding N value is $N_1(z, t)$, say. Then

$$N(z,\ t) = N_1(z, t) + \Delta N(z, t), \tag{11}$$

where $\Delta N(z, t)$ is the change in electron density due to the change in solar zenith distance. Assuming $\Delta N \ll N$, we find from (10),

$$\Delta N(z, t) = e^{-w(z,t)} \int_{c_1}^{t} e^{w(z,t)}\, p(t) \Delta q(z, t)\, dt, \tag{12}$$

where

$$w(z, t) = \int_{c_2}^{t} 2\alpha N_1(z, t)\, \mathrm{d}t \tag{13}$$

$$= \int_{c_2}^{t} \frac{\mathrm{d}t}{\tau_1(z, t)} \tag{14}$$

and $\Delta q(z, t) = q(z, t) - q_1(z)$.
Relation (12) clearly demonstrates the effect of the relaxation time

$$\tau_1 = 1/2\alpha N_1\ (z, t).$$

Remembering that the change in Δq with time is much smaller than in N_1, we obtain from (12) the following second-order approximation:

$$\Delta N(z, t) = \Delta N^{(2)} = N_1(z, t) \frac{\Delta q(z, t)}{2q(z)} \left[1 - \frac{1}{2\alpha N_1} \frac{N_1'}{N_1} + \frac{1}{(2\alpha N_1)^2} \left\{ \frac{N_1''}{N_1} - \left(\frac{N_1''}{N_1}\right)^2 \right\}\right], \tag{14}$$

where $N' = \dfrac{\partial}{\partial t} N_1(z, t)$, etc.

From this relation (14) the change in layer shape during the eclipse caused by the change in solar zenith distance can be computed.

N_1, N_1', and N_1'' can be obtained from the relations of RYDBECK, and RYDBECK and WILHELMSSON. It should be sufficient in this connection, however, to write down only N_1'/N_1 for second contact, as the expressions are fairly complex for other values of time. One has

$$\left\{\frac{N_1'(z, t)}{N_1(z, t)}\right\}_{\text{2nd cont.}} \cong - \left(\frac{2N_0(z)\alpha t_0}{3\pi^2}\right)^{1/3} \cdot \frac{1}{t_0}. \tag{15}$$

If, for the moment, we satisfy ourselves using only the first-order approximation, we have

$$\{\Delta N^{(1)}(z, t)\}_{\text{2nd cont.}} = N(0, z) \frac{\Delta q(z, t)}{2q_1} \underbrace{\left\{1 + \frac{1}{(\sqrt{32}\pi\alpha N(0,\ z) t_0)^{2/3}}\right\}}_{\xi(z)}. \tag{16}$$

where $\xi(z)$ does not differ much from unity.

The electron-density curve with first-order corrections thus becomes at second contact (beginning of totality)

$$N''(0, z) = N(0)_{\max} e^{\frac{1}{3}(1-\Delta z-e^{-\Delta z})} \underbrace{\left\{1 - \frac{\xi(z)}{2} + \frac{\xi(z)}{2} e^{-\mu e^{-\Delta z}}\right\}}_{\psi(z)}, \tag{17}$$

where

$$\mu = \frac{\cos \chi_1}{\cos \chi_2} - 1, \tag{17a}$$

and χ_2 is the solar zenith distance at second contact (maximum phase).

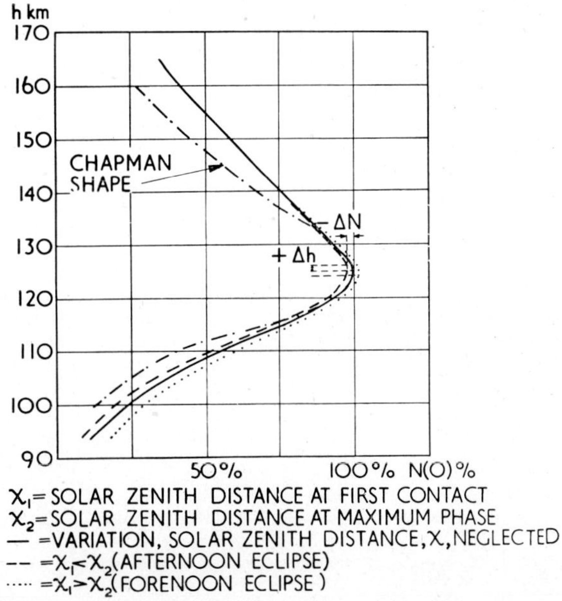

Fig. 4. The change in E-layer shape at second contact (beginning of totality) due to variable solar zenith distances for forenoon and afternoon eclipses.

The correction factor $\psi(z)$ will displace the electron-density maximum a certain distance Δh_d and will change the maximum electron density by a small amount ΔN_d. Expansion of (17) around the $h = h_m$ level yields (neglecting higher powers in μ, which for the 1954 eclipse at Onsala (Gothenburg) was only 0·06 [$\chi_1 \simeq 38°$; $\chi_2 \simeq 34°$])

$$\Delta h_d^{(1)} \simeq \tfrac{3}{2}\, \mu \xi H, \tag{18}$$

and

$$\Delta N_d^{(1)} \simeq -\frac{\mu\xi}{2} N(0)_{\max} \tag{19}$$

For a forenoon eclipse $\chi_1 > \chi_2$, and μ is negative. This means that the layer is lowered and the electron density increased. For an afternoon eclipse (like the one at Onsala, 1954) $\chi_1 < \chi_2$, and conditions are reversed.

The actual first-order correction in layer shape is depicted in Fig. 4. The change is of the same order of magnitude as the experimental error.

3.

It is desirable to extend (5) even to partial eclipses. RYDBECK (1954) has shown that this can be done quite accurately.

If the minimum electron density at the total eclipse is $N_{\min}$, and N_0 is the electron density at first contact, the minimum electron density $N_{\min_M}$ of a partial eclipse of magnitude M (and exposure coefficient p) becomes

$$N_{\min_M} \cong [N_0{}^4 p^2 + N^4_{\min}\{1 - (1 - M)^2\}]^{1/4} \tag{20}$$

For the corresponding critical frequencies we have*

$$f_{c_{\min_M}} \cong f_{c_0}\left[p^2 + \left(\frac{f_{c_{\min}}}{f_{c_0}}\right)^8 \{1 - (1 - M)^2\}\right]^{1/8} \tag{21}$$

Fig. 5 demonstrates the nature of relations (20) and (21). When the normal

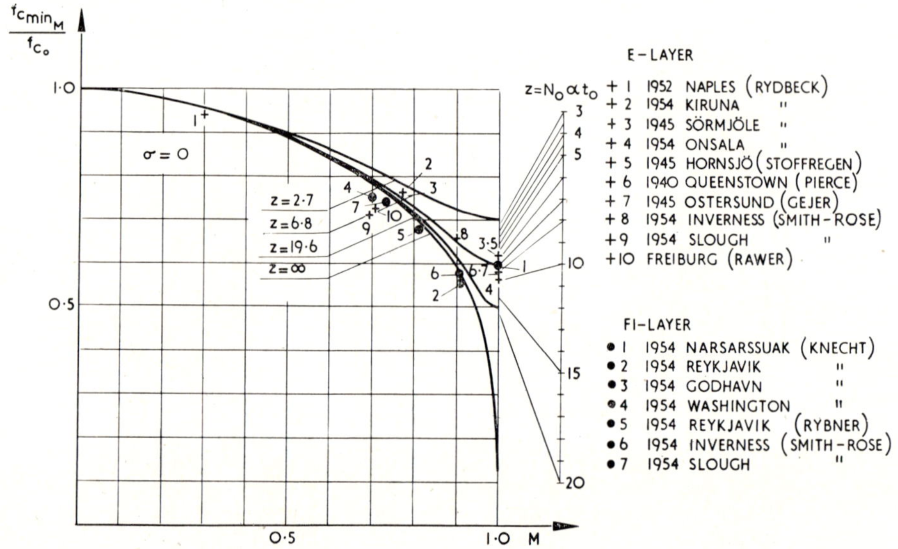

Fig. 5. The reduction of E- and F-layer critical frequencies during the maximum phase of a partial eclipse of magnitude M.

ionospheric experimental errors are considered, it is clearly shown by Fig. 5 that it is hardly worth while to observe eclipses of smaller magnitude than $M = 0{\cdot}8$.

We have in Fig. 5 plotted a number of eclipse observations. It is clearly demonstrated that most *F1*-layer values are grouped in such a manner that, when comparing them with E layer values, one is led to assume that the average $\alpha N_0 t_0$ is at least three to four times greater than for the E layer. Fig. 5 also demonstrates that the accuracy of most ionospheric measurements, especially of the E layer, is not good enough to be used for a detailed check of E- and *F1*-layer eclipse behaviour.

REFERENCES

RYDBECK, O. E. H. 1946 *Trans. Chalmers Univ. Tech.*, No. 53.

RYDBECK, O. E. H., and WILHELMSSON, H. 1954 *Trans. Chalmers Univ. Tech.*, No. 149.

* For a relation between M and p, see RYDBECK and WILHELMSSON (1954).

Solar Eclipses and the Ionosphere (*Special Suppl. Vol. 6 to J. Atmosph. Terr. Phys.*), *Pergamon Press, London*, 1956

An Ionospheric Layer during a Solar Eclipse

F. Lied
Norwegian Defence Research Establishment, Kjeller

Abstract—The changes in electron density and nondeviative absorption for an ionospheric layer have been calculated for the eclipse of 30 June 1954 at Kjeller. The recombination coefficient and the collisional frequency have been taken from published results. A comparison is made between the normal approximative method of analysing eclipse observations and the more detailed treatment given here.

1. Introduction

It is customary when analysing ionospheric observations taken during a solar eclipse to assume that the maximum electron density is given by:

$$\frac{\mathrm{d}N}{\mathrm{d}t} + \alpha N^2 = q_0 \cos \chi \, p(t) \tag{1}$$

where α is the recombination coefficient, $p(t)$ is the unobscured fraction of the sun's disk, q_0 is a measure of the ion production at the height of the layer maximum before the eclipse, and χ is the sun's zenith distance. When analysing critical frequency observations by comparing observed N/N_0 with N/N_0 calculated from (1) for different values of α, it is assumed that the layer maximum remains at the same height during the eclipse, and also that horizontal and vertical diffusion can be disregarded.

When considering ionospheric absorption, it is customary to write

$$\int_{-\infty}^{+\infty} N \nu \, \mathrm{d}h \sim N_{\max} , \tag{2}$$

an expression based upon the assumption that the layer retains its form during the eclipse.

It was felt that these assumptions should be tested by considering a particular layer in detail.

2. An Ionospheric Layer during an Eclipse

As an example of a complete analysis, a layer having

$$N_0 = 10^4 \, e^{1/2\,(1 - z - 1\cdot 29 e^{-z})} \qquad \text{electrons/cm}^3 \tag{3}$$

where $z = \dfrac{h - 90}{9}$ (h in km) was chosen. This value of N_0 corresponds to a reasonable D layer at Kjeller (60° N, 11° E) in the summer. It has been assumed, in accordance with Nicolet (1953), that

$$\nu = 3\cdot 96 \times 10^{12} \, e^{-\frac{h}{5\cdot 75}} \tag{4}$$

and that α is a function of height, as given by Mitra and Jones (1954).

The eclipse of 30 June 1954 was chosen for this investigation, and $p(t)$ was determined for a height of 120 km, where the maximum obscuration was 96 per cent. The time of first contact was 1226, maximum eclipse being at 1340, and fourth contact at 1451 M.E.T.

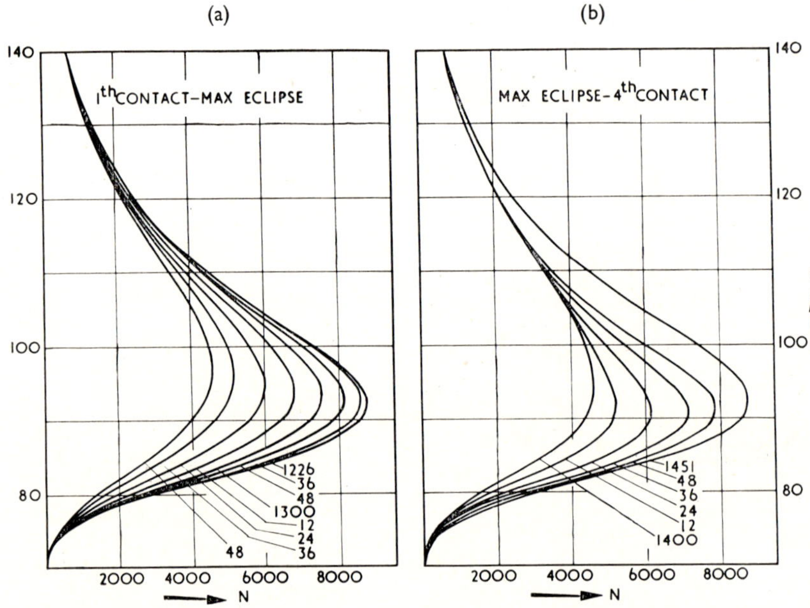

Fig. 1. Electron-density profiles during the eclipse.

In this case the electron density at any height, when diffusion is disregarded (RYDBECK, 1946), is given approximately by

$$\frac{dN}{dt} + \alpha(h)N^2 = q_0 p(t), \tag{5}$$

where $q_0 = N_0{}^2\alpha(h)$, and N_0 is determined from equation (3), which gives the electron density at first contact ($p(t) = 1$).

With these assumptions it is possible to calculate the change in form and density of the ionized layer during the eclipse, and also to compute the non-deviative absorption of a radio wave propagated twice (up and down) through the layer.

2.1. *Change of density and form*

In Fig. 1 the electron-density profiles at different periods during the eclipse are shown.

There is a very pronounced assymmetry in the curves a and b, showing that the layer disappears most rapidly at the bottom, where αN_0 is great, and showing also that the layer re-forms from the bottom, where $\alpha N_0{}^2$ is great. The height of maximum electron density follows a curve shown in Fig. 2.

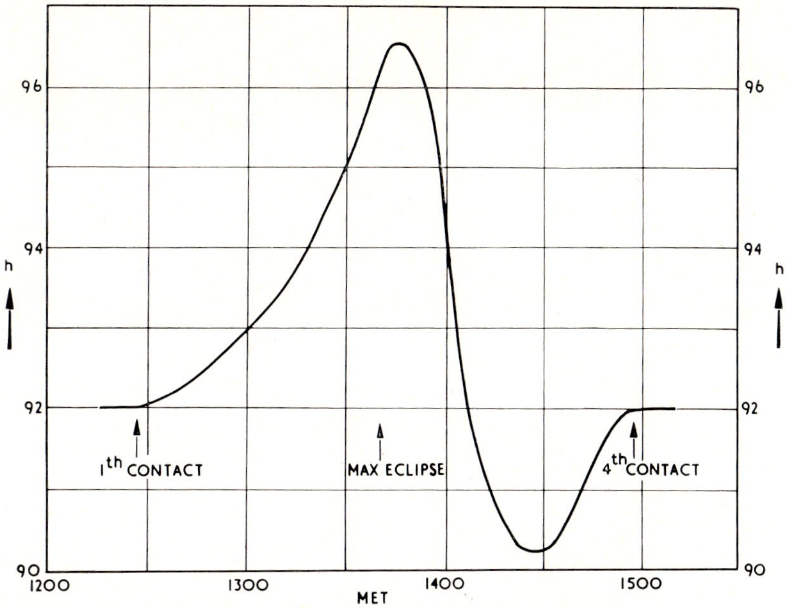

Fig. 2. Height of maximum electron density.

2.2. *Nondeviative absorption*

The total nondeviative absorption for a given angular frequency $\omega(= 2\pi f)$ determined by integration twice throughout the layer is given by

$$S = 2 \frac{e^2 10^5 2\pi}{cm} 8{\cdot}69 \int_{-\infty}^{+\infty} \frac{N\nu \, dh}{\nu^2 + (\omega + \omega_L)^2} \text{ decibel}$$

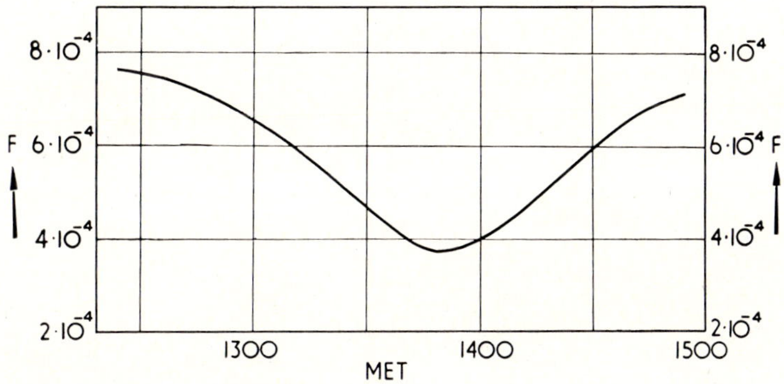

Fig. 3. Nondeviative absorption for an ionized layer during a solar eclipse.

where h is in km and $\omega_L = 2\pi f_H \cos \theta$, f_H and θ having their usual significance. The total nondeviative absorption is therefore given by

$$S = 46 \,.\, 10^3 \,.\, 2 \int_{-\infty}^{+\infty} \frac{N\nu \, dh}{\nu^2 + (\omega + \omega_L)^2} \text{ decibel}$$

$$= 46 \,.\, 10^3 \,.\, F \text{ decibel (say).}$$

Results obtained by solving the above equation for $f = 2$ Mc/s are shown in Fig. 3.

3. Discussion

In order to obtain an idea of the inaccuracies involved when the normal procedure is applied to ionospheric observations made during an eclipse, curves are given in Fig. 4 showing:

(i) F/F_0,

(ii) N/N_0 for the height of maximum electron density at first contact ($\alpha N_0 = 0{\cdot}7 \times 10^4$ in our case),

(iii) N_{max}/N_0 as a function of time. N_{max} is the maximum electron density at each instance.

The suffix 0 indicates values of a variable at first contact.

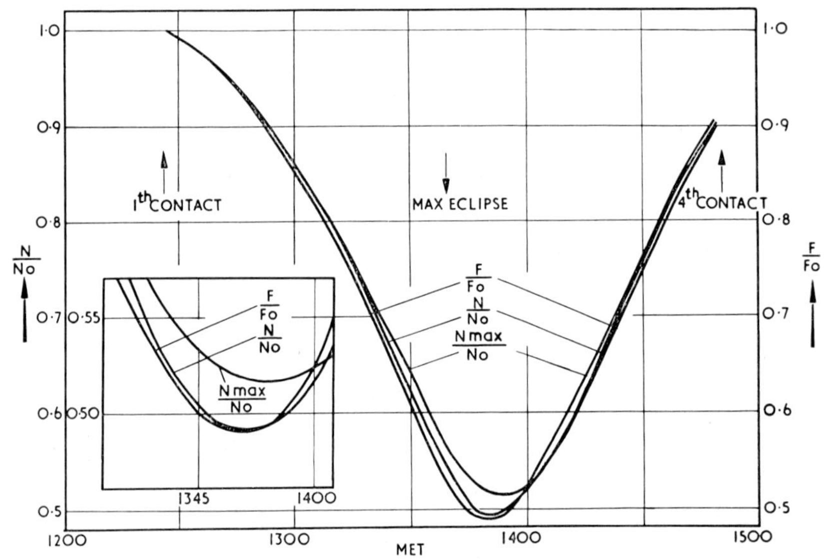

Fig. 4. Comparisons of F/F_0, N/N_0, and N_{max}/N_0.

A slight disagreement is evident between the different curves; in particular, values of absorption increase more rapidly after maximum eclipse than do values of N/N_0 and N_{max}/N_0. This is due to the re-forming of the layer at a lower height, where the number of collisions is greater. Furthermore, the error of interpreting N_{max}/N_0 as N/N_0 (comparing N_{max}/N_0 with curves for constant $N_0\alpha$) is illustrated.

The layer has been chosen to give approximately 35-db nondeviative absorption on a frequency of 2 Mc/s. The recombination coefficient α varies rather rapidly throughout the layer, and the curves given in Fig. 4 are consequently rather accentuated. The error involved in applying the standard method for interpreting *E*- and *F1*-layer results is probably negligible. Absorption measurements, however, require more careful treatment.

Acknowledgement

The author is indebted to the computing section of the N.D.R.E. for most able assistance.

References

Mitra, A. P., and Jones, R. E. 1954 *J. Geophys. Res.*, **59**, 391.
Nicolet, M. 1953 *J. Atmosph. Terr. Phys.*, **3**, 200.
Rydbeck, O. E. H. 1946 *Trans. Chalmers Univ. Tech.* No. 53.

Solar Eclipses and the Ionosphere (Special Suppl. Vol. 6 to J. Atmosph. Terr. Phys.), Pergamon Press, London, 1956

Some Observations on the Influence of a Solar Eclipse upon the Ionosphere

P. DOMINICI

Istituto Nazionale di Geofisica, Rome

Abstract—A general analysis to a first approximation of the effect of a solar eclipse on the ionosphere is given and applied to data obtained at Rome during the partial eclipse of 30 June 1954, together with data from some previous eclipses. It is concluded that the rate of electron loss varies as if the process of electron removal were recombination in the E and $F1$ layers and attachment in the $F2$ layer, the recombination and attachment coefficients decreasing with altitude. In the $F2$ layer the decrease of electron density is, for similar conditions, more marked in winter eclipses than in summer eclipses, and it is suggested that this may be associated with $F1$-$F2$ superposition. Secondary minima of electron density observed shortly before and after the optical eclipse may be explained in terms of coronal occultation, while minima further removed from the optical eclipse are ascribed to corpuscular eclipse effects. Definite eclipse effects in the sporadic E layer (decrease in fEs) and the appearance of a sub-stratification between the E and $F1$ layers, are sometimes observed, and may be associated with the descent of banks of charges from the $F1$ layer.

Attention is called to the need for common standards and methods in future eclipse work.

IT is well known that a solar eclipse offers important information on the phenomena controlling the development of the ionospheric layers.

The most general equation of the electron equilibrium in the ionosphere,

$$\frac{dN}{dt} = q(t) - l(t)$$

where N is the electron density, $q(t)$ and $l(t)$ are the rates of electron production and of electron loss respectively, and t is the time, is generally written in one of the two following forms:

$$\frac{dN}{dt} = q(t) - \alpha N^2 \, ; \qquad \frac{dN}{dt} = q(t) - BN$$

if one assumes as the principal process for the electron loss ionic recombination (α is the effective recombination coefficient) or attachment to neutral atoms and molecules (B is the attachment coefficient).

Assuming that (a) the principal ionizing agent in the upper atmosphere is the ultra-violet solar radiation; (b) the ionizing power of the sun is uniformly distributed over its visible disk, we write:

$$q(t) = q_M S \cos \chi$$

where χ is the zenith angle of the sun, S is the visible fraction of the area of the solar disk, q_M is the value of q when $\chi = 0$ and $S = 1$.

Assuming one knows the astronomical "magnitude" m (eclipsed fraction of the solar diameter) of the eclipse, the minimum value d of the distance between the centres of the disks of the sun and of the lunar shadow is $d = 1 - m$, and this distance, measured in solar diameters, is, during the eclipse:

$$D = \sqrt{d^2 + \left(\frac{t_2 - t}{t_2 - t_1}\right)^2 (1 - d^2)}$$

where t_1 is the commencement time of the eclipse and t_2 the time of greatest occultation. The visible fraction of the solar disk is:

$$S = \frac{2}{\pi} (\text{arc} \sin D + D \sqrt{1 - D^2}) = \frac{4}{\pi} D \left(1 - \frac{D^2}{6} - \frac{D^4}{40} - \ldots\right).$$

We now consider the ionospheric conditions about the maximum of the eclipse. The electron equilibrium in the ionosphere is not instantaneous, and the maximum electron density N of a particular layer attains its minimum N_m at the time $(t_2 + \Delta t)$ instead of t_2.

We indicate with the suffixes $n,1,2,m$ respectively the values of the "normal" conditions (i.e., in absence of eclipse), conditions at the time t_1, at the time t_2, and at the minimum of electron density. If $(dN/dt)_n \simeq 0$ when $t = t_2$, we may write, for example in the recombination case:

$$\begin{cases} \alpha_m N_m{}^2 = q(t_2 + \Delta t) \simeq q_M \left[S_2 \cos \chi_2 + (\cos \chi_1 - S_2 \cos \chi_2) \dfrac{\Delta t}{t_2 - t_1}\right] \\ \alpha_n N_n{}^2 \simeq q_M \cos \chi_2 \end{cases}$$

Putting: $k = S_2 + \left(\dfrac{\cos \chi_1}{\cos \chi_2} - S_2\right) \dfrac{\Delta t}{t_2 - t_1}$ and if $\alpha_m = \alpha_n$, we have:

$$1 = k \bigg/ \left(\frac{N_m}{N_n}\right)^2 \text{[recombination]}$$

Analogously for the attachment case:

$$1 = k \bigg/ \left(\frac{N_m}{N_n}\right) \text{[attachment]}$$

We may write at the instant t_2:

$$\begin{cases} -\dfrac{N_2 - N_m}{\Delta t} \simeq q_M S_2 \cos \chi_2 - \alpha_m N_2{}^2 \\ 0 = q_M \cos \chi_2 - \alpha_n N_n{}^2 \end{cases}$$

and we have

$$\alpha_n \simeq \frac{N_2 - N_m}{\Delta t (N_2{}^2 - S_2 N_n{}^2)}$$

Analogously for the attachment case:

$$B_n \simeq \frac{N_2 - N_m}{\Delta t (N_2 - S_2 N_n)}$$

We have applied this simple first-approximation analysis to the ionospheric data obtained at Rome during the partial solar eclipse of 30 June 1954 (Fig. 1), and moreover to data obtained by others in some past eclipses* which occurred in calm geomagnetic conditions and for which $(dN/dt)_n \simeq 0$ when $t \simeq t_2$.

The results appear in Fig. 2 and in Fig. 3. In Fig. 2 the ratios $k/(N_m/N_n)$ and

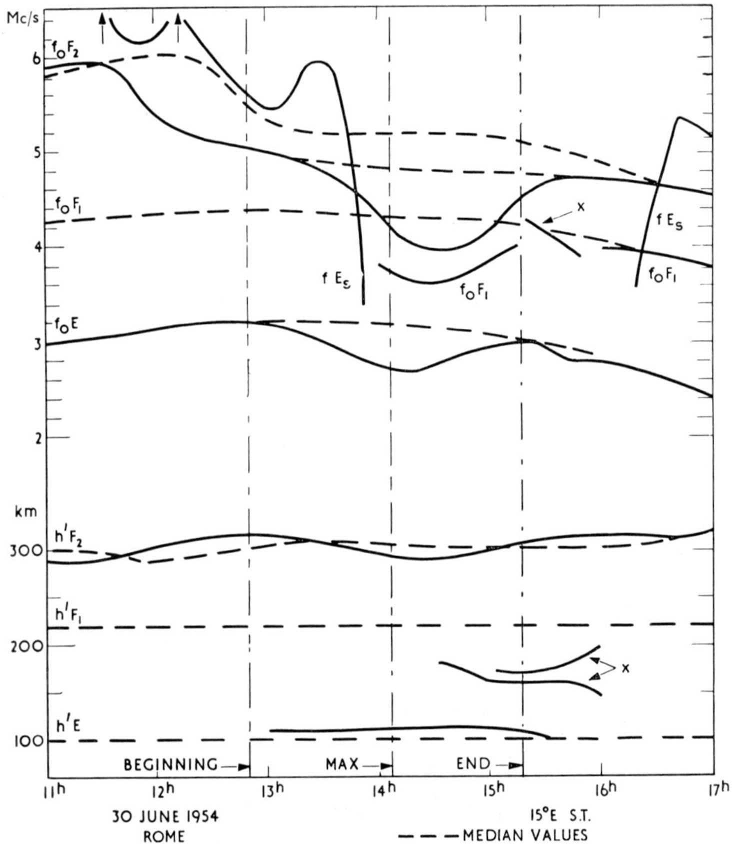

Fig. 1.

$k/(N_m/N_n)^2$ are plotted with the zenith angle of the sun; the encircled values, which are strangely out of place, are the values calculated for the $F2$ layer from data obtained in high-latitude stations during the eclipse of 9 July 1945, for which we suppose an anomaly of local origin.

The examination of these figures leads to the following conclusions:

(a) In the E and $F1$ layers the rate of electron loss varies with the second power of the electron density, as if the process of electron removal were ionic recombination.

(b) In the $F2$ layer the rate of electron loss varies with the first power of the

* "The effect of solar eclipses upon the ionosphere," Bibliography by Netherlands Postal and Telecommunications Services, No. 768 (1954).

electron density, as if the process of electron removal were attachment to neutral atoms and molecules.

(c) In the *F2* layer the decrease of electron density during a solar eclipse is, for identical conditions, more important in winter eclipses than in summer eclipses.

(d) The recombination and attachment coefficients decrease with height.

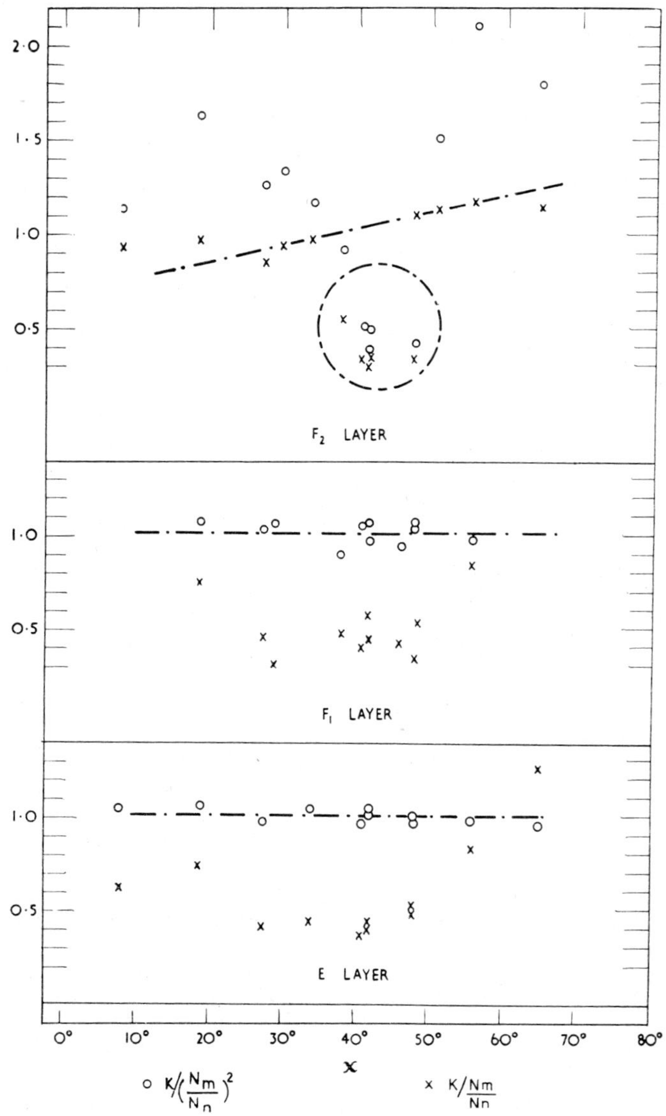

Fig. 2.

The seasonal anomaly of the eclipse-effect in the *F2* layer can be explained, perhaps, with a superposition hypothesis of the two layers of the *F* region, as suggested many years ago by L. V. BERKNER (FLEMING, 1939). The cooling of the ionosphere during the eclipse has the result of raising the level of maximum electron

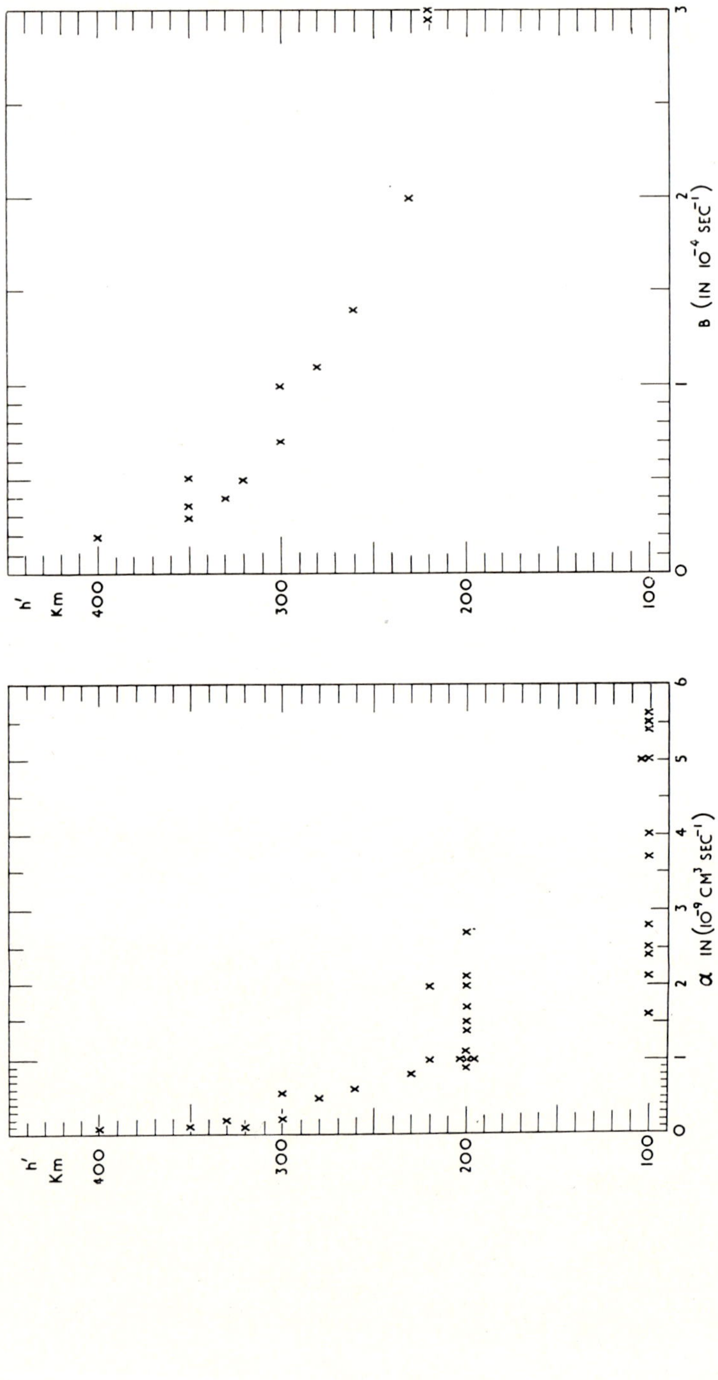
h'
Km
400
300
200
100
0
1
2
3
B (IN 10⁻⁴ SEC⁻¹)
h'
Km
400
300
200
100
0
1
2
3
4
5
6
α IN (10⁻⁹ CM³ SEC⁻¹)

Fig. 3

density in each layer towards zones with lower electron loss; the raising is greater for a smaller vertical density-gradient of the ionizable molecules, i.e., the *F2* layer tends to separate itself from the *F1* layer. If the critical frequency f_0F2 is due to the superposition zone between the *F1* and the "true" *F2*, one understands that the eclipse-effect upon this frequency depends on the vertical movements of the two layers and also on their "degree of superposition," and is stronger in conditions of large superposition, i.e., in winter months. Unfortunately, we are not able to test this hypothesis with the data obtained at Rome during the eclipse of 25 February 1952 (ZANOTELLI, 1953), since the *F1* data are missing due to interferences, and we hence leave the question for further study.

Referring to Fig. 1, we may note three interesting phenomena in the ionospheric eclipse of 30 June 1954 observed at Rome: the existence of secondary minima of critical frequency outside the limits of the optical eclipse in the *E*, *F2*, and perhaps *F1* layers; a marked eclipse-effect upon the sporadic *E* ionization; the appearance of a sub-stratification (*X* in Fig. 1) between the *E* and *F1* layers.

In order to explain the first observation, it is necessary to consider the eclipse of the "ionizing" sun instead of the "visible" sun. The "ionizing" sun, which includes the corona and the corpuscular emissions, is larger than the disk of the "visible" sun, and is not uniform, strong ionizing sources being localized in particular active zones of the solar surface and atmosphere. The secondary minima of electron density observed shortly before and after the optical eclipse may be explained in terms of coronal occultation (WELLS and SHAPLEY, 1946), and the minima observed far outside the limits of the photon eclipse in terms of corpuscular eclipse. The occultation of active zones of solar surface may explain the irregularities sometimes observed in ionospheric characteristics during a solar eclipse. These questions are all of great interest, and it is hence necessary systematically to integrate in the next eclipses the ionospheric data with heliophysical data.

Further to the second observation, we may note that, in the eclipses of 14 February 1934 in Southern Seas (MINOHARA *et al.*, 1934), of 19 June 1936 in Japan (MAEDA and ISAGAWA, 1936), and of 9 July 1945 in Northern Sweden (RYDBECK, 1946), a strong decrease of the sporadic *E* critical frequency has been observed. These facts and some regularities in the seasonal and diurnal behaviour of sporadic *E* lead one to suppose some relation between the formation of sporadic *E* and the activity of the normal layers: for example, by electron photodetachment processes from negative ions (CHATTERJEE, 1953), and/or, as suggested by MARTYN (1951) by the descent of banks of charges from the *F1* layer to the *E* level. The decrease in the sporadic *E* ionization sometimes observed during a solar eclipse may be consequently due to a real eclipse-effect.

The third observation, viz., the appearance of a sub-stratification between the *E* and *F1* layers during the eclipse of 30 June 1954, may be ascribed, perhaps, to the phenomenon suggested by MARTYN, i.e., the formation of the sporadic *E* layer also by descent of charges from the *F1* layer. One sees in Fig. 1 that this short-lived stratification has the form of two close thin layers, one disappearing by descent to *E* level, the other by ascent to *F1* level. Something like this has been observed in other eclipses (LEDIG *et al.*, 1946), viz., the appearance of a new layer between the *F1* and the *F2* layers. This is perhaps explicable as a particular case of partial merging of the two layers; in fact, if the maximum electron density of the superposition zone has a value comprised between the maximum value of the *E1* and "true" *F2* densities, one can observe three critical frequencies in the *F* region.

Concluding Remarks

If one examines the available data on ionospheric eclipses, one is, in general, able to deduce conclusions only of very general character, already known, in part, in other ways, such as: (a) that the principal ionizing agents of the upper atmosphere are electromagnetic radiations from the sun; (b) that the rate of electron removal is proportional to the second power of electron density in the *E* and *F1* layers, to the first power in the *F2* layer. The actual demands of ionospheric physics, on the contrary, require a more stringent investigation of the experimental facts. For example, we, in practice, have no information on the behaviour of the level of maximum electron density of each ionospheric layer during a solar eclipse, and hence we cannot study effects due to hypothetical vertical movements of the ionosphere during an eclipse.

We think that work in the future could be improved in order to establish:

(i) A common standard for obtaining and presenting the ionospheric data regarding a solar eclipse;

(ii) A common method, at least to a first approximation, for the reduction of the experimental *h'f* curves to curves of electron density/geometric height, in order to study at various sites and in comparable conditions the eclipse effects at the heights at which they effectively occur.

(iii) A common method, at least to a first approximation, for the correlation between ionospheric data and some heliophysical magnitudes during a solar eclipse.

References

CHATTERJEE, B. 1953 *J. Atmos. Terr. Phys.*, **3**, 229.

FLEMING, J. A. 1939 *Terrestrial Magnetism and Electricity*, p. 462.

LEDIG, P. G., *et al.* 1946 *Terr. Mag. Atmos. Elect.*, **51**, 411.

MAEDA, K., and ISAGAWA, Y. 1936 *Rep. Rad. Res. Jap.*, **6**, 91.

MARTYN, D. F. 1951 *Mixed Comm. Ionosph. Proc.*, 2nd Mtg., p.44

MINOHARA, T., ITO, Y., and MAEDA, K. 1934 *Rep. Rad. Res. Jap.*, **4**, 51

NAKATA, Y. 1950 *Rep. Ion. Res. Jap.*, **4**, 21.

RYDBECK, O. E. H. 1946 *Trans. Chalmers Univ. Tech.*, No. 53.

WELLS, H. W., and SHAPLEY, A. H. 1946 *Terr. Mag. Atmos. Elect.*, **51**, 401.

ZANOTELLI, G. 1953 *Ann. Geof.*, **6**, 367.

Solar Eclipses and the Ionosphere (Special Suppl. Vol. 6 to J. Atmosph. Terr. Phys.), Pergamon Press, London, 1956

SECTION III

IONOSPHERIC ECLIPSE RESULTS

Report on Recent Japanese Ionospheric Observations during Solar Eclipses

Y. NAKATA AND T. YONEZAWA
Radio Research Laboratories, Kokubunji, Tokyo

Abstract—Results are considered from observations made during the annular eclipse of 9 May 1948 at Wakkanai and the partial eclipses of 12 September 1950 and 14 February 1953 at Tokyo and Akita. The appearance of a new stratification in the *F* layer appears to be a not-unusual phenomenon, although it has often been overlooked in the past. The processes of electron dissipation in the *F1* and "new *F*" layers are considered. During the 1953 eclipse *Es* layer ionization observed at Tokyo was found to show marked parallelism with the intensity of 200 Mc/s solar radio emission.

THE following facts have been revealed by our ionospheric observations made during the annular eclipse of 9 May 1948 at Wakkanai [45·4° N, 141·7° E] (NAKATA, 1950) and the partial eclipses of 12 September 1950 and 14 February 1953 at Tokyo [35·7° N, 139·5° E] (NAKATA, 1954) and at Akita [39·7° N, 140·1° E] (NAKATA, 1955).

(i) Recent observations have made it clear that a new stratification other than the familiar *F1* and *F2* layers appears during a solar eclipse, but it appears that hitherto this has been limited to observations in equatorial regions (LEDIG *et al.*, 1946; ESTRABAUD, 1952; PIGGOTT, 1953; MINNIS, 1955). Our observations, however, showed that this phenomenon could occur at stations in middle latitudes such as Wakkanai and Akita. During the 1948 eclipse at Wakkanai a new layer was observed to split from the *F2* layer immediately after first contact and to descend until, after about 15 min, it was obscured by the lower *F1* layer. This layer reappeared again for about 30 min before the last contact (Fig. 1). A similar behaviour of a new stratification was evident during the 1953 eclipse at Akita, but in this case the new layer could be observed for the greater part of the eclipse. It must be emphasized that we can detect, on closer examination, the appearance of the new layer on some of the ionospheric records taken during earlier solar eclipses, though this seems to have been overlooked hitherto. Conventional analyses of earlier eclipse data may thus not always be reliable, because in some cases this new stratification was erroneously taken as the *F2* (or *F1*) layer. In future, in analysis of eclipse data we must be very cautious of this point.

(ii) A simple method has been developed to distinguish between the two types of electron disappearance, i.e., the recombination type and the attachment type. It is concluded from the analysis of eclipse data that electrons in the *F1* layer are removed by a recombination-type process, the rate coefficient being about 4×10^{-9} cm^3 sec^{-1}, while disappearance of electrons in the new stratification seems to be effected by an attachment-type process. With regard to the *F2* layer, definite conclusions cannot be obtained owing to its very complicated behaviour.

(iii) The sporadic E-layer ionization observed at Tokyo during the 1953

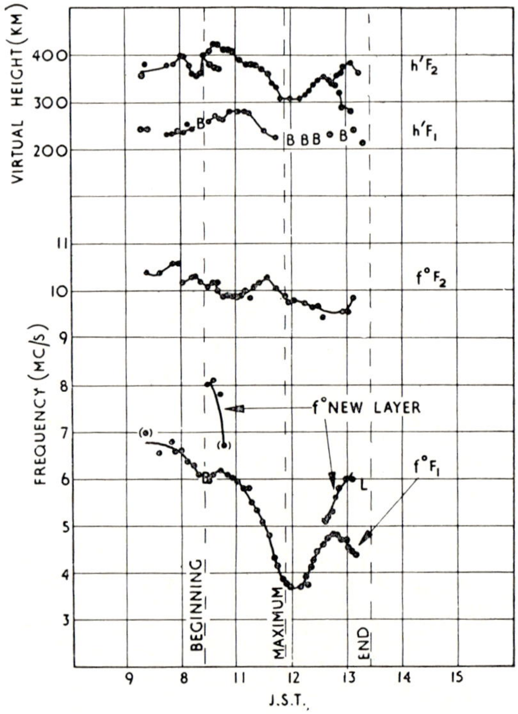

Fig. 1

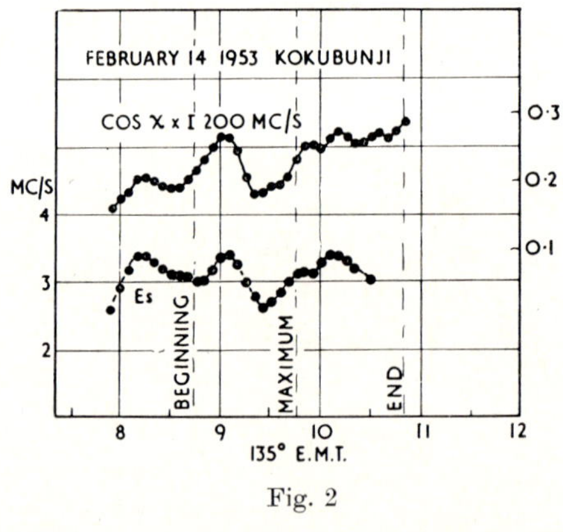

Fig. 2

eclipse was found to vary in marked parallelism with the intensity of 200 Mc/s solar radio emission (observed at Hiraiso, 36·4° N, 140·6° E) multiplied by the cosine of the solar zenith distance (Fig. 2). Both quantities reached a maximum at about 15 min after the beginning of the eclipse and fell to a minimum at about

20 min before the maximum of the eclipse. In particular, it may be noted that a general decrease in *fEs* and solar-radio noise began simultaneously at about 45 min before the eclipse. These facts might be taken as indicating that the 200 Mc/s solar radio emission is in some way related to the radiation responsible for the production of the sporadic-*E* layer and the solar effective radius for these radiations is larger than that of the visible disk by about 45 per cent.

References

Estrabaud, S. 1952 *Compt. Rend.*, **235,** 1521.

Ledig, P. G., Jones, M. W., Giesecke, A. A., and Chernosky, E. T. 1946 *Terr. Mag. Atmos. Elec.*, **51,** 411.

Minnis, C. M. 1955 *J. Atmosph. Terr. Phys.*, **6,** 19.

Nakata, Y. 1950 *Rep. Ionosphere Res. Japan*, **4,** 21.

1954 *J. Radio Res. Labor.*, **1,** No. 2.

1955 *Proc. Semi-Ann. Meet. Radio Res. Labor.*, No. 8 (Preliminary report in Japanese).

Piggott, W. R. 1953 *Proc. 3rd Meet. Mixed Comm. Ionosphere* (Canberra) p. 42.

Solar Eclipses and the Ionosphere (Special Suppl. Vol. 6 to J. Atmosph. Terr. Phys.), Pergamon Press, London, 1956

The Ionosphere above Lindau during the Solar Eclipse on 30 June 1954

W. Becker

Institut für Ionosphärenforschung in der Max-Planck-Gesellschaft, Lindau/Harz

Abstract—Due to the high transmitting power of the routine sweep-frequency equipment used (50 kW), the ordinary *E1* critical frequencies of the *E1* layer could be determined up to the maximum of the solar eclipse in spite of sometimes very intense *Es* echoes. They clearly demonstrate a drop of the f^0E1 values before the optical eclipse in the *E1* layer. The sporadic *Es* ionization and dynamical processes within the *E1* layer allowed no deduction of any recombination coefficients. Stormlike phenomena within the *F* layer forbade a deduction of the *F1* or *F2* recombination coefficients here too.

The observations at Lindau (51°39′ N, 10°7′30″ E) were carried out with the routine sweep-frequency equipment. The characteristics were:

Peak pulse power 50 kW; frequency range 1–16 Mc/s logarithmically divided into eight steps; sweep-time for each step 58 sec; two rhombic aerial systems both for transmitting and receiving.

According to U.R.S.I. conventions, records were taken every ten minutes from sunrise to sunset, beginning with 25 June. The last day of this observational period was 5 July.

Fig. 1 shows the ordinary critical frequencies of the *E1*, *F1*, and *F2* layers for 30 June (full lines), and for comparison the median values of the other ten days, the control days (dotted lines). The two thin lines enveloping the dotted f^0E1 curve represent the highest and the lowest value ever observed during the control days at the respective time. Shortly before the eclipse maximum the f^0E1 curve is broken; this means that the respective values are a bit uncertain because of partial *Es* blanketing. h' denotes the minimum virtual heights of the different layers, *A* means no value because of *Es* blanketing, *K* indicates stormlike phenomena, *p* measures the rate of obscuration of the sun's disk by the moon, and *i* the distance from the sun's limb to the moon's limb in units of the visible sun's diameter for a height of 100 km above Lindau. In order to demonstrate the reliability of the deduced data, Figs. 2, 3, and 4 are given; they represent some of the records taken.

It may be mentioned that always, also in the low-frequency band, $2 \times E1$ reflections have been observed. This was due to the high peak-pulse power of the equipment. Thus it was possible to read the *E1* critical frequencies at cusps (marked by arrows in Figs. 2 and 3), though these could not always be detected amongst the first-order *E1* and frequently very intense *Es* reflections.

It is striking that f^0E1 drops before the onset of the optical solar eclipse in the *E1* layer. At this time the f^0E1 values in Fig. 1 are well below the variation-range of f^0E1 of the ten control days. But it seems to be impossible to fix the beginning of the drop exactly, and so also the effective solar diameter comprising the radiation of the corona, by making use of the *i* values. This is because some unknown dynamical and temporary *Es* ionization processes had apparently influenced the *E1* layer at that time. Evidence is given in Fig. 2. Thus record 1207^h shows a slight stratification within the *E1* layer; the same cannot be detected in the records from 1217–1237^h, but it is seen again in the records from 1247–1317^h MET.

More or less intense *Es* echoes can be seen on every record. For the same reasons

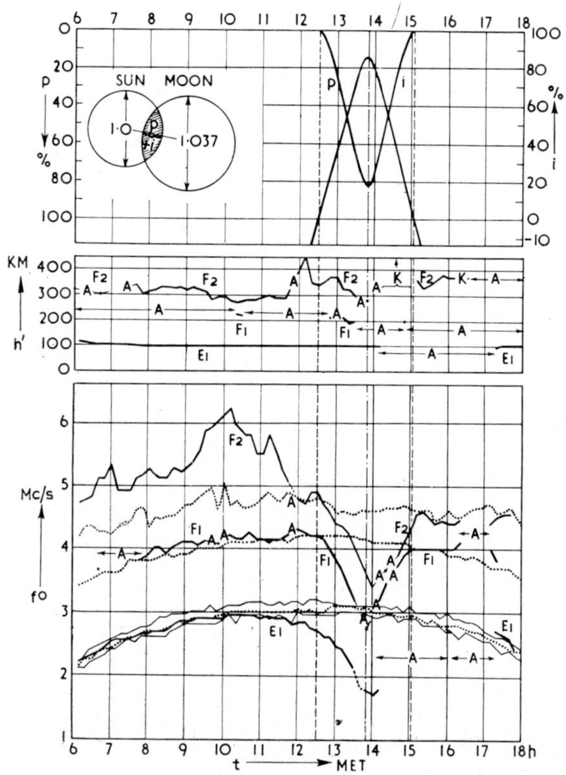

Fig. 1. f^0 means the ordinary critical frequencies and h' the virtual height of the *E1*, *F1*, and *F2* layers during the solar eclipse on 30 June, 1954, represented by full thick curves; the median values of the ten control days are shown by dotted curves. The full thin curves represent the highest and lowest f^0E1 values of the control days; p measures the percentage of the obscured sun's disk at a height of 100 km above Lindau; i is expressed in percentage of the visible solar diameter.

we do not try here to deduce any recombination coefficient of the *E1* layer for the first part of the eclipse.

The uncertainty of the f^0E1 values before and shortly after the eclipse maximum (Fig. 3) prevents us from attempting a determination of the *E1* recombination coefficient here too.

Stormlike phenomena within the *F2* region (Fig. 4) forbid the same attempt being made for the *F1* or *F2* layer.

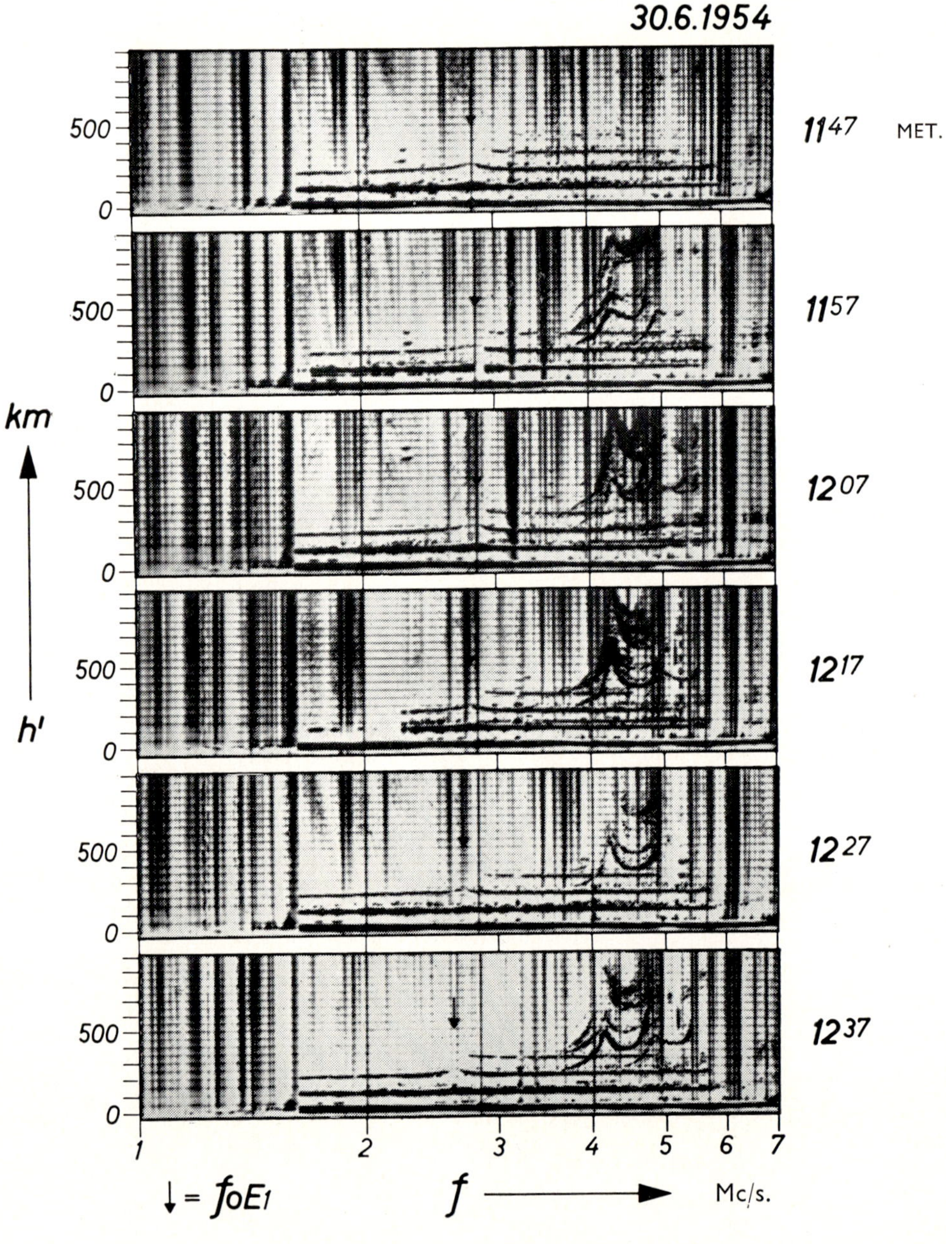

Fig. 2. Sweep-frequency records taken at Lindau around the beginning of the solar eclipse on 30 June, 1954.

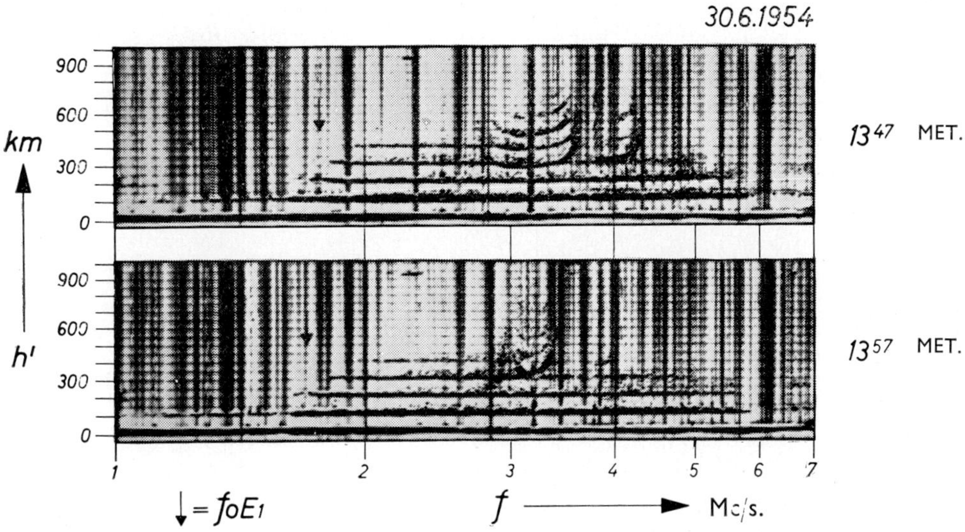

Fig. 3. Sweep-frequency records taken at Lindau at solar eclipse maximum on 30 June, 1954.

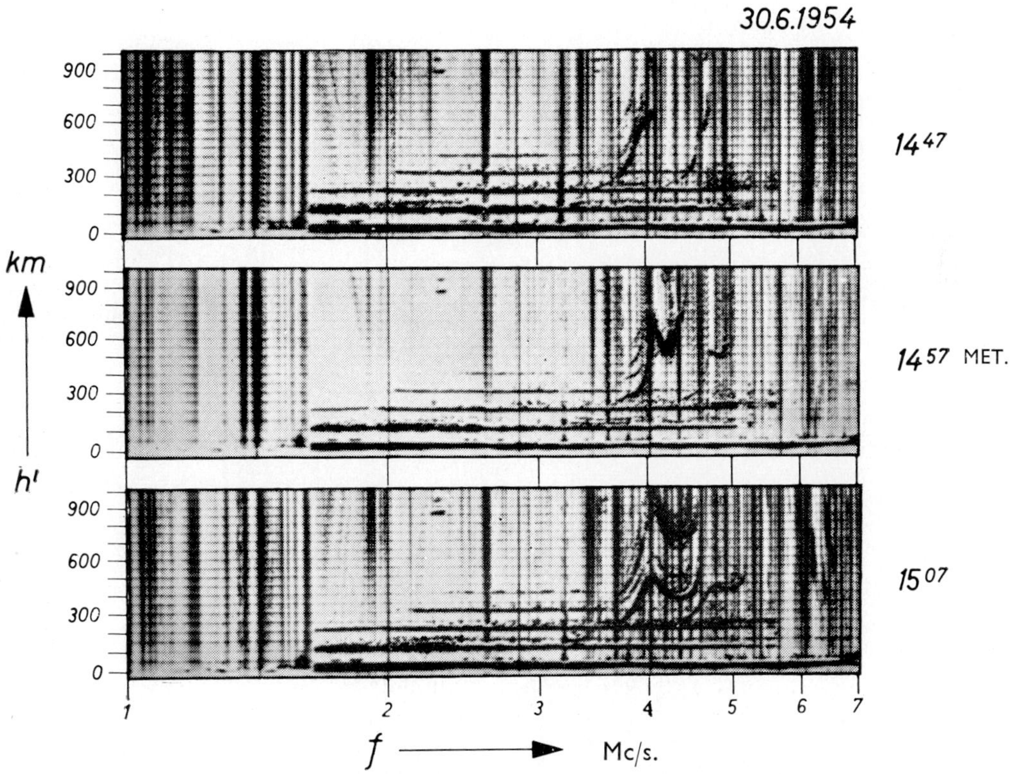

Fig. 4. Sweep-frequency records taken at Lindau showing stormlike phenomena in the *F2* layer after solar eclipse maximum.

Solar Eclipses and the Ionosphere (Special Suppl. Vol. 6 to J. Atmosph. Terr. Phys.), Pergamon Press, London, 1956

Ionospheric Observations in Canada during the Solar Eclipse of 30 June 1954

K. DAVIES

Radio Physics Laboratory, Defence Research Telecommunications Establishment, Ottawa.

Abstract—During the solar eclipse of 30 June 1954, height/frequency ionosphere sweeps were recorded at six Canadian stations. Three of these stations lay in the auroral zone, and so a special study has been made to ascertain the influence of a corpuscular eclipse. There is some evidence for such an effect occurring some three to four hours before the visible eclipse. At stations outside the auroral zone, normal eclipse effects were observed in the *F2*, *F1*, and *E* layers.

Ionospheric absorption measurements on 2·0 Mc/s measured at Fort Chimo indicate a definite effect due to solar ultra-violet radiation, and there is also some evidence for a corpuscular effect some four hours before the visible eclipse.

1. INTRODUCTION

DURING the solar eclipse of 30 June 1954 a special programme of ionospheric sounding measurements was carried out at six Canadian ionosphere stations. Three of these stations are situated in the so-called auroral zone, and the others lie to the south of this zone. The three auroral-zone stations are Baker Lake, Churchill, and Fort Chimo, and the remaining three stations are St. John's, Ottawa, and Winnipeg. The locations of these six stations with regard to the auroral zone are shown in Fig. 1.

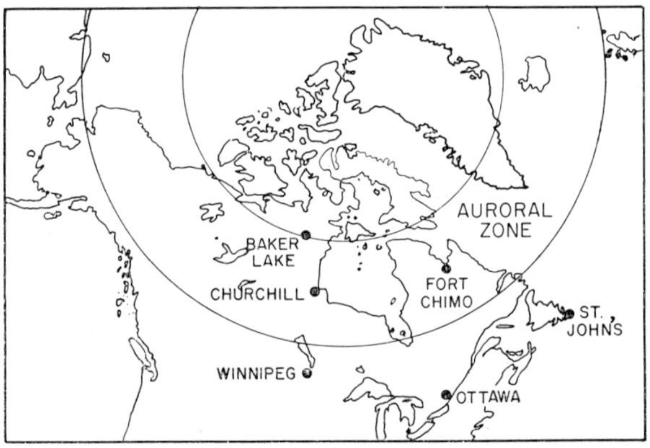

FIG. 1. Locations of ionospheric stations with respect to the auroral zone.

Special attention was paid to the observations made in the auroral zone to determine the effect, if any, of a corpuscular eclipse. The auroral regions should be particularly suitable for such an investigation, as it is believed that the Aurora Borealis is caused by streams of electrically-charged solar corpuscles which are deflected to the auroral zone by the dynamo action exerted by the earth's magnetic field. CHAPMAN (1932) has shown that, for a stream of neutral solar corpuscles, the corpuscular eclipse should occur about two hours before the optical eclipse. Furthermore, the central line of the corpuscular eclipse track would be about two thousand miles to the east of the optical eclipse track. No such calculations have been made with regard to a charged corpuscular eclipse. The paths of totality of the optical eclipse at different altitudes are shown in Fig. 2.

In addition to studying the effect of the eclipse on the *E*, *F1*, and *F2* layers by conventional *h'f* techniques, ionospheric absorption measurements were made at Fort Chimo to investigate the behaviour of the *D* region. For the sake of clarity, the critical frequency data for the *E*, *F1*, and *F2* layers will be considered separately.

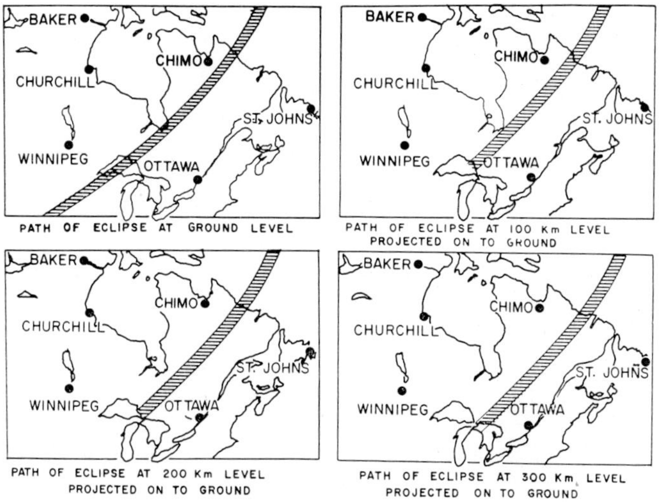

FIG. 2. Path of totality of the eclipse of 30 June 1954 across Canada at four levels.

2. Equipment

The equipments used for recording the variation of equivalent height with frequency were either type LG17 or type C2 automatic ionosphere recorders. The performance characteristics of these two types of equipment are similar. Their peak output powers are 10 kW, contained in pulses of 50μsec duration. The pulse repetition rate is 30 per sec. For the eclipse observations, 100-km height markers and frequency markers every 1·0 Mc/s were used. Vertical delta antenna systems, terminated with 600-ohm resistors, were used throughout.

At the three southern stations the critical frequencies were estimated to the nearest 0·05 Mc/s, but accuracies greater than about 0·2 Mc/s were not possible under the ionospheric conditions which exist in the auroral zone.

Because of the rapidly varying characteristics of the ionosphere in high latitudes, fast-frequency sweeping is desirable. Hence, sweep speeds of 15 sec (C2) and 16 sec (LG17) were used. On the ten control days (five days before and five days after 30 June 1954) sweeps were taken once every 5 min, from three hours before the start to two hours after the end of the optical eclipse. On the day of the eclipse continuous sweeping was employed over the same period. Sweep records were obtained photographically on 35-mm films, which were used for detailed study, and panoramic displays were photographed on 16-mm film. The latter were studied with the aid of a motion-picture projector.

3. DATA FOR ST. JOHN'S, OTTAWA, AND WINNIPEG

For the sake of convenience, the data for the three stations outside the auroral zone are considered together. The values of the critical frequencies of the ordinary E-layer trace (f_0E), the $F1$ trace (f_0F1), and the $F2$ trace (f_0F2) on the day of the

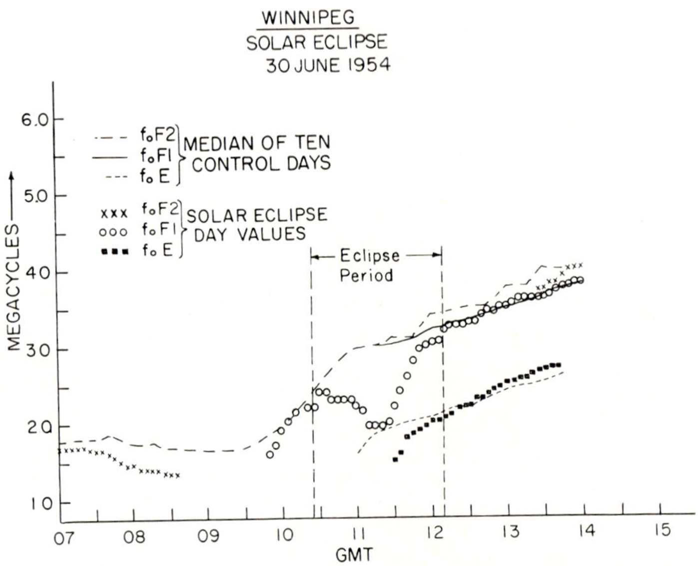

FIG. 3. Variation of critical frequencies at Winnipeg during the solar eclipse of 30 June 1954.

eclipse are plotted in Figs. 3, 4, and 5, together with the median values of these characteristics for the ten control days. Some relevant data on the optical eclipses at these stations are presented in Table 1.

Table 1

Station	Eclipse		Maximum obscuration per cent			Expected decrease in N_{max} per cent			Observed decrease in N_{max}			Sunrise G.M.T.
	Begins	Ends	0 km	100 km	200 km	E	$F1$	$F2$	E	$F1$	$F2$	
St. John's	1014	1224	72	74	76	49	51	52	40	49	37	0808
Ottawa	1010	1206	86	88	90	65	68	69		62	37	0918
Winnipeg	1024	1210	89	86	84	63	60	58		58		0958
Baker Lake	1043	1228	69	67	65	43	41	40	30			0727
Churchill	1034	1221	77	75	73	50	48	47				0900
Fort Chimo	1026	1228	95	94	92	75	72	70				0810

3.1. *The Eclipse in the* E *layer*

The eclipse at Winnipeg (see Fig. 3) started so shortly after local ground sunrise that the E layer was not observed until after the eclipse had reached its maximum. The observed effect was, therefore, merely to delay the appearance of the E layer

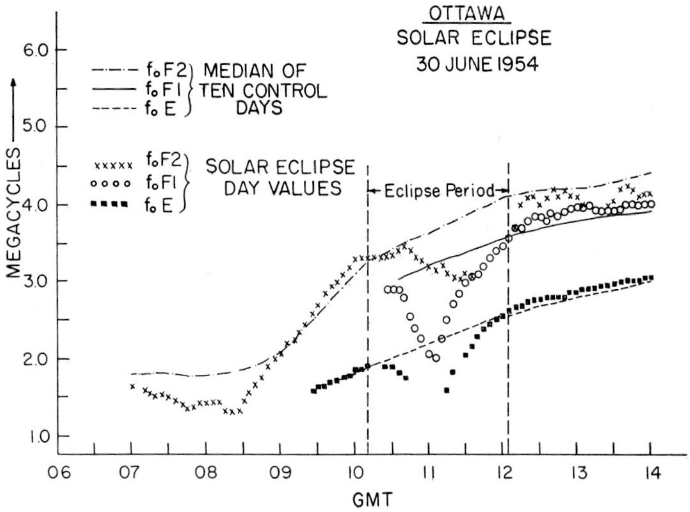

FIG. 4. Variation of critical frequencies at Ottawa during the solar eclipse of 30 June 1954.

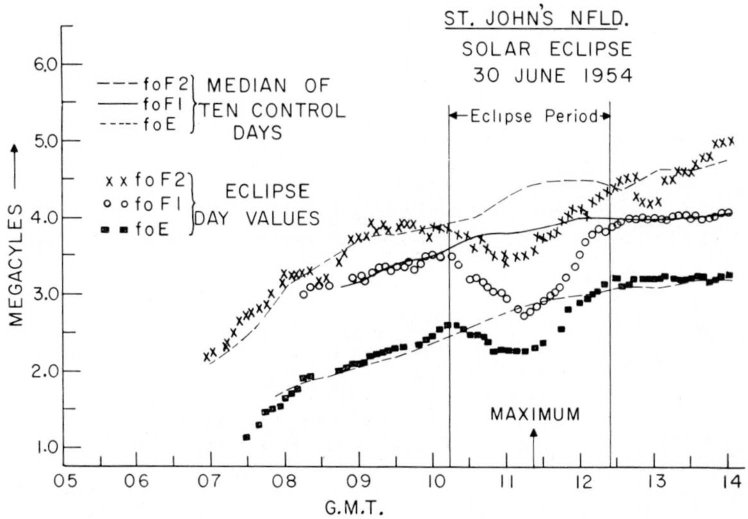

FIG. 5. Variation of critical frequencies at St. John's during the solar eclipse of 30 June 1954.

by about 30 min. It should be noted that, due to interference caused by broadcasting stations, the lower-frequency limit of the equipment was effectively 1·5 Mc/s.

The minimum value of f_0E was obscured at Ottawa also, but there is no doubt as to the effect produced by the eclipse.

At St. John's the eclipse started about two hours after local ground sunrise, so that the E layer was well developed at that time. The departures of f_0E values from the control day median values are plotted in Fig. 6. It will be noticed that the minimum value of f_0E was attained about ten minutes before maximum obscuration. It may be relevant to note that, at all three stations, after the end of the eclipse f_0E attained a somewhat higher value than the median for the control days. Similar phenomena have been observed by other workers (MINNIS, 1955; MCLEISH, 1948).

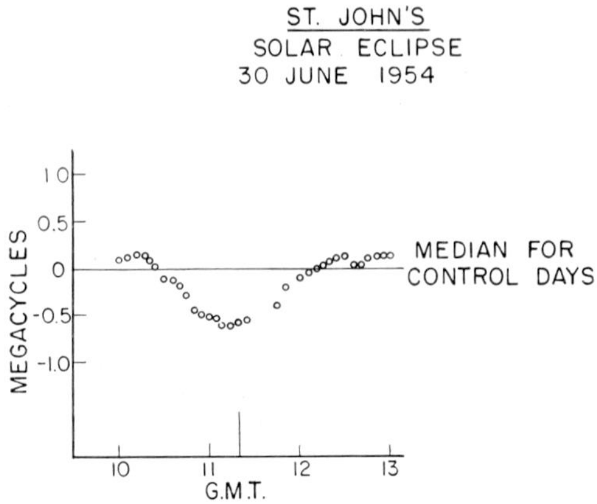

FIG. 6. Departures of f_0E from median values for control days.

3.2. *The Eclipse in the* F1 *Layer*

The effects of the eclipse on the maximum frequencies reflected from the *F1* layer show this layer to be the most sensitive to the decrease in solar ionizing radiation. Indeed, the decreases in N_{max} are nearly equal to the values calculated on the assumptions of a uniformly radiating solar disk and a simple recombination law of electron dissipation, i.e.,

$$\frac{dN}{dt} = q - \alpha N^2 \tag{1}$$

where N = electron density, α = recombination coefficient, and q = rate of production of electrons.

The times of minimum f_0F1 are almost coincident with maximum obscurations. It is evident that, at Ottawa, the *F1* layer had not assumed a separate identity until after first contact. At Winnipeg, the night *F* layer (generally identified with the *F2* layer) actually develops into the day *F1* layer near dawn, and the *F2* layer appeared later in the day (at about 1330 G.M.T.). The latter phenomenon is quite common over Canada.

3.3. *The* F2 *Layer*

Figs. 4 and 5 show that the eclipse has a definite effect on the *F2* layer. It will be seen from Table 1 that the decrease in N_{max} is less than would be expected if a simple recombination law of electron dissipation were operative. At St. John's, f_0F2 appears to have reached its minimum value a few minutes before maximum obscuration, whereas at Ottawa the minimum was not reached until about 30 min after maximum obscuration.

4. Data for Baker Lake, Churchill, and Fort Chimo

Ionospheric sweeps in the neighbourhood of the auroral zone are very complex, due to the presence of "spread" echoes and rapidly moving clouds of sporadic *E* ionization. Thus any eclipse effects which may be produced in the ionosphere

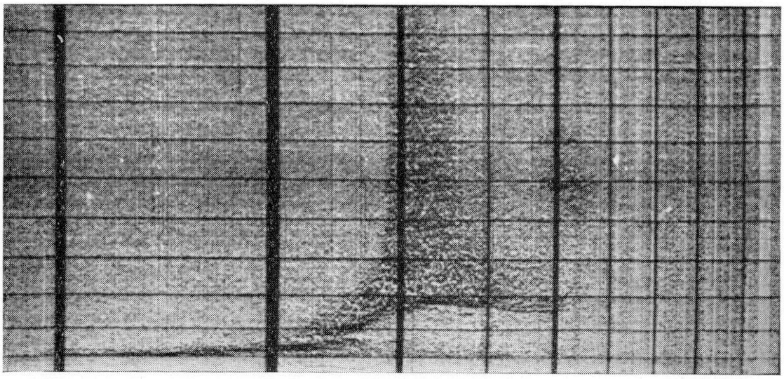

Fig. 7. Typical ionospheric sweep at Churchill, 1144 G.M.T., 30 June 1954.

are likely to be obscured. Nevertheless, an attempt has been made to scale the *h′f* curves in conventional terms, and the results are presented below.

4.1. *Effects in the* E *Layer*

The *E* layer in the auroral zone differs greatly from the *E* layer elsewhere, in that it can exist when the sun is absent. This layer is sometimes called "nite *E*."

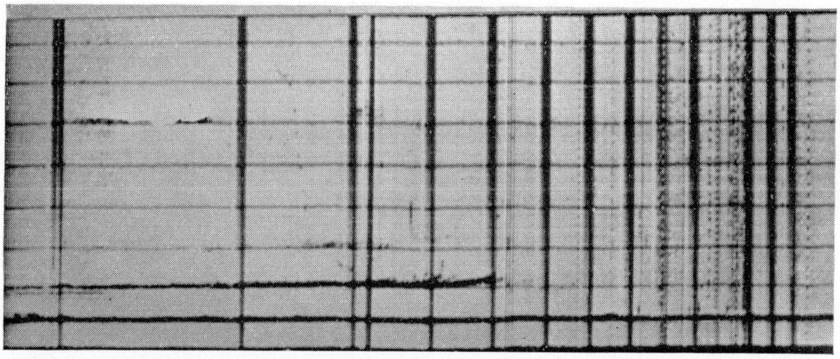

Fig. 8. Typical ionospheric sweep at Fort Chimo, 1141 G.M.T., 30 June 1954.

The nite *E* layer differs from what is usually termed sporadic or nocturnal *E* by the possession of the following characteristics:

(i) it is not transparent,
(ii) it exhibits retardation near the critical frequency,
(iii) it causes retardation of echoes reflected from the *F* region,
(iv) it merges into the normal daytime *E* layer when the latter is formed.

Typical sweeps taken at Churchill and at Fort Chimo during the eclipse are shown in Figs. 7 and 8. It is important to appreciate the difficulties involved in the interpretation of ionospheric records obtained in and around the auroral zone. The existence of different types of sporadic E makes the scaling of critical frequencies rather uncertain. Nevertheless, it is possible to define traces (as has been done above) and follow the course of given characteristics with time.

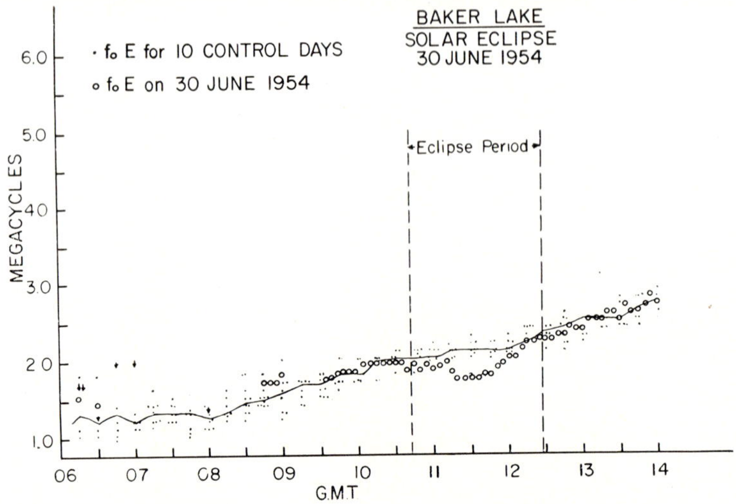

FIG. 9. Variation of f_0E at Baker Lake during the solar eclipse of 30 June 1954.

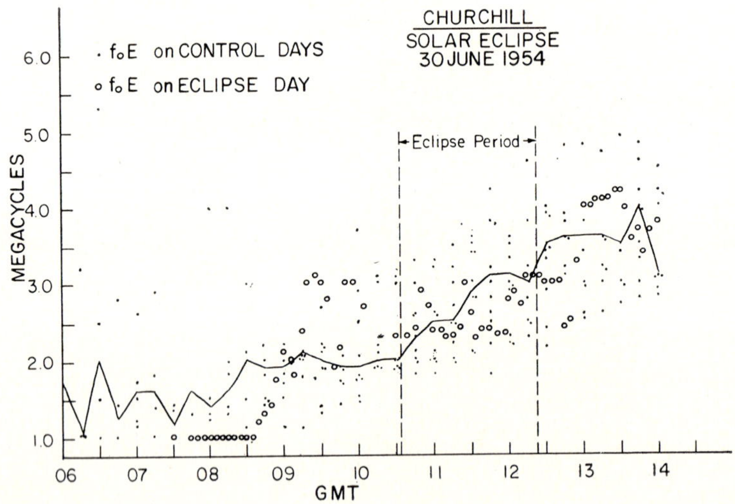

FIG. 10. Variation of f_0E at Churchill during the solar eclipse of 30 June 1954.

The E-layer critical frequencies thus obtained are plotted in Figs. 9, 10, and 11 for Baker Lake, Churchill, and Fort Chimo respectively. The eclipse at Baker Lake (see Fig. 9) began 3 h 16 m after sunrise, so that a considerable fraction of the electron density in the E layer would be due to ultra-violet radiation. Comparison of f_0E during the eclipse with the individual control-day values show that there is a definite optical eclipse effect with a minimum of f_0E near maximum obscuration. The sudden appearance of the E layer at about 0840 G.M.T. may be significant, especially as f_0E both at Churchill and at Fort Chimo increased rapidly at about

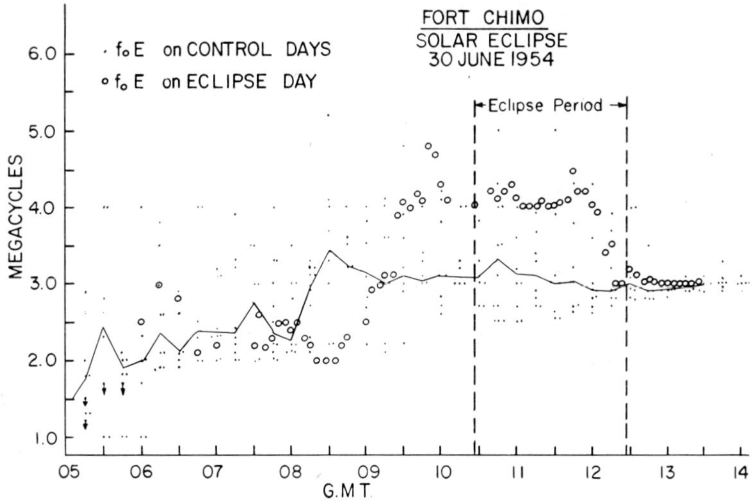

FIG. 11. Variation of f_0E at Fort Chimo during the solar eclipse of 30 June 1954.

this time. This phenomenon may well be brought about at the end of a corpuscular eclipse, but the times of fourth contacts at the three stations were so close that this conclusion is uncertain. There is little evidence of optical eclipse effects at Churchill and at Fort Chimo.

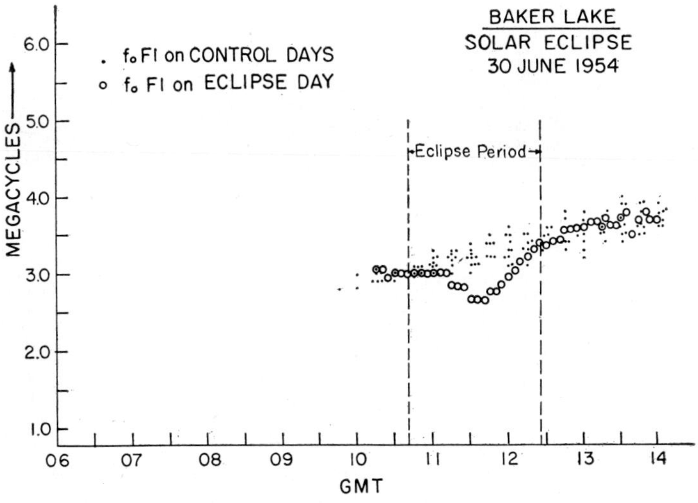

FIG. 12. Variation of f_0F1 at Baker Lake during the solar eclipse of 30 June 1954.

4.2. *The* F1 *Layer*

Values of f_0F1 during the eclipse, together with available values for the control days, are shown in Figs. 12 and 13 for Baker Lake and Churchill respectively.

The data for the control days are limited by the blanketing effect of the E layers. The effect of the optical eclipse at Baker Lake is quite pronounced, but blanketing by the E layer at Churchill prevented observations of f_0F1 being made at the start of the visible eclipse. The $F1$ layer was not observed at Fort Chimo during the eclipse.

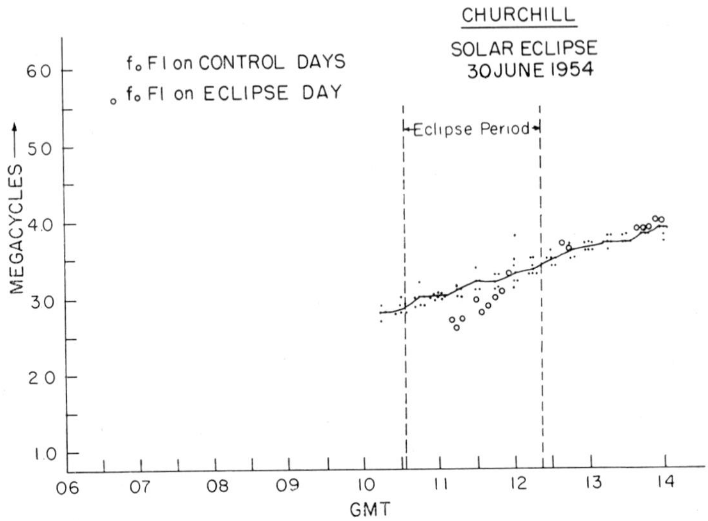

FIG. 13. Variation of f_0F1 at Churchill during the solar eclipse of 30 June 1954.

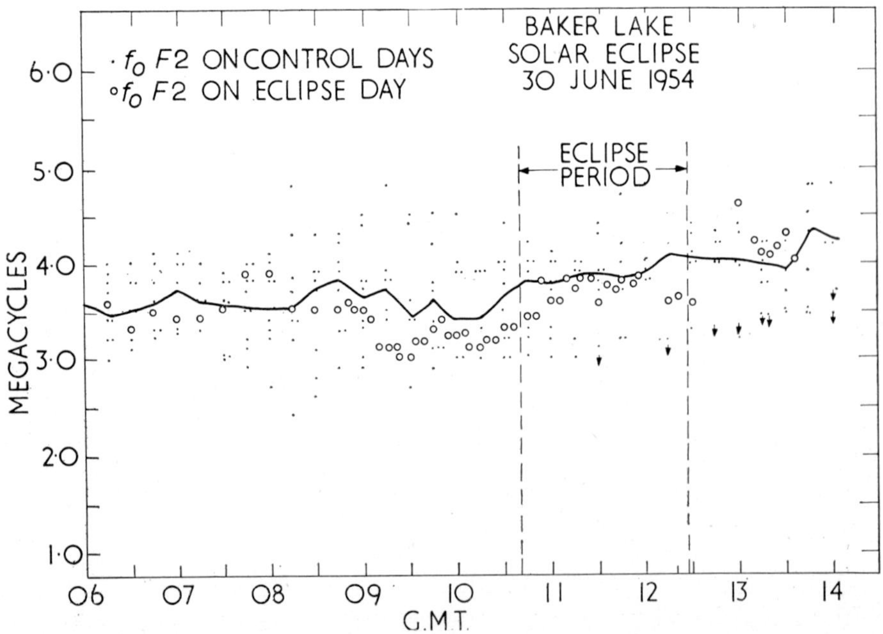

FIG. 14. Variation of f_0F2 at Baker Lake during the solar eclipse of 30 June 1954.

4.3. *The* F2 *Layer*

Considerable difficulties were experienced in scaling f_0F2, especially at Churchill, because of the spread of the echoes (see Fig. 7), and the $F2$ layer at Fort Chimo was not observed at all during the eclipse because of blanketing by the lower layers. The values of f_0F2 obtained at Baker Lake (Fig. 14) and at Churchill (Fig. 15) cannot be said to show significant eclipse effects.

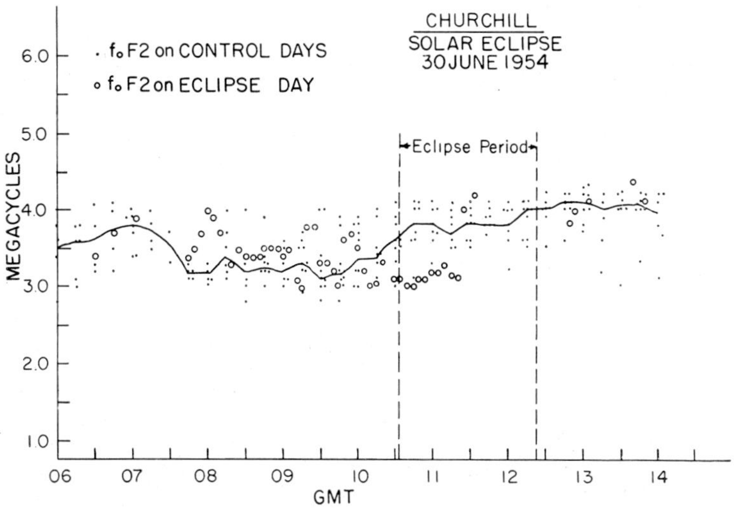

FIG. 15. Variation of f_0F2 at Churchill during the solar eclipse of 30 June 1954.

5. Ionospheric Absorption Measurements at Fort Chimo

5.1. *Introduction*

There is some evidence to show that corpuscular ionizing agencies bring about increased absorption of radio waves in the D region (Cox and Davies, 1954).

As part of the Canadian eclipse programme, it was decided to study the influence of the solar eclipse on the D region in the auroral zone. Fort Chimo was chosen for this study for the following reasons:

(i) it is located in a region of maximum auroral intensity,

(ii) the percentage obscuration of the sun due to the eclipse was greater at Fort Chimo than at any other Canadian ionosphere station.

5.2. *Equipment*

The equipment consisted of a pulse-sender of approximately 1 kW peak output power, together with a commercial receiver (suitably modified for pulse reception) with a gain control in the intermediate frequency amplifier. Auxiliary display equipment was included. The repetition frequency was 60 per sec. The antennae were two quarter-wave horizontal dipoles cut for a frequency of 2·0 Mc/s and orientated at right angles to each other.

5.3. *Method of Measurement*

To study the absorption of radio waves, it was decided to measure "the apparent reflection coefficient" ρ. This quantity is defined in terms of the relative amplitudes of the once, twice, thrice, etc., reflected signals (MITRA, 1952).

Let A_1 = the relative amplitude of the once-reflected signal,
A_2 = the relative amplitude of the twice-reflected signal,

and so on for A_3, etc.
Then

$$\rho = \frac{2A_2}{A_1} = \frac{3A_3}{2A_2}, \quad \text{etc.} \tag{2}$$

It is possible to obtain ρ from the relative amplitude A_1 of the once-reflected signal and its virtual height of reflection h using the expression:

$$\rho = BA_1h \tag{3}$$

where B is a quantity, characteristic of the equipment, which can be determined empirically from equations (1) and (2) when multiple echoes are present. Since, for most of the time, only the first reflected signal is large enough to measure with any degree of accuracy, ρ was generally calculated from equation (2), although on every possible occasion A_2 and A_3 were also measured.

The total ionospheric absorption, in decibels, is $-20 \log_{10} \rho$.

The relative amplitudes (A_1, A_2, A_3, etc.) were obtained by measuring visually the gain required to keep the output of the receiver constant. Observations were made from 0730 to 1430 G.M.T. on each day from 26 June to 5 July 1954, inclusive. Measurements were made every 10 sec and, in order to reduce the errors due to fading, these measurements were averaged over intervals of 10 min. To estimate the degree of accuracy of the average of such a sample, sets of measurements were carried out in which 400 amplitude measurements, each separated by a 10-sec interval, were made near noon. This period was chosen since, on the average, it was expected that during this interval the ionospheric absorption remained fairly constant. It is assumed that during the period of the 400 measurements the true ionospheric absorption remained sensibly constant, and that the mean value of the 400 measurements furnished the correct value of absorption.

To determine the average error of a sample of n consecutive amplitude measurements, the whole group of 400 measurements was divided into samples of sizes 1, 5, 10, 20, 40, 80, 160, and 320 consecutive values. For each sample the deviation of the mean from the grand mean (of 400 values) was determined. In this way a series of running averages of amplitude measurements was obtained for each group, and the standard deviation (i.e., the r.m.s. value) of these averages, expressed as a percentage of the grand mean, was thus obtained. This operation was performed on all the above samples, and for the following days: June 22, 23, 24, 25, 26, 30, July 3. The variation of the percentage standard deviation with the size of the sample is shown in Fig. 16. This shows that the mean of a sample of fifty measurements is likely to be in error by about 30 per cent. It is of some interest to compare this result with the sampling measurements of BEYNON and DAVIES (1954). These workers found that, in Britain, the mean of a sample of fifty successive

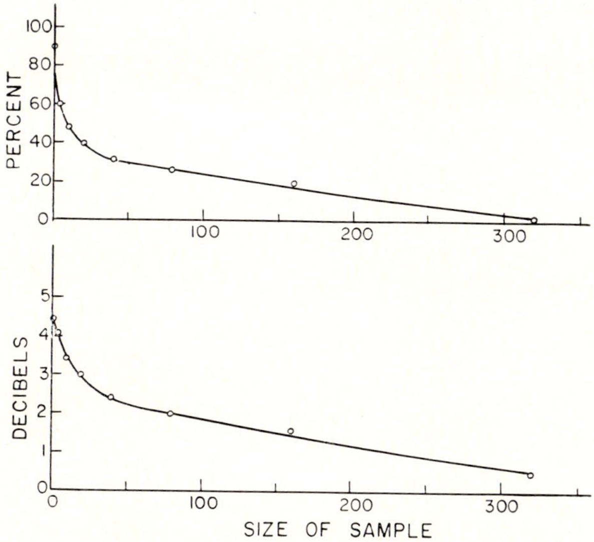

FIG. 16. Sampling of amplitude measurements. Variation in percentage standard deviation with size of sample. Measurements taken at 10-sec intervals.

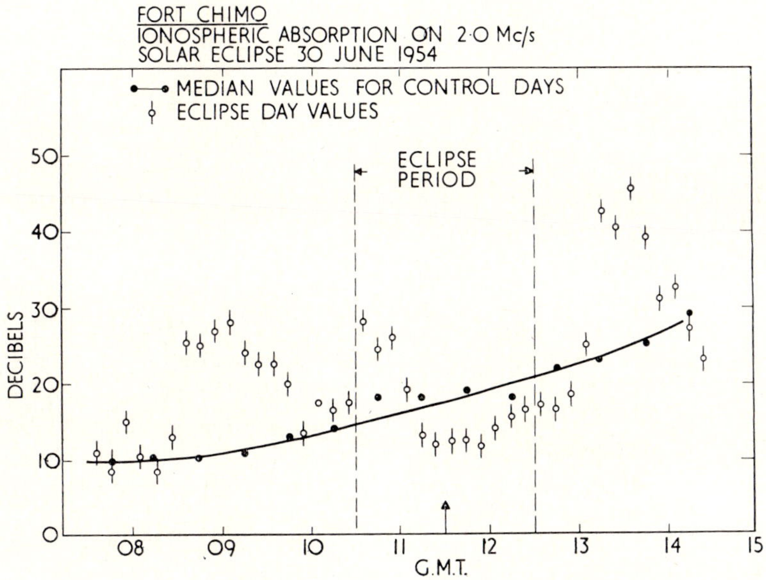

FIG. 17. Variation of ionospheric absorption on 2·0 Mc/s at Fort Chimo during the solar eclipse of 30 June 1954.

amplitude measurements taken at intervals of 10 sec is likely to be in error by about 20 per cent. This difference between the accuracies of measurements at the two places is probably due to the greater complexity of the ionosphere in the auroral zone.

5.4. *Eclipse Data*

The absorption results during the eclipse are plotted in Fig. 17. The following features may be worth noting:

(i) the large increase in absorption (about 15 db) near 0830,

(ii) the decrease in absorption during the visible eclipse (near 1100),

(iii) the minimum of absorption is attained about 20 min after the maximum obscuration.

The increase in $-\log_{10} \rho$ near 0830 starts at about the same time as the increase in f_0E (see Fig. 11). This increase may be due to corpuscular eclipse effects taking place some $3\frac{1}{2}$ to 4 h before the visible eclipse. However, this conclusion is rather uncertain in view of the fact that the variations of $-\log_{10} \rho$ on the control days are of the same order of magnitude.

There is no doubt that the visible eclipse had a marked effect on the ionospheric absorption. It is difficult to determine the percentage decrease in absorption as the general level of absorption, before and after the eclipse, appears to be considerably greater than on the control days. The relaxation time of 20 min is close to that obtained from studies of the diurnal variation of absorption (Davies and Hagg, 1955).

6. Discussion

In general, the observations at the stations outside the auroral zone are in agreement with those of other workers. The effect of the eclipse is most pronounced in the *F1* layer, where the changes of ionization are practically in phase with the eclipse, and is least in the *F2* layer.

The only unusual feature of the observations on the *E* layer (outside the auroral zone) is the fact that at St. John's f_0E attained its minimum value a few minutes before maximum obscuration. An explanation of this, on the basis of non-uniformity in the distribution of active centres on the solar disk, seems unlikely, as the only solar activity at the time of the eclipse was some slight coronal activity near the west limb. Since, however, a similar phenomenon was not observed in the *F1* layer at St John's, too much importance should not be attached to this observation.

The observations on the *F1* layer appear to be in agreement with those of most other workers, and indicate that it behaves as a Chapman-type layer.

The behaviour of the *F2* layer is rather more complex. Thus, at St. John's the minimum value of f_0F2 appears to have occurred shortly before maximum obscuration, due allowance being made for the diurnal variation. On the other hand, at Ottawa the phase-lag was of the order of 30 min. Such wide variations in the phase-lag in the *F2* layer have been observed on many occasions.

The eclipse observations in the auroral zone can be summarized by saying that the effect of the optical eclipse was limited to Baker Lake (*E* and *F1*) and Churchill (*F1*). Critical frequency and absorption data suggest the possibility of a corpuscular eclipse effect which took place some 4 h before the optical eclipse. It may be relevant to note that, on 30 June 1954, f_0F2 at Ottawa showed a somewhat similar effect near 0830 G.M.T.

There was a definite decrease in ionospheric absorption during the optical eclipse, with a phase-lag comparable in value with that observed from studies of the diurnal variation.

7. Acknowledgements

The author wishes to express his thanks to the Defence Research Board for permission to publish this work, and the Canadian Department of Transport for carrying out the experimental programme. The measurements were made with the assistance of Mr. E. E. Stevens, and the scaling of the records was done with the assistance of Mr. E. L. Hagg, both of the Radio Physics Laboratory.

References

Beynon, W. J. G., and Davies, K. 1954 *J. Atmosph. Terr. Phys.*, **5**, 273.
Chapman, S. 1932 *Monthly Not. Roy. Astron. Soc.*, **92**, 413.
Cox, J. W., and Davies, K. 1954 *Canad. J. Phys.*, **32**, 743.
Davies, K., and Hagg, E. L. 1955 *J. Atmosph. Terr. Phys.*, **6**, 18.
McLeish, C. W. 1948 *Canad. J. Res.*, **A26**, 137.
Minnis, C. M. 1955 *J. Atmosph. Terr. Phys.*, **6**, 91.
Mitra, S. K. 1952 *The Upper Atmosphere*, 2nd edition, p. 249.

Solar Eclipses and the Ionosphere (Special Suppl. Vol. 6 to J. Atmosph. Terr. Phys.), Pergamon Press, London, 1956

Ionospheric Measurements in Norway during the Total Solar Eclipse of 30 June 1954

B. LANDMARK

Norwegian Defence Research Establishment, Kjeller

Abstract—Continuous $h'f$ recordings and absorption measurements were made at various Norwegian stations during the total eclipse of 30 June 1954, a time of extremely low sunspot activity. It is found that no good fit is possible between experimental and theoretical results for any of the stations for the E or $F1$ layers on the assumption of uniform radiation from the solar disk. A solar model in which about 70 per cent of the ionizing agency is uniformly distributed over the disk, with about 15 per cent limited to the western limb, 5 per cent to the eastern limb, and about 10 per cent spread over the central part, is found to give good agreement for the E and $F1$ layers. Values obtained for the recombination coefficient α are in the range $(3{\cdot}0\text{–}4{\cdot}0) \times 10^{-8}$ for the E layer and $(1{\cdot}5\text{–}2{\cdot}0) \times 10^{-8}$ for the $F1$ layer.

The $F2$ region shows a marked solar control at a period around maximum eclipse, and may also exhibit a corpuscular eclipse effect.

The absorption results, interpreted in terms of a "Chapman" D layer, fit better to the solar model than to a uniform sun, with a value of $\alpha N_0 = 13 \times 10^{-4}$.

DURING the solar eclipse of 30 June 1954 ionospheric measurements were made by the geophysical section of the Norwegian Defence Research Establishment at three different stations as follows:

Station	Geographical co-ordinates	Max. obscuration at ground (%)
Tromöya	58° N, 9° E	99·5
Kjeller	60° N, 11° E	99·0
Tromsö	70° N, 19° E	75·4

At all three stations $h'f$ recordings were taken, the Mark-II NPL recorders being used. Control measurements were made each quarter of an hour five days before and five days after the day of the eclipse, and on 30 June the equipments were run continuously between 0900 MET and 1700 MET, one record being obtained approximately each third minute. Absorption measurements were made at Kjeller and at Tromöya. Again, control measurements were made on three days before and after the eclipse. During the control period the value of the absorption coefficient was recorded as a median value over 15-min intervals, and on the day of the eclipse absorption values for the period 1130 to 1700 MET were deduced over 6-min intervals.

In addition, at Kjeller and at Tromsö the signal field strength from the Rugby transmitter was also recorded. These measurements were made as part of a programme drawn up by the radio group at the Cavendish Laboratory, and we have not made a detailed analysis of the results.

In interpreting the observed results from the $h'f$ observations for the E and the $F1$ layers, the equation of balance between production and loss of free electrons has been assumed to be

$$\frac{dN}{dt} = p(t)\, q_0 \cos \chi - \alpha N^2, \tag{1}$$

where N is the number of free electrons per unit volume, $q = q_0 p(t) \cos \chi$ is the number of free electrons produced per unit volume per unit time, α is the effective recombination coefficient, χ the zenith angle of the sun, and $p(t)$ is the fraction of ionizing radiation unobscured.

For the E and the $F1$ layers, equation (1) will approximately be valid for the maximum electron density in the layers, with a constant value of the recombination coefficient α. If one assumes therefore that the ionizing radiation is uniformly distributed over the sun's disk, it becomes possible from equation (1) to calculate the variations of the critical frequencies one should expect for different values of α.

From our observations, we have concluded that a good fit between theoretical and observed curves cannot be obtained for any value of α, either for the E or for the $F1$ layer for any of the stations, on the assumption of uniform radiation from the sun's disk. This result is interesting because the solar eclipse of 30 June 1954 occurred at a time of extremely low solar activity.

In order to obtain a better fit between observed and calculated curves, a "solar model" has been adopted which assumes a nonuniform distribution of the ionizing agency over the sun's disk; and on the assumption that all ionizing agency is limited to the solar disk, quite a good agreement between observed and theoretical curves has been obtained both for E and $F1$ layers for all three stations. In the solar model adopted, about 70 per cent of the ionizing agency is uniformly distributed over the solar disk, and superimposed on this about 15 per cent is limited to the western limb of the sun, about 5 per cent to the eastern limb, and about 10 per cent is spread over the central part of the sun's disk.

With regard to the $F2$ layers, it is found that at a period around maximum eclipse the observations can only be explained in terms of a marked solar control of the ionizing agency. A smaller dip of the critical frequencies observed at all three stations, some time before the optical eclipse, may possibly be explained in terms of a corpuscular eclipse.

When interpreting the absorption measurements, it has been assumed that the absorption of the waves is proportional to the maximum electron density in a D layer, and that this maximum electron density varies according to equation (1). The absorption results have been analysed both in terms of a uniform distribution of the ionizing agency over the sun's disk and in terms of the solar model adopted in connection with the studies of the E and $F1$ layer results. It is found that a better agreement is obtained when the solar model is used than when a uniform distribution is assumed. A value of $\alpha N_0 = 13 \times 10^{-4}$ is deduced for the D layer from the measurements.

In the table on p. 56 are listed the values for the recombination coefficient α, and the total rates of generation of electrons q_0, as deduced from our observations. The values given in the table show good agreement with values found by other investigators from eclipse observations, but they are quite high compared with the generally accepted values.

A more detailed account of the investigation will be published elsewhere.

Station	*E* layer		*F1* layer	
	$\alpha \times 10^8$	q_0	$\alpha \times 10^8$	q_0
Tromöya	3·2	420	1·8	1000
Kjeller	4·0	530	2·0	1100
Tromsö	3·0	400	1·5	800

Solar Eclipses and the Ionosphere (Special Suppl. Vol. 6 to J. Atmosph. Terr. Phys.), Pergamon Press, London, 1956

The Solar Eclipse of 30 June 1954: Preliminary Report

W. STOFFREGEN
Ionosphere Observatory, Upsala

Abstract—Ionospheric measurements at two stations in Sweden during the eclipse of 30 June 1954 are summarized. The behaviour of the *Es* layer was similar to that of the *E* and *F* layers during the eclipse, but some hours before the eclipse the *Es* ionization was high. Field-strength measurements for long and medium waves indicated a short, sharp increase near the eclipse maximum. The noise-level of atmospherics was also greater during eclipse.

DURING the solar eclipse of 30 June 1954, the following programme was carried out by the Upsala Ionosphere Observatory in Sweden:

UPSALA 59° 48·1′ N: 17° 36′ E

1. Ionospheric records 1·4–17 Mc/s, sweep time 6 min.
2. Field-strength record of Rugby, 16 kc/s.
3. Field-strength records of Kalundborg, 245 kc/s, and Warsaw, 227 kc/s.
4. Atmospherics recorded by a modified atmoradioscope.

EKENÄS 56° 29′ N; 16° 11′ E

1. Ionospheric records 1·4–17 Mc/s, sweep time 3 min.
2. Field-strength records of Göteborg, 980 kc/s, and Motala, 191 kc/s.
3. Field-strength measurements of a number of stations on the medium- and long-wave range.
4. Atmospherics recorded by a modified atmoradioscope.

The full results will be published later when detailed studies of the material have been made. This report includes only some of the essential results which may be of interest for comparison with other investigations of this solar eclipse.

Figs. 1 and 2 show the characteristic behaviour of the *E*, *Es*, *F1*, and *F2* layers at Upsala and Ekenäs respectively.

The following points are of interest:

1. The variation of the *Es* layer at Upsala and Ekenäs is very similar to the *Es* variation recorded during the solar eclipse of 9 July 1945 at Hörnsjö in the north of Sweden. In both cases high ionization densities of *Es* were recorded some hours before the eclipse. During the eclipse the *Es* ionization decreases in the same manner as the other layers, and increases again after the eclipse. It is important to notice that this special type of *Es*, existing during the summer at temperate latitudes, is really controlled by the ultra-violet light of the sun.

2. The decrease of the ionization density of the *E*, *F1*, and *F2* layers seems to be great compared with earlier eclipses. This may be explained by the low sunspot activity in 1954. This eclipse occurs near the minimum of the eleven year sunspot period. The sun was very quiet during the eclipse, and large decreases in ionization were measured both at Ekenäs and Upsala. Of special interest is the large decrease of *NF2*. Table 1 gives preliminary values of the decrease in

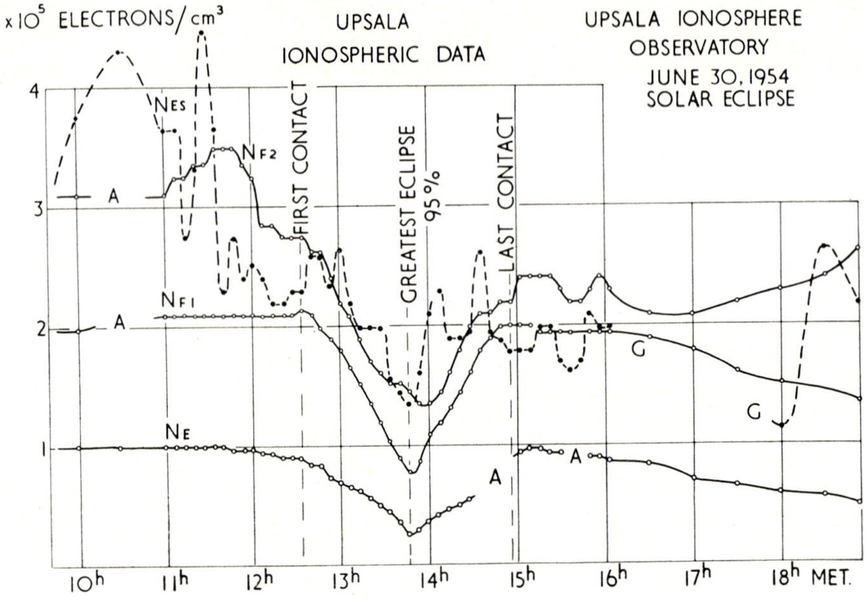
x10⁵ ELECTRONS/cm³
UPSALA
IONOSPHERIC DATA
UPSALA IONOSPHERE
OBSERVATORY
JUNE 30, 1954
SOLAR ECLIPSE
FIRST CONTACT
GREATEST ECLIPSE
95 %
LAST CONTACT
NES
NF2
NF1
NE
A
G
4
3
2
1
10h
11h
12h
13h
14h
15h
16h
17h
18h MET.

Fig. 1.

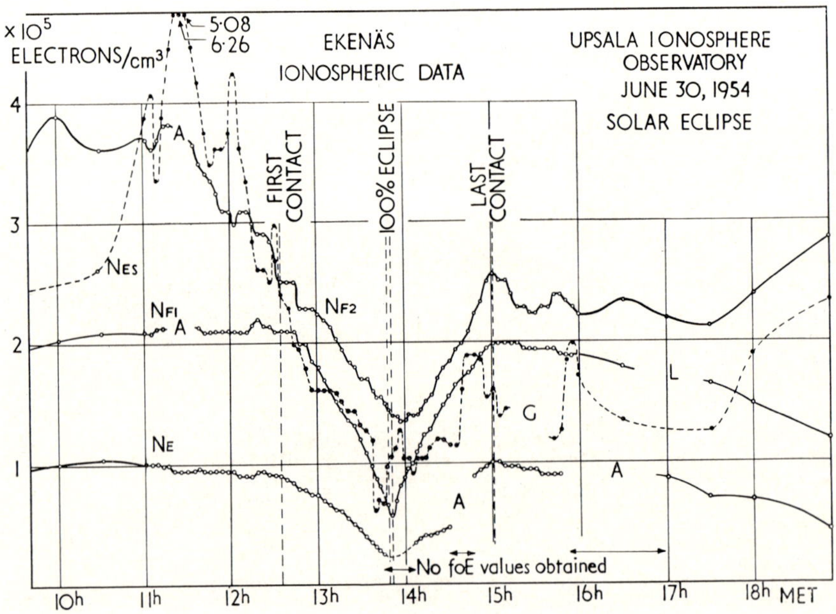
x10⁵
ELECTRONS/cm³
5·08
6·26
EKENÄS
IONOSPHERIC DATA
UPSALA IONOSPHERE
OBSERVATORY
JUNE 30, 1954
SOLAR ECLIPSE
FIRST CONTACT
100% ECLIPSE
LAST CONTACT
NES
NF1
NF2
NE
A
G
L
No foE values obtained
4
3
2
1
10h
11h
12h
13h
14h
15h
16h
17h
18h MET

Fig. 2.

ionization and of the effective recombination coefficient α. (Values of α obtained during the 1940 and 1945 eclipses are also included in the Table.)

Table 1

Layer	N_{min}/N_0	α
	Ekenäs, 1954. 100 per cent eclipse	
E	0·25	$5{\cdot}1 \times 10^{-8}$
F1	0·27	$1{\cdot}7 \times 10^{-8}$
F2	0·51	$>3 \times 10^{-9}$
	Upsala, 1954. 95 per cent eclipse	
E	0·28	
F1	0·37	
F2	0·49	
	Brazil, 1940 (GILLILAND)	
E		$>2 \times 10^{-8}$
F1		1×10^{-8}
F2		$(\sim 1 \times 10^{-10})$
	Sörmjöle, 1945 (RYDBECK)	
E	0·37	$1{\cdot}1 \times 10^{-8}$
F1	0·38	$0{\cdot}6 \times 10^{-8}$
F2	0·74	$\sim 3 \times 10^{-9}$
	MCLEISH, 1945	
E		$1{\cdot}6 \times 10^{-8}$
F1		$1{\cdot}4 \times 10^{-8}$
F2		$1{\cdot}0 \times 10^{-9}$

The delay of the variation of *E* and *F1* in relation to the eclipse is of the order of about 2–3 min. The delay is somewhat greater for *F2* than for the *E* and *F1* layers.

3. A short time after the maximum of the eclipse at about 1410, the increase of the *E* and *F1* densities at both stations is more rapid than expected. For the interpretation of this abnormality it will be of interest to study other observations.

Fig. 3 shows the original recordings of signals from Rugby on 16 kc/s, and the curve shown in Fig. 4 is based upon the same record. The field strength of Rugby is recorded by means of an oscillograph in order to avoid interference by atmospherics, and the curve in Fig. 4 shows the variation, together with the eclipse curve at the contact points. The field strength of the received signal shows a symmetrical eclipse variation. The increase is about 44 per cent. This compares with an increase at night of about 91 per cent. It is remarkable that the variation of the field strength of Rugby is limited to the time 1316–1410 M.E.T. when the eclipse varies between 60 per cent—total—50 per cent obscuration. Before and after the increase a slight decrease in field strength is observed. This seems to be typical on other field-strength records too. It will be of interest to compare this record with records of Rugby taken at other places during this eclipse. It is also of interest to note that corresponding night values, when the increase is similar to the maximum increase during the eclipse, are not reached before 1 h 50 min after sunset and 1 h 20 min before sunrise. This corresponds to the height of the earth-shadow at about 25 to 30 km.

Records of Rugby (16 kc/s) and other long- and medium-wave stations during

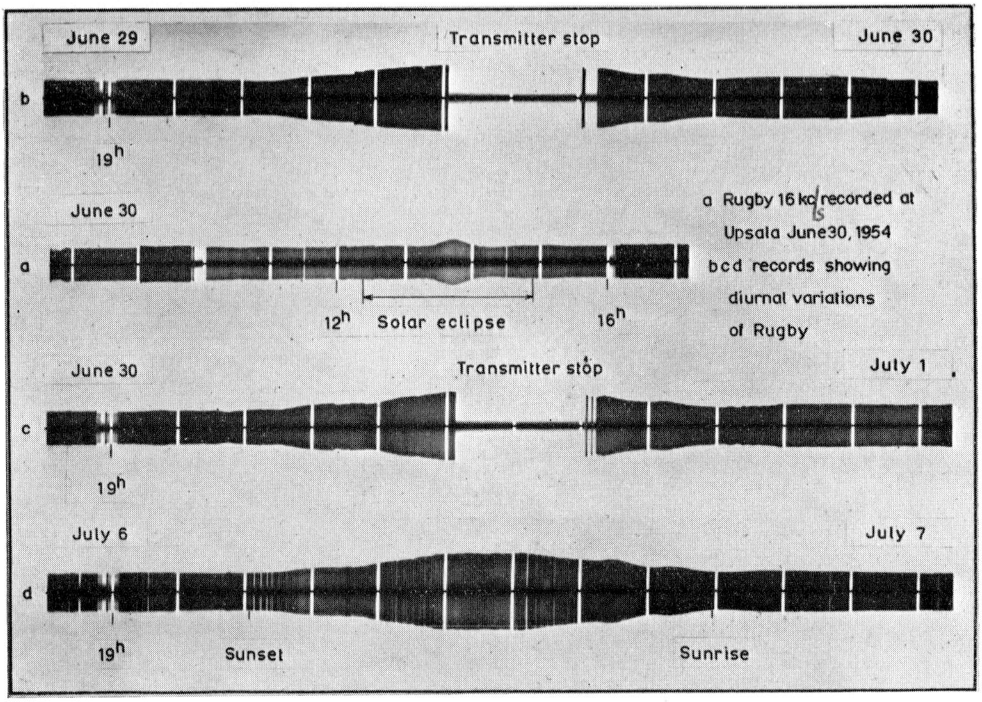

Fig. 3.

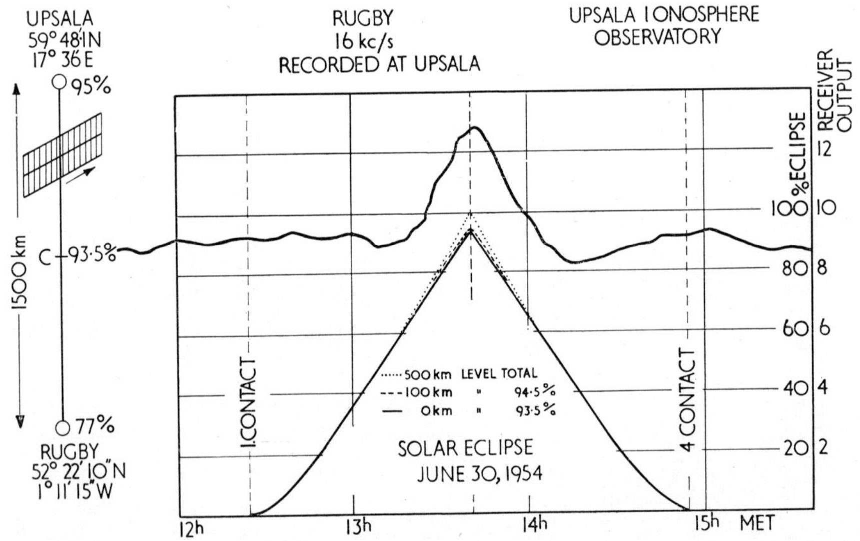

Fig. 4.

the eclipse are taken principally to study the variation of absorption in the lower part of the ionosphere. Depending on the path-length, the position of the path in relation to the totality-zone of the eclipse, and the frequency of the transmission, different effects are to be expected. Some of the recorded stations are represented by Figs. 5–10 and Table 2. It is remarkable that the increase in field strength of Göteborg (980 kc/s) is very high on the path Göteborg–Ekenäs, whilst Oslo (218 kc/s) under the same conditions (100 per cent totality) shows only a small effect. Not one of the recorded long-wave stations shows a pronounced effect. The high field-strength increase of Göteborg (980 kc/s), 34 db, measured at Ekenäs, and of Kalundborg II (1061 kc/s), 33 db, measured at Stockholm (Table 2), may be due to the fact that these relatively high frequencies are readily

Table 2.

Station	Records made at	f (kc/s)	Eclipse (%) (centre-point of path)	Total distance of path (km)	Beginning of field-strength variations eclipse (%)	End of field-strength variations eclipse (%)	Maximum increase of field strength
Kalmar	Ekenäs	1562	100	21	—	—	no increase
Szczecin	Ekenäs	1259	96	360	—	—	×1·8?
Hörby	Ekenäs	1178	98·5	170	~75	~30	×5·47
Tallinn	Ekenäs	1034	97	590	—	—	×1·43
Göteborg	Ekenäs	980	100	285	~70	~45	×47·6
Frankfurt AFN	Ekenäs	872	89	860	—	—	×1·46
Deutschl. send.	Ekenäs	782	94	480	—	—	no increase
Stockholm	Ekenäs	773	99·5	330	~70	~60	×2·1
Hof RIAS	Ekenäs	683	92	670	—	—	no increase
Vilna	Ekenäs	665	100	590	—	—	no increase
Potsdam	Ekenäs	611	94	480	—	—	no increase
Riga	Ekenäs	575	99·9	490	—	—	×1·08
Moskva II	Ekenäs	263	96	1320	—	—	no increase
Lahti	Ekenäs	254	95	745	—	—	no increase
Kalundborg I	Ekenäs	245	97	320	—	—	no increase
Warszawa	Ekenäs	227	97	570	—	—	no increase
Oslo	Ekenäs	218	100	490	(65)	~70	×1·15
Motala	Ekenäs	191	100	230	—	—	no increase
Deutschl. send.	Ekenäs	185	94	480	—	—	no increase
Kalundborg I	Upsala	245	100	590	~60	~25	×1·15
Warszawa	Upsala	227	100	870	~70	~30	×1·27
Rugby	Upsala	16	93·5	1500	~80	(50)	×1·44
Kalundborg II	Stockholm	1061	100	570	~80	~60	×45

affected when the absorption decreases. On the lower frequencies the decrease of absorption is apparently not great enough to cause a pronounced effect. From this point of view it is interesting that Rugby (16 kc/s) shows such a large increase. This may be due to the fact that the level of reflection for such signals is very low, about 70–80 km.

Determination of the atmospheric-noise level is effected by recording the voltage of a condenser with a high-leakage resistor, which is connected to an amplifier by means of a diode. Because of the high time-constant of the system, these records give smoothed values of the atmospheric noise.

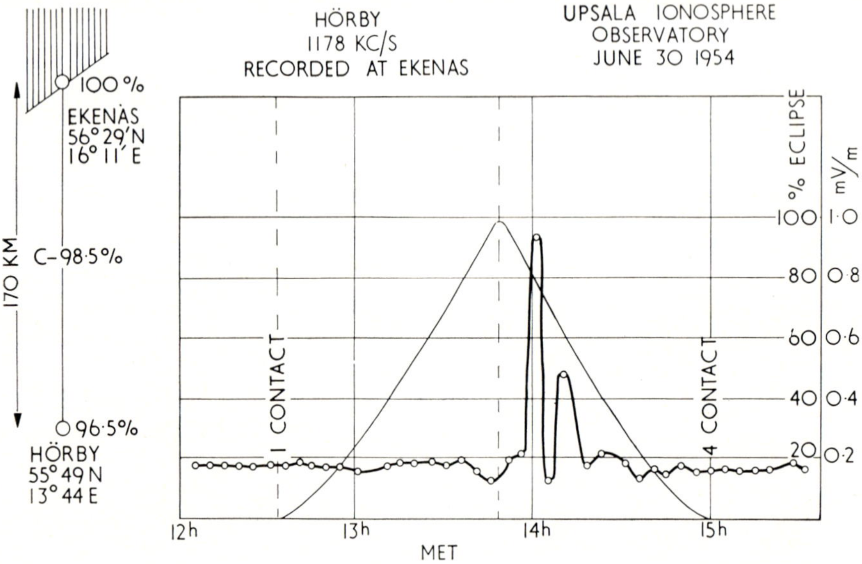

Fig. 5.

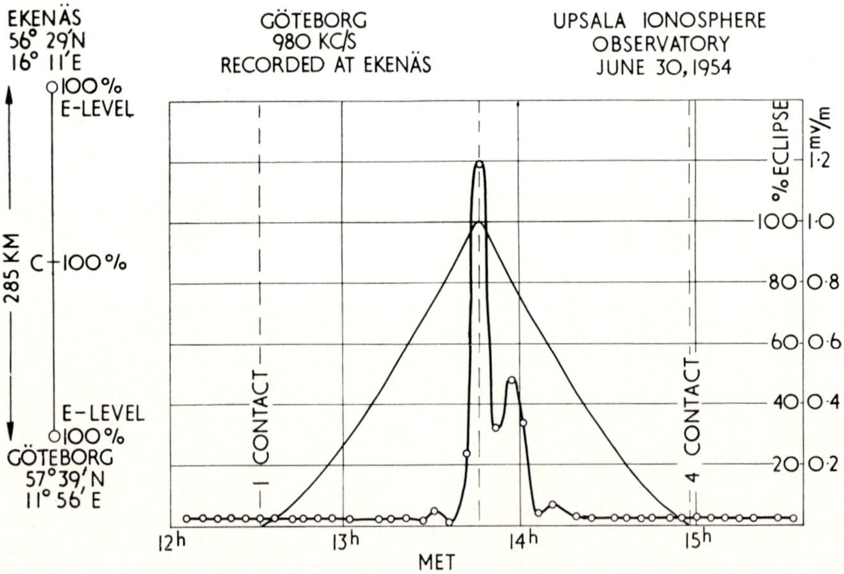

Fig. 6.

Atmospherics from all directions are recorded at Upsala. The situation on the day of the eclipse was quite favourable in that the main part of thunderstorm activity was in the southern direction and most of the atmospherics had to cross

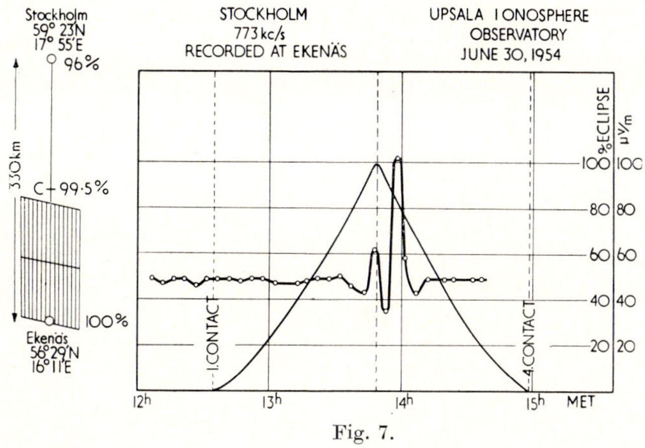

Fig. 7.

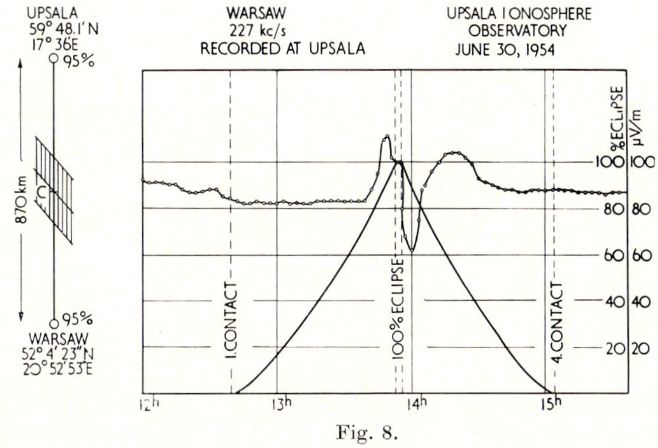

Fig. 8.

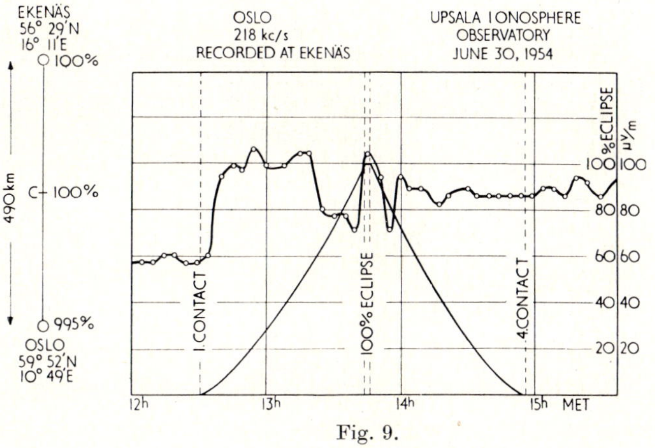

Fig. 9.

the totality zone. A slight increase is observed at Upsala: Fig. 11 shows the course of the activity, together with that on six control days. The effect measured at Ekenäs was very small, as may be expected. The increase at Upsala is of

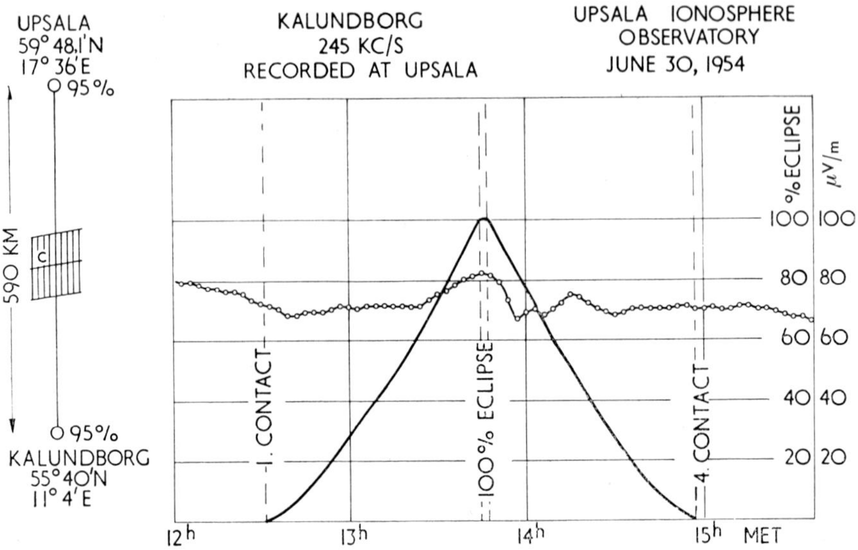

Fig. 10.

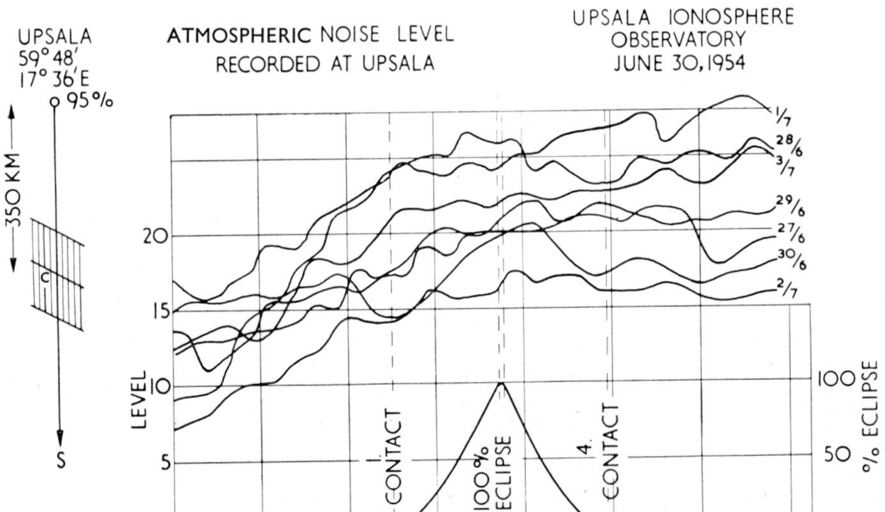

Fig. 11.

the order of 15 per cent. This observation is valuable, in that it is complementary to the other field-strength measurements, since it covers the lowest frequency range, 5–10 kc/s.

Solar Eclipses and the Ionosphere (Special Suppl. Vol. 6 to J. Atmosph. Terr. Phys.), Pergamon Press, London, 1956

Experimental Ionospheric Observations of the Solar Eclipse of 20 June 1955 at Baguio City, Philippine Islands

J. J. HENNESSEY and J. S. TORRES
Ionospheric Station, Manila Observatory, Baguio City

Abstract—Studies of parameters for all the chief ionospheric layers are given for the eclipse of 20 June 1955 in the Philippine Islands, which reached its maximum phase (96 per cent obscuration) near noon. Values of f_{min} showed pronounced maxima before the first contact and after the last contact. Stratification of the *F* layer was observed on the control days as well as on the eclipse day, and an intermediate layer between the *F1* and *F2* layers formed in the second half of the eclipse. There was no certain evidence of a corpuscular eclipse.

A SOLAR eclipse which reaches its maximum at a time when the sun is near the zenith should present an excellent opportunity for checking the ionizing influences responsible for the ionosphere. The eclipse over Baguio of 20 June 1955, beginning with first contact at 1047 and with last contact at 1349, had a maximum of about 96 per cent at ground level at 1220. Very noticeable changes in ionospheric parameters occurred on the eclipse day as compared with the characteristics scaled on the five days preceding and the five days following the eclipse day. By using a C-2 ionosonde (CARROLL, 1952) on loan from the United States National Bureau of Standards, ionograms were taken every five minutes on the ten control days beginning at 0900 and going to 1630. Prior to and following this, interval records were taken every quarter of an hour, and one minute after each hour. On the eclipse day a similar programme was followed, with additional records being taken from 1210 to 1245. All times are local mean time (120° east).

IONIZING AGENCY BEFORE AND AFTER THE ECLIPSE

On the eclipse day before first contact and after last contact the presence of an intense ionizing agency was noticeable, not only in the absorbing region, but also

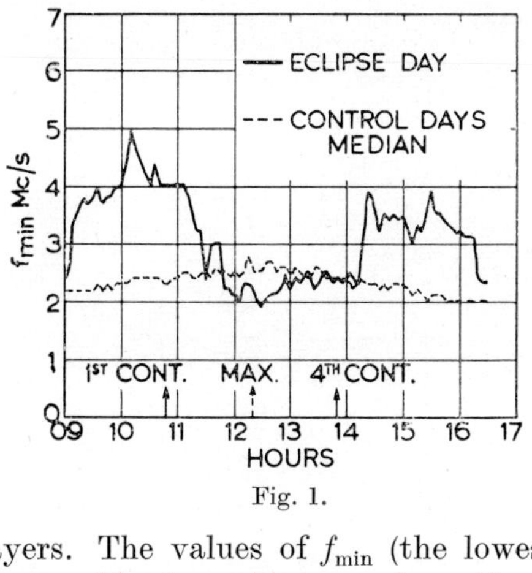

Fig. 1.

in the *F1* and *F2* layers. The values of f_{min} (the lowest frequencies found on the ionogram, and presumably dependent on absorption in the *D* region) began to increase at 0900 (Fig. 1). At 1000 the f_{min} value was so high that malfunctioning of the receiver was suspected. A check of the receiver resulted in the loss of the

1015 sweep, but showed that the receiver was responding properly. The 1010 ionogram showed an f_{min} of 5 Mc/s. From 1010 f_{min} dropped to lower values, and at 1130, i.e., forty-three minutes after first contact, it was down to the level of the values on the control days. During the period of greatest eclipse f_{min} decreased below the values on control days. The lowest f_{min} was recorded at 1230. After the eclipse, this characteristic increased beyond the normal value. It is

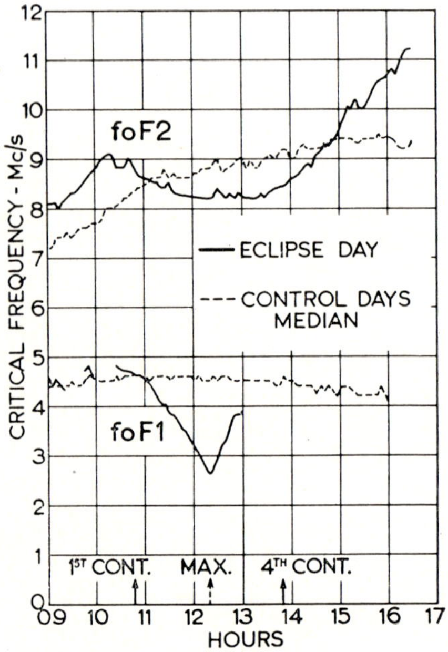

Fig. 2.

interesting to note that the largest value of f_{min} occurred thirty-seven minutes before first contact and a corresponding maximum occurred about thirty-six minutes after last contact. These time intervals may be in error by ± 5 min. The f_{min} resumed normal values before 1700.

The f_0F2 characteristic showed higher values before first contact and after last contact than the median of the control days (Fig. 2). At 1220 the ionization in the $F2$ layer dropped to 86 per cent of the control-day value (Fig. 3). This relative drop would have been greater, or the recombination of ions would have been more pronounced, if the f_0F2 characteristic prior to first contact had not been so high.

The f_0F1 characteristic before first contact also showed a value higher than that for the control days (Fig. 2). After last contact the *Es* layer blanketed the $F1$ layer. Starting at first contact, f_0F1 dropped uniformly with the advance of the eclipse. At 1220 f_0F1 reached a minimum value of 2·6 Mc/s. Such a low value has not been reported at this station either in the hours after sunrise or before sunset. The ionization of the $F1$ layer at the maximum of the eclipse was reduced to 33 per cent of the value on the control days.

The normal E layer began to appear through the previous heavy absorption at about 1150. The value of f_0E decreased with the progress of the eclipse and increased after the maximum of the eclipse had passed. At about 1320 f_0E was restored to the value which obtained on the control days. Though the sporadic E layer was present on the control days, on 20 June no sporadic E was apparent from 0900 on. Later, Es of the sequential type became evident at about 1300.

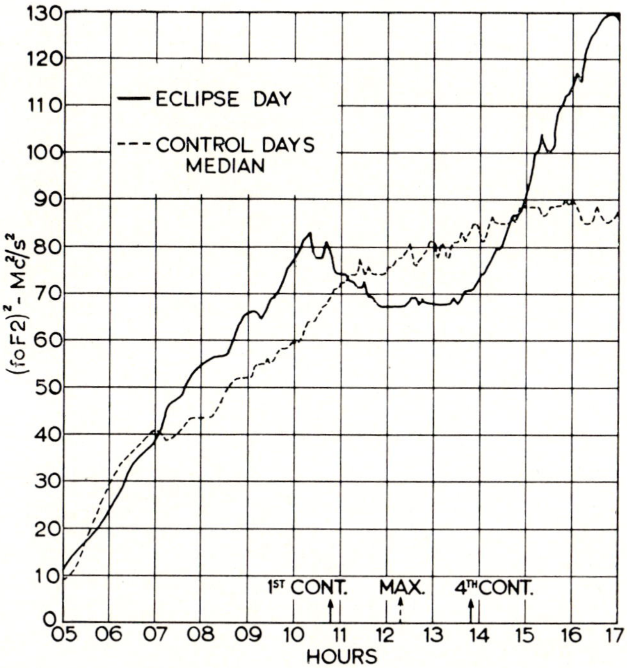

Fig. 3.

In explaining the variations in the ionosphere which were observed on the eclipse day the significance of the time of the eclipse should not be minimized. At this local time might not the moon reflect a considerable amount of ultra-violet light into the earth-region of the eclipse? The records seem to demand an intensification of the ionizing agency before and after the eclipse.

Corpuscular Eclipse

A careful study of the records gives no certain evidence of a corpuscular eclipse two or more hours before the start of the optical eclipse (cf. Gupta and Mitra, 1954). A slight dip in the $F2$ ionization density about 20 min before first contact seems to be better explained by the eclipsing of ultra-violet radiation from some region in the solar atmosphere, if extraterrestrial sources are sought for this variation.

Stratification Phenomena

The ionograms on the control days both before and after the eclipse day showed pronounced stratification in the $F2$ layer. Indeed, at certain times, the stratification was such as to require a really different layer, the *G layer*. On each of the control days this stratification started at about 0900 and continued until about three hours after local noon. On the eclipse day the same type of stratification began at about 0900 and continued to form until 1025. From then until 1125 the stratification started to coalesce into a single layer. At 1155 the *F2* layer formed, in contradistinction to this stratification, in a strong single layer with clear sharp turn-ups at the beginning of the *F2* trace. On a linear sweep the *F2* layer trace would have the approximate appearance of a parabola with a vertical axis. Due to the eclipsing of the solar radiation, the incipient "*G* layer" evidently was subjected to rapid ionic recombination. At least, sufficient recombination took place in this layer to reduce its density of ionization below that of the *F2* layer. Alternatively, the ionization of the higher layer or stratum moved down in virtual height into the region of the lower layer. This would increase the density of ionization, an apparent phenomenon observed daily near sunset at this station when the *F1* and *F2* layers coalesce. On the eclipse day, the ionograms suggest that both recombination of ionization in the *G* and *F2* layers, and the penetration of the ionization of the virtually higher layer into the lower layer, took place.

Though the sequential *Es* blanketed the *F1* layer at about 1300, at the same time a distinct layer began to be formed between the *F1* and the *F2* layers. The progress of the development of this intermediate layer showed that it was distinct from the *F1* layer.

We are indebted to the U.S. National Bureau of Standards for the loan of the equipment.

REFERENCES

CARROLL, J. M. 1952 *Electronics*, **25**, 128.
GUPTA, P. K. S., and MITRA, S. N. 1954 *Nature*, **173**, 814.

Solar Eclipses and the Ionosphere (*Special Suppl. Vol. 6 to J. Atmosph. Terr. Phys.*), *Pergamon Press, London*, 1956

Temperature-controlled Variations of the Ionosphere during an Eclipse

O. BURKARD
Institut für Meteorologie und Geophysik, University of Graz

Abstract—Hitherto the changes in f_0E and f_0F1 in an eclipse have been explained by the relation

$$\frac{\mathrm{d}N}{\mathrm{d}t} = D \,.\, q_0 \cos \chi - \alpha' N^2$$

where D is the fraction of unobscured sun's disk and α' a constant.

But now we know that the diurnal variation of f_0E is approximately proportional to $\cos^{\frac{1}{3}} \chi$, and for f_0F1 the diurnal variation is represented by $f_0F1 \propto \cos^m \chi$, where the values of m change approximately between 0·15 and 0·28. Such a variation can be explained by the assumption that there is a linear increase of temperature with height and a recombination coefficient $\alpha' = \alpha'_0 \,.\, p^k \,.\, T^n$ (p = pressure; k, n = indeterminate constants). An attempt to explain the changes in electron density at Khartoum during the eclipse of 25 February 1952 has shown that, using the same relations, there is no necessity to assume any nonuniform distribution of the sources on the sun's disk as calculated by MINNIS. It is necessary to neglect the term $\mathrm{d}N/\mathrm{d}t$, but this may be a smaller error than to calculate with a constant recombination coefficient (i.e., constant temperature). The term $\mathrm{d}N/\mathrm{d}t$ can be neglected in the diurnal changes in f_0E and f_0F1. The temperature at the height of maximum electron density decreases shortly after the beginning of the eclipse, and increases rapidly after the eclipse maximum.

THE behaviour of the E layer as well as of the $F1$ layer during a solar eclipse has been explored many times. Usually the simple formula

$$\frac{\mathrm{d}N}{\mathrm{d}t} = q - \alpha N^2$$

has been taken as a basis for investigation of the balance of the ionization. There is no success in fitting the observational material to this equation, for it has been found that the ionization after the maximum of the eclipse increases much faster than this formula allows. But there are other arguments against the application of this simple equation.

It is an accepted fact that the diurnal variation of the critical frequency of the E layer changes as $(\cos \chi)^{1/3}$, not as $(\cos \chi)^{1/4}$, as the above-mentioned equation requires. The conditions of the $F1$ layer are quite similar: the critical frequencies are proportional $(\cos \chi)^m$, where m changes with the seasons between 0·15 and 0·28.

Fig. 1 shows an arbitrary example where $\log f_0$ is plotted against $\log \cos \chi$ (the July 1954 median values for Slough). This diagram demonstrates that, for the E layer, f_0E is proportional to $(\cos \chi)^{1/3}$, and for the $F1$ layer f_0F1 is proportional to $(\cos \chi)^{1/6}$.

For a better understanding of the actual situation during a solar eclipse it is useful to consider the dependence of $\log f_0$ on $\log D$ (where D = fraction of unobscured sun's disk). Fig. 2 shows this procedure for the $F1$ layer on the occasion of the eclipse of 25 February 1952 at Khartoum. The curves show lower values before the eclipse and higher values after the eclipse. This behaviour is actually quite contrary to the above-mentioned balance equation. If we write this equation as

$$N^2 = \frac{qD}{\alpha} - \frac{1}{\alpha}\frac{\mathrm{d}N}{\mathrm{d}t}$$

and consider that $\mathrm{d}N/\mathrm{d}t$ is negative *before* the maximum obscuration and positive *after* the eclipse, the critical frequency should be higher before the eclipse maximum than afterwards for the same degree of obscuration (because in our case cos χ varies only slightly). The observation shows the contrary behaviour.

Recently Mr. MINNIS has tried to eliminate these difficulties by the assumption

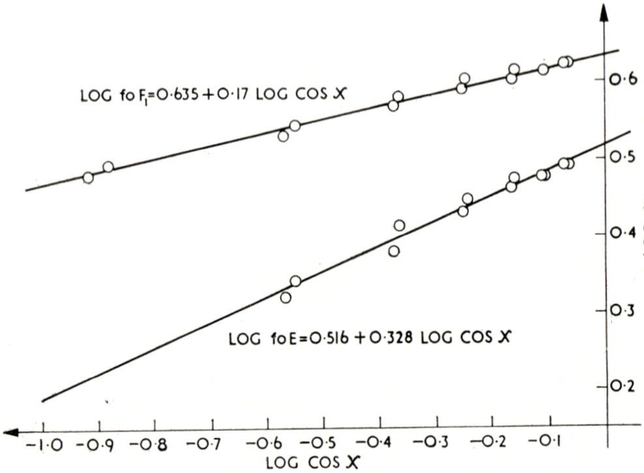

Fig. 1. Log f_0E and log f_0F1 as a function of log cos χ, Slough, July 1954.

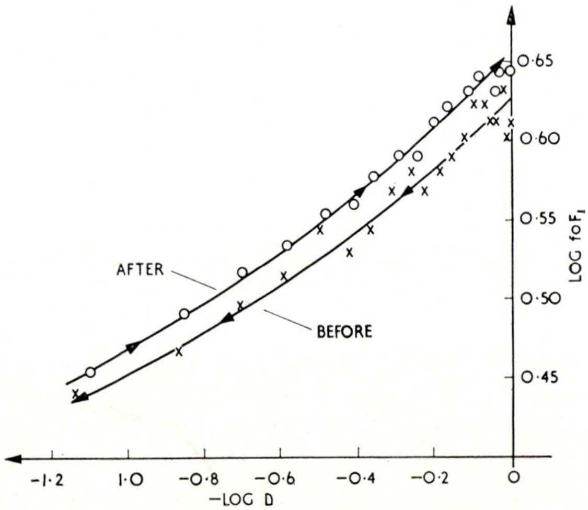

Fig. 2. Log f_0F1 before and after the eclipse of 25 February 1952 at Khartoum.

that the different parts of the sun's disk radiate inhomogeneously and that certain areas radiate more intensely than the rest of the sun's disk. This is a tempting method to overcome the difficulties mentioned, but the hypothesis does not consider that the normal undisturbed behaviour of both layers cannot be explained by the formula

$$\frac{dN}{dt} = q - \alpha N^2$$

In my paper (*J. Atmosph. Terr. Phys.* **8,** 83–90, 1956) I have tried to explain the observed linear connection between $\log f$ and $\log\cos\chi$, usually written as

$$\log f_0F1 = \log C + m \,.\, \log\cos\chi$$

By assuming (as usual) a flat surface of the earth, monochromatic ionizing radiation, and so on, we can represent the production of ions or electrons by

$$q = q_0 \cdot \frac{p}{T} \cdot e^{-\frac{Ap}{g\cos\chi}}$$

(after the well-known deduction from CHAPMAN), where the pressure p, the absolute temperature T and $\cos\chi$ are variables, and the other terms are constant, without considering a variation of the intensity of the ionizing radiation. Against this ionization production, q, operates the process of recombination of electrons which is given by $\alpha \,.\, N^2$. (It is not quite clear if this is a real or only an apparent recombination.) As Fig. 1 shows, the observations for the morning are scarcely distinguishable from the afternoon observations, and therefore $\mathrm{d}N/\mathrm{d}t$ may be neglected. It is seen that

$$q_0 \frac{p}{T} e^{-\frac{Ap}{g\cos\chi}} = \alpha N^2$$

where no assumption is made concerning α. Written in logarithmic form, this becomes:

$$2\log N = \log q_0 + \log p - \log(\alpha T) - \frac{Ap}{g\cos\chi}$$

This equation applies first of all for any height. A comparison with the measured values of the critical frequency is possible if the maximum concentration of electrons is known. This requires

$$\frac{\mathrm{d}}{\mathrm{d}p}\log N = 0$$

Then we obtain the additional equation

$$\frac{Ap}{g\cos\chi} = 1 - \frac{p}{\alpha T}\frac{\mathrm{d}(\alpha T)}{\mathrm{d}p}$$

Since $N = 1{\cdot}24 \times 10^{-8} \,.\, f^2 = \sigma \,.\, f^2$, it is seen that

$$4\log f_0 = \log\frac{q_0 \cdot g}{\sigma^2 \,.\, A} + \log\cos\chi + \log\left[1 - \frac{p}{\alpha T}\frac{\mathrm{d}(\alpha T)}{\mathrm{d}p}\right] - \log(\alpha T) - 1 + \frac{p}{\alpha T}\frac{\mathrm{d}(\alpha T)}{\mathrm{d}p} \tag{1}$$

On the other hand, we get from the observational material the simple equation:

$$4\log f_0 = 4\log C + 4m\log\cos\chi \tag{2}$$

It is easy to show how to transfer the complicated function (equation 1) to equation (2). It is sufficient to express the coefficient of recombination as $\alpha = \alpha_0 \cdot p^k \,.\, T^n$, where α_0, k, and n are constants. The values of these constants may

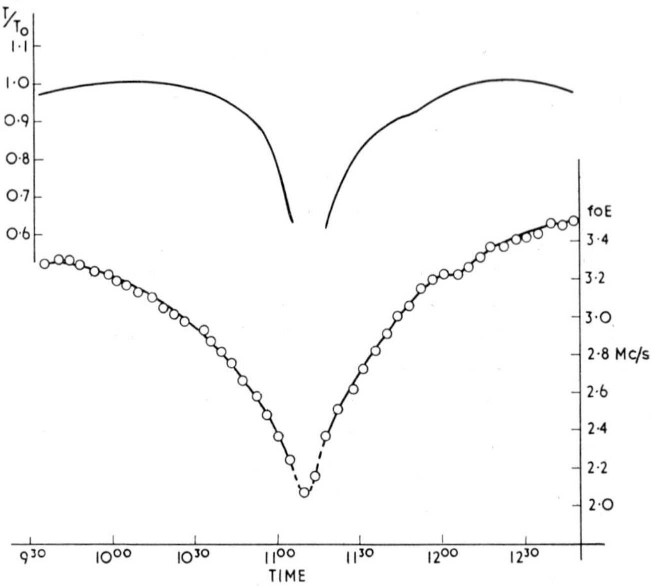

Fig. 3. Calculated temperature, *E*-layer, KHARTOUM, 25 February 1952.

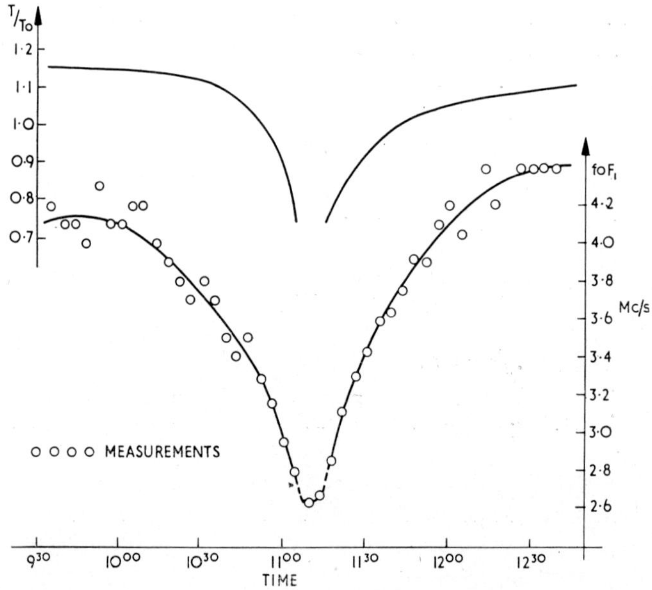

Fig. 4. Calculated temperature, *F1*-layer, KHARTOUM, 25 February 1952.

be simple numbers, like 0, ± 1, ± 2, $\pm \frac{1}{2}$, $\pm \frac{1}{3}$, and so on. Only the value $n = -1$ is not possible.

It should be added that we have assumed the temperature gradient $\mathrm{d}T/\mathrm{d}h$ is constant (valid only for the range of the region concerned). A variation of the index m (equation 2) results from a variation of the temperature gradient $\mathrm{d}T/\mathrm{d}h$, since

$$4m = 1 - k + (1 + n)R \frac{\mathrm{d}T}{\mathrm{d}h}$$

where R is the gas constant.

It is natural to investigate the situation during an eclipse, and for this purpose the observations of Khartoum were used. By computing values during an eclipse it is found that the temperature decreases before total obscuration and increases rapidly after the eclipse, these changes corresponding to the obscuration of the sun's disk. The upper curve of Fig. 3 shows the ratio of the temperature on the eclipse-day to the temperature on the control-days (the temperatures refer to the maximum of the layer). The lower curve shows the variation of the critical frequency as computed from the temperature variation. This diagram shows that it is not necessary to assume an unsymmetrical radiation of the sun's disk. Moreover, the calculations are made using equations which are valid for undisturbed conditions. The assumptions involved in these calculations are as follows:

(1) The relation $\log f_0E = 0{\cdot}556 + 0{\cdot}330 \log \cos \chi$ for the control-days.

(2) The constant $n = 1$. (Another value would not change the results appreciably.)

(3) $\mathrm{d}T/\mathrm{d}h =$ constant during the whole eclipse, having the value observed on the control-days. Such an assumption is suggested by the fact that $\mathrm{d}T/\mathrm{d}h$ does not change during the day.

(4) $\mathrm{d}N/\mathrm{d}t$ may be neglected if $q \,.\, D > \mathrm{d}N/\mathrm{d}t$. This is not valid for a very large obscuration.

The same conditions were used for the *F1* layer, except the assumption (1) now becomes $\log f_0F1 = 0{\cdot}677 + 0{\cdot}210 \log \cos \chi$. The results are shown in Fig. 4, where the upper curve shows the variation of relative temperature and the lower curve the computed variation of the critical frequency f_0F1.

The exact analysis of further observations during eclipses will decide if the above interpretation is valid or not.

Solar Eclipses and the Ionosphere (Special Suppl. Vol. 6 to J. Atmosph. Terr. Phys.), Pergamon Press, London, 1956

The Behaviour of the *E1* Layer during the Solar Eclipse of 25 December 1954

M. E. SZENDREI and M. W. MCELHINNY
Physics Department, Rhodes University, Grahamstown

Abstract—Two methods are attempted to explain deviations in the maximum electron density of the *E1* layer observed during the eclipse from the expected effects from a solar disk of uniform intensity. Some success is met by postulating more intense regions on the solar surface and correlating these with astronomical observations. In the second method, a temperature variation with time of day is considered; while this indicates a reasonable diurnal variation on a normal day, it points to a rise in temperature during the middle part of the eclipse instead of an expected drop, and hence an explanation in terms of temperature variation must be rejected.

1. INTRODUCTION

IONOSPHERIC readings at vertical incidence were taken in Grahamstown during an annular eclipse of the sun which occurred on 25 December 1954. The centre of the belt of annularity passed very near to a height of 100 km above Grahamstown, the various phases of the eclipse at this height being shown in Table 1. During annularity, 85·8 per cent of the solar disk was covered.

Table 1. Times of contact at 100 km.
South African Standard Time = U.T. + 2 h

1st contact	2nd contact	3rd contact	4th contact
0646	0800	0807	0935

A specially designed and completely automatic ionosphere sounder was used, covering the range from 1·5 to 15·0 Mc/s. The equipment delivered pulses of 100-μs duration at an average output of 2 kW. The variation in frequency was obtained by a method described by WADLEY (1949): this was done by beating a fixed oscillator at 30 Mc/s with a variable oscillator operating between 31·5 and 45·0 Mc/s. The latter also served as the first oscillator of a double conversion receiver, so that it was automatically tuned to the transmitter at all frequencies. Thus both transmitter and receiver covered the full frequency range by rotation of a single condenser. From the *h'f* records, frequencies could be read to an accuracy of 0·02 Mc/s, and virtual heights to the nearest 2 km.

2. OBSERVATIONS AND PRELIMINARY RESULTS

Readings were taken over thirty control days, centred on the eclipse day, every quarter of an hour from 0500 to 1000 S.A. Standard Time, and at hourly intervals for the rest of the day. On 25 December, for the duration of the eclipse, readings were taken at 2½-min intervals. All records were scaled for true heights by the method described by KELSO (1952). Although this method is satisfactory for the *F* layers, no significant change in the true height or thickness of the *E1* layer could be detected, owing to absorption at the low frequencies. It was gratifying to note from data supplied by the Magnetic Observatory, Hermanus, that the period under consideration was magnetically quiet.

From the critical frequency, f_0, the maximum electron density N of the *E1* layer was calculated from:

$$N = 1{\cdot}24 \times 10^4 f_0^2 \tag{1}$$

where f_0 is expressed in Mc/s. Mean values of N as a function of time of day were calculated for the *E1* layer for the thirty control days.

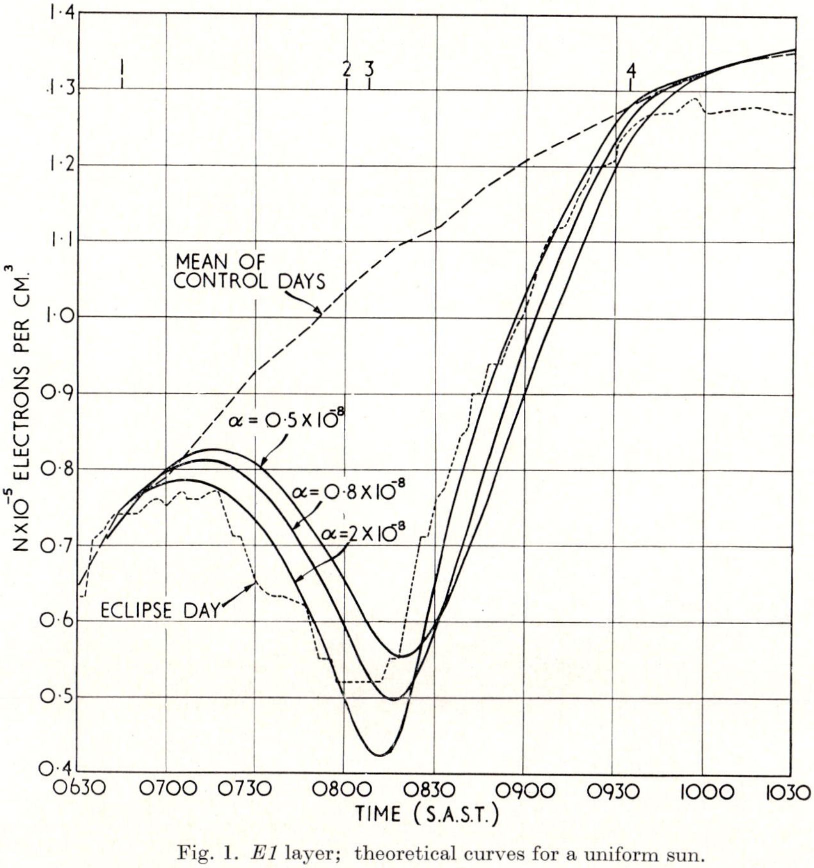

Fig. 1. *E1* layer; theoretical curves for a uniform sun.

Some preliminary results for this eclipse are being published elsewhere (Szendrei and McElhinny, 1955). It was possible to show that the maximum electron densities may be approximately explained by assuming that the solar radiation responsible for the *E1*, *F1*, and *F2* layers comes uniformly from the solar disk. The method is illustrated in Fig. 1, where a number of theoretical curves for various values of the recombination coefficient α are drawn and compared with the experimental data for the *E1* layer. It will be observed that, while the general shape of the theoretical curves agrees with the experimental observations, no perfect fit could be obtained. In all cases the minima occur too late, and none of them accounts for the rapid drop in the maximum electron density observed during the early phases of the eclipse. By taking the value of the minimum as a criterion for determining α, it appears that for the *E1* region the recombination coefficient is about 7×10^{-9} cm³ sec⁻¹. Similar calculations for the *F1* and *F2* layers gave recombination coefficients of 5×10^{-9} and 5×10^{-11} respectively.

3. Effect of a Non-uniform Solar Disk

In interpreting ionospheric eclipse data it is usual to assume that, on a normal day,

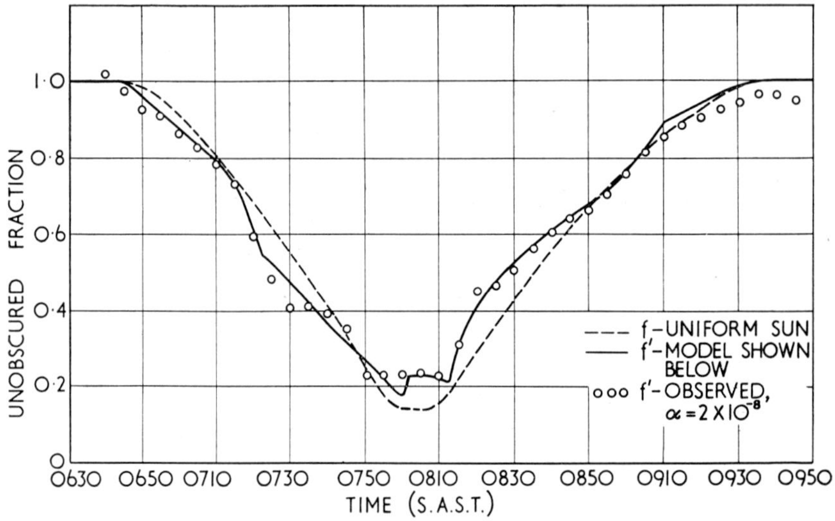

Fig. 2. Unobscured fractions of the sun.

the rate of increase of electron density can be expressed in the form:

$$\frac{dN}{dt} = q - \alpha N^2 \tag{2}$$

where q is the rate of ion production per cm³. If the radiation from the sun is uniform over the whole solar disk, then, during an eclipse, equation (2) must be modified to

$$\frac{dN}{dt} = fq - \alpha N^2 \tag{3}$$

in which f represents the uncovered fraction of the sun's disk, and may be calculated from astronomical data. The theoretical curves shown in Fig. 1 were obtained by using equations (2) and (3).

In order to obtain a better fit to the experimental results, a method similar to that outlined by Minnis (1955) was applied. If the radiation does not come uniformly from the solar disk, equation (3) will take the form:

$$\frac{dN}{dt} = f'q - \alpha N^2 \tag{4}$$

where f' is the unobscured fraction of the solar intensity and in general will not coincide with the astronomical value f. Values of N and dN/dt were obtained from the experimental eclipse data, while q can be obtained for any assumed value of α from the mean curve for the control days. On account of the difficulty of reading dN/dt accurately from the experimental values, and in order that

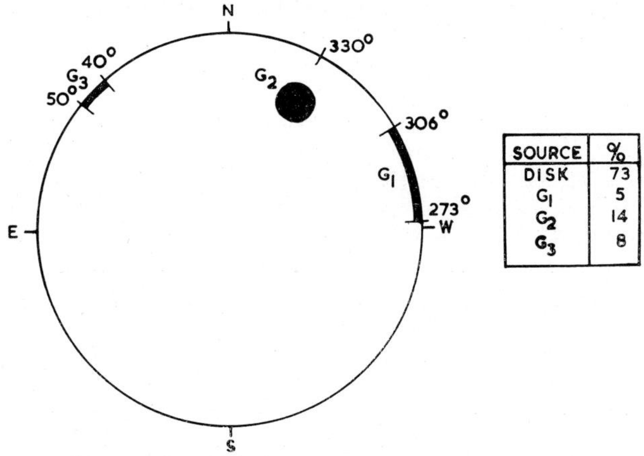

Fig. 3. Positions and intensities of bright regions on the solar surface.

f' be equal to unity at the beginning and end of the eclipse, a value of $\alpha = 2 \times 10^{-8}$ cm^3 sec^{-1} was chosen for calculating f'. It must be pointed out that the value of α used does not materially affect the relative values of f', which are shown

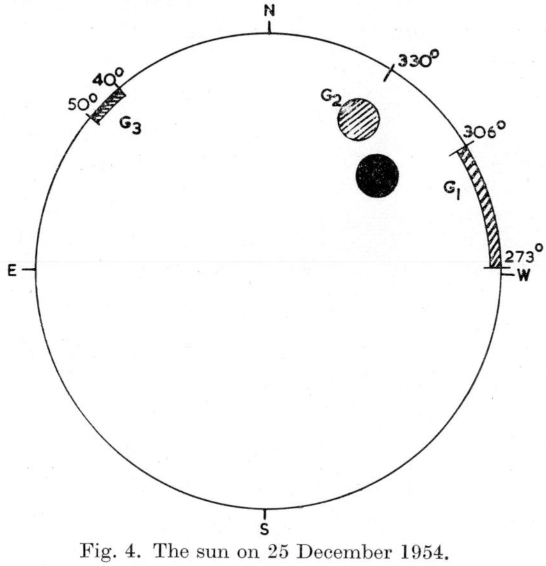

Fig. 4. The sun on 25 December 1954.

in Fig. 2, together with the values of f calculated from astronomical data. At the end of the eclipse f' does not rise to the value of unity, because on the eclipse day the experimental curve did not rise to the normal value of N after the eclipse (Fig. 1).

The best fit to the f' values was found by postulating the existence of three more intense regions on the solar surface. The positions and intensities of these regions are shown in Fig. 3. Since eclipse data cannot supply information as to whether these more intense regions occur on the northern or southern hemisphere, in Fig. 3 they are all shown as occurring on the northern hemisphere. The curve for f' according to this model is also shown in Fig. 2.

Comparison with sketches made with the aid of a spectroheliscope and prominence spectroscope showed that:

(i) the source *G3* coincides exactly with the position of a solar prominence;

(ii) the source *G2* is slightly displaced from a solitary sunspot surrounded by a bright $H\alpha$ region;

(iii) there is no correlation with the source *G1* as the western limb of the sun was completely inactive.

These results are illustrated in Fig. 4. Unfortunately, data on the distribution of the coronal green line have not yet been received.

4. Effect of a Possible Temperature Variation

It seems remarkable that in most eclipse measurements on the *E1* layer, such as those by PIERCE (1948) on the eclipse of October 1940, HARANG (1945) on the eclipse of July 1945, MINNIS (1955) on the eclipse of February 1952, and again MINNIS on the eclipse of June 1954 (private communication), the rapid drop in N at the beginning of the eclipse appears to indicate, on the previous explanation, a brightening of the solar western edge. As from these measurements there is no corresponding brightening of the eastern edge, a coronal effect seems unlikely.

Although temperature variation in the ionosphere is a debatable point (GERSON, 1951), it represents a possibility which should not be overlooked. Such a variation would affect the value of N both through a change in q and a change in α.

According to Chapman's theory, if χ is the solar zenith distance and q_0 is the rate of ion production for an overhead sun,

$$\frac{dN}{dt} = q_0 \cos \chi - \alpha N^2 \tag{5}$$

Equation (5) indicates that under quasi-equilibrium conditions

$$N \propto (\cos \chi)^{1/2} \tag{6}$$

If the temperature T varies in the course of the day, we may assume that

$$N \propto (\cos \chi)^n \tag{7}$$

since a variation in temperature will affect the value of q_0. In this case, in equation (5), q_0 must be replaced by $q_0 \, . \, T_0/T$ and α by $\alpha(T/T_0)^r$, T_0 being the temperature when the sun is directly overhead. Hence equation (5) becomes

$$\frac{dN}{dt} = q_0 \frac{T_0}{T} \cos \chi - \alpha \left(\frac{T}{T_0}\right)^r . \, N^2 \tag{8}$$

Equation (7) also implies that

$$\frac{\mathrm{d}N}{\mathrm{d}t} = q_0 (\cos \chi)^{2n} - \alpha N^2 \tag{9}$$

and, from equations (8) and (9),

$$T = T_0 \,.\, (\cos \chi)^{\frac{1-2n}{1+r}} \;. \tag{10}$$

The value of n obtained from the control days by plotting log N against log cos χ was 0·6. Values of r of 0·5, -1, and -3 have been suggested (GERSON, 1951; SEATON, 1948; BARAL and MITRA, 1950). To satisfy the common-sense condition that if a change of temperature occurs at all, it must be such that T rises during the middle part of the day, from equation (10) it appears that $r = -3$. Hence equation (10) becomes

$$T = T_0 \,.\, (\cos \chi)^{0\cdot 1} \;. \tag{11}$$

In order to make $T = 300°$ K at local noon, the temperature at 0630 and 1800 should be 264°K, i.e., a reasonable rise of 36°K during the daytime period.

Values of $\alpha = 5\cdot 7 \times 10^{-9}$ cm³ sec⁻¹ and $q_0 = 100$ cm⁻³ were found to give the best fit from equation (9) to the control curve.

If $\mathrm{d}N/\mathrm{d}t = 0$, by putting $r = -3$ in equation (8) we have

$$\left(\frac{T}{T_0}\right)^2 = \frac{\alpha N^2}{q_0 \cos \chi} \tag{12}$$

and during the eclipse, replacing q_0 by fq_0,

$$\left(\frac{T}{T_0}\right)^2 = \frac{\alpha N^2}{fq_0 \cos \chi} \;. \tag{13}$$

This applies to the minimum value of N during the eclipse, and, from the eclipse data, this would indicate a rise of approximately 110°K during the middle of the eclipse!

We must conclude that the observed deviation from the expected effects for a solar disk of uniform intensity during the eclipse cannot be explained by a change in temperature. This seems to confirm that the conclusions of BATES (1951) to this effect are also applicable to the *E1* layer.

5. ACKNOWLEDGEMENTS

The authors take this opportunity of thanking the Council of Scientific and Industrial Research for a substantial grant towards the building of the automatic equipment and a scholarship to one of them; Prof. J. A. GLEDHILL for several discussions; Dr. A. K. DAS of the Astrophysical Observatory, Kodaikanal, for supplying the astronomical sketches and photographs of the sun; and Dr. A. M. van Wijk, of the Magnetic Observatory, Hermanus, for communicating the magnetic data.

References

Baral, S. S., and Mitra, A. P.	1950	*J. Atmosph. Terr. Phys.*, **1,** 95.
Bates, D. R.	1951	*Proc. Phys. Soc.*, **B64,** 805.
Gerson, N. C.	1951	*Rep. Prog. Phys.*, **14,** 316.
Harang, L.	1945	*Terr. Mag. Atmos. Elec.*, **50,** 307.
Kelso, J. M.	1952	*J. Geophys. Res.*, **57,** 357.
Minnis, C. M.	1955	*J. Atmosph. Terr. Phys.*, **6,** 91.
Pierce, J. A.	1948	*Proc. Inst. Radio Engrs.*, **36,** 8.
Seaton, S. L.	1948	*J. Meteorol.*, **5,** 204.
Szendrei, M. E., and McElhinny, M. W.	1955	*J. Atmosph. Terr. Phys.* (in press).
Wadley, T. L.	1949	*J. Inst. Elect. Engrs.*, Pt. III, **96,** 483.

Solar Eclipses and the Ionosphere (Special Suppl. Vol. 6 to J. Atmosph. Terr. Phys.), Pergamon Press, London, 1956

The *F2* Layer during Eclipses of 1952, 1954, and 1955

C. M. Minnis
Radio Research Station, Slough

Abstract—It is desirable that the electron density as a function of height in the *F2* layer should, where possible, be computed, because the changes in f_0F2 alone during an eclipse can give a misleading impression of the response of the layer to the drop in incident radiation intensity.

At Khartoum in 1952, f_0F2 remained almost constant for two hours during the eclipse, although there were marked changes in electron density at constant height at all levels. Assuming that these changes can be accounted for by an attachment process, values of the attachment coefficient have been derived which are similar to those obtained by Savitt at Bocayuvo. The great height of the layer at Khartoum has allowed values to be computed for heights up to 400 km.

The small difference between f_0F1 and f_0F2 at Inverness and Slough in 1954 has made it difficult to deduce the shape and true height of the *F2* layer. After allowing for the retardation due to the *F1* layer, it is concluded that the observed fall in f_0F2 was not accompanied by any appreciable change in the height or thickness of the *F2* layer. The attachment coefficient determined for the peak of the *F2* layer was considerably smaller than the values obtained at Bocayuvo and Khartoum for the same height.

1. Introduction

The changes which occur in the *E* and *F1* layers during a solar eclipse are, in general terms, explicable in terms of Chapman layers and rapid changes in the intensity of the incident radiation. In contrast, the response of the *F2* layer is complex, and it is not usually possible to predict the changes in electron density and layer height which are found to occur. Rydbeck (1946) has drawn attention to systematic differences between the changes which take place in f_0F2 during eclipses at high- and low-latitude stations, and the results obtained in recent years tend to confirm that the change in f_0F2 at low-latitude stations may be negligibly small during an eclipse which does not coincide with a time of day when f_0F2 is changing rapidly.

A closer examination of the *h'f* records made during an eclipse shows that, even when the maximum electron density does not change, there may still be important changes in the shape and true height of the *F* layer as a whole. In view of this, when analysing the effects of an eclipse on the *F* layer, it is desirable to attempt to estimate the distribution of electron density as a function of true height. The conclusions based on analyses of this kind for the 1952 and 1954 eclipses are summarized in the sections which follow.

2. February 1952 Eclipse: Khartoum

Although the eclipse was total at Khartoum, there was no perceptible eclipse effect on f_0F2, which remained constant, within a range of $\pm 0{\cdot}1$ Mc/s, for two hours before rising slightly near the end. Kelso's (1952) method was used to determine the electron density as a function of true height, and the results show that, at the maximum phase of the eclipse, the peak of the *F2* layer had risen to a height about 100 km above its normal level. At least partly as a result of this, the density at constant height fell at all levels by a considerable amount and there were, in addition, simultaneous changes in the shape of the layer (Fig. 1). After totality, a temporary ledge, which has been referred to as the $F1\frac{1}{2}$ layer, appeared just above the *F1* layer and eventually merged into the *F2* layer (Fig.2). This

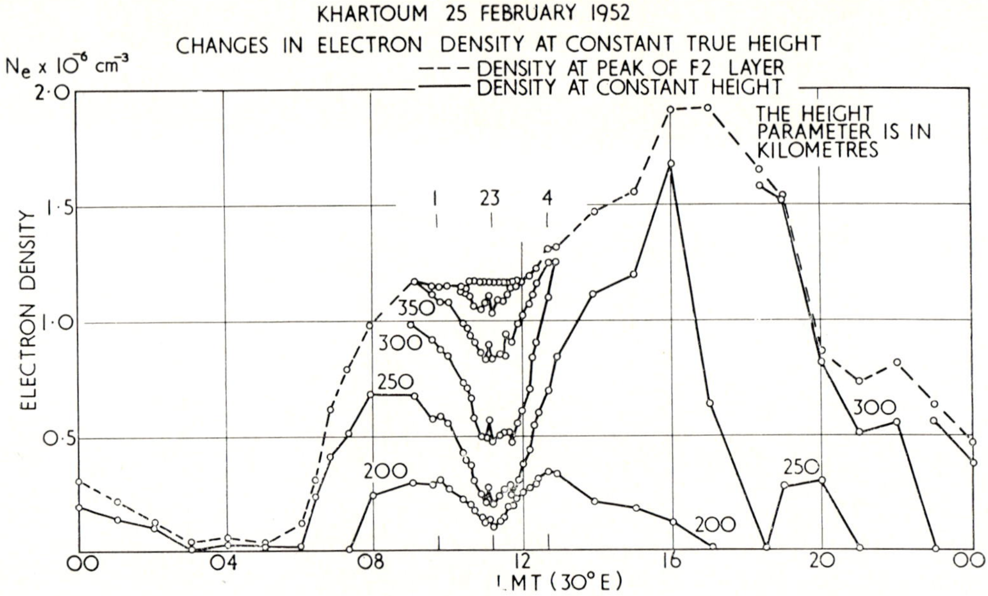
KHARTOUM 25 FEBRUARY 1952
CHANGES IN ELECTRON DENSITY AT CONSTANT TRUE HEIGHT
$N_e \times 10^{-6}$ cm^{-3}
--- DENSITY AT PEAK OF F2 LAYER
—— DENSITY AT CONSTANT HEIGHT
THE HEIGHT PARAMETER IS IN KILOMETRES
ELECTRON DENSITY
2·0
1·5
1·0
0·5
1 23 4
350
300
250
200
00 04 08 12 16 20 00
LMT (30° E)
Fig. 1.

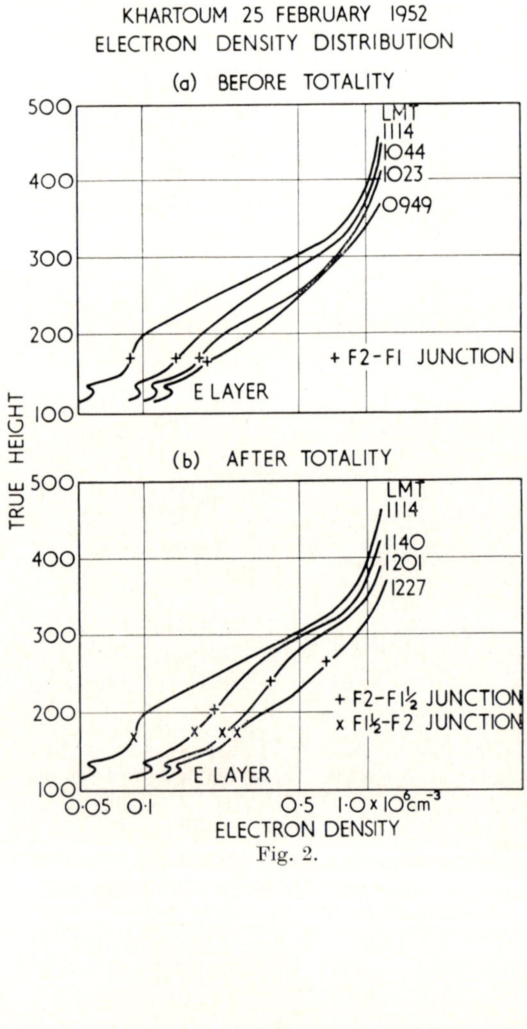
KHARTOUM 25 FEBRUARY 1952
ELECTRON DENSITY DISTRIBUTION
(a) BEFORE TOTALITY
LMT
1114
1044
1023
0949
+ F2-F1 JUNCTION
E LAYER
(b) AFTER TOTALITY
LMT
1114
1140
1201
1227
+ F2-F1½ JUNCTION
x F1½-F2 JUNCTION
E LAYER
TRUE HEIGHT
500
400
300
200
100
0·05 0·1 0·5 1·0 x 10^6 cm^{-3}
ELECTRON DENSITY
Fig. 2.

ledge is probably analogous to the "new $F2$ layer" which has been reported by LEDIG *et al.* (1946), ESTRABAUD (1952), and PIGGOTT (1952) during eclipses which occurred in the morning while f_0F2 was still rising rapidly.

The mechanism of the $F2$ layer is not sufficiently well understood to allow a quantitative explanation to be given for such changes, but they appear to be consistent with a rapid fall with increasing height in the rate of disappearance of free electrons and with the assumption that electrons generated in the $F1$ layer may move upwards to contribute to the ionization in the $F2$ layer (MINNIS, 1955).

SAVITT (1950) has used the results of the 1947 eclipse at Bocayuvo to estimate for various heights the value of b in the expression

$$\frac{dN}{dt} = q - bN .$$

The method used was to find a value of b which made the calculated and observed minimum values of N equal; differences in the times of occurrence of the minima were ignored. A similar analysis of the Khartoum results leads to values which are quite close to those quoted by SAVITT (Table 1), but, in view of the possibility of vertical movement, the detailed interpretation of such figures requires caution.

Table 1. Estimated values of b in the F layer

h (km)	$b \times 10^4$ (sec^{-1})			
	Bocayuvo 1947	Khartoum 1952	Inverness 1954	Slough 1954
200	—	5·5	—	—
220	6·5	—	1·5	2·5
250	5·4	5·0	—	—
300	3·5	2·5	—	—
320	2·8	—	—	—
350	—	0·9	—	—
400	—	0·6	—	—

3. JUNE 1954 ECLIPSE: INVERNESS AND SLOUGH

During this eclipse the maximum electron densities in the $F1$ and $F2$ layers differed, for much of the time, by only 20 per cent. For this reason it is difficult to compute the changes in the height and shape of the $F2$ layer owing to the very large group retardation due to the $F1$ layer. Nevertheless it seemed to be worth trying to deduce what the virtual height of the $F2$ layer would have been in the absence of the $F1$ layer, and then to compute the true height and thickness from the derived $h'f$ curves using the Appleton-Beynon (1940) method. This has been done, and, although great accuracy cannot be claimed for the results, it seems reasonable to conclude that there were no important changes in the height or thickness of the $F2$ layer at either station.

Both at Inverness and Slough the maximum electron density fell to about half its normal value, and, if it is assumed that the height of the layer remained constant, values of b for the height of maximum ionization can be computed as explained earlier. The values of b for Inverness and Slough are smaller than those found for Khartoum and Bocayuvo (Table 1). If, however, the layer remained at a constant height at the high-latitude stations, it would seem reasonable to expect a lower value of b, since at low-latitude stations b would presumably contain an additional term arising from the upward movement of the layer.

4. June 1955 Eclipse: Singapore

Only the values of f_0F2 are available so far, but it is evident that the effect of the eclipse (which was about 45 per cent total at Singapore) on the maximum density in the $F2$ layer was negligible. This is consistent with the Khartoum results and with Rydbeck's classification, and it remains to be seen what changes occurred in the vertical distribution of the ionization.

5. Conclusions

It is not yet possible to account quantitatively for the observed effects of eclipses on the $F2$ layer. The results obtained during recent eclipses appear to point to considerable differences between the response of the layer at high- and low-latitude stations, which are probably due to differences in the height of the layer and the dip of the earth's magnetic field. It also seems possible that at low latitudes some of the $F1$-layer ionization can contribute to that in the $F2$ layer. Estimated values, obtained at Khartoum, of a coefficient which reflects the electron attachment rate in the $F2$ layer, are in reasonable agreement with those measured at Bocayuvo, but exceed the values for Inverness and Slough.

Acknowledgement

The work described above was carried out as part of the programme of the Radio Research Board and is published by permission of the Director of Radio Research of the Department of Scientific and Industrial Research.

References

Appleton, E. V., and Beynon, W. J. G. 1940 *Proc. Phys. Soc.*, **52**, 518.
Estrabaud, S. 1952 *Compt. Rend.*, **235**, 1521.
Kelso, J. M. 1952 *J. Geophys. Res.*, **57**, 357.
Ledig, P. G., Giesecke, A. A., Jones, M. W., and Chernosky, E. J. 1946 *Terr. Mag. Atmos. Elec.*, **51**, 411.
Minnis, C. M. 1955 *J. Atmosph. Terr. Phys.*, **6**, 91.
Piggott, W. R. 1952 *Proc. Mixed Commission on the Ionosphere*, August 1952, 42.
Rydbeck, O. E. H. 1946 *Trans. Chalmers Univ. Tech.*, No. 53.
Savitt, J. 1950 *J. Geophys. Res.*, **55**, 385.

Solar Eclipses and the Ionosphere (Special Suppl. Vol. 6 to J. Atmosph. Terr. Phys.), Pergamon Press, London, 1956

Comparaison des Résultats des Sondages Ionosphériques à Bangui, Ibadan, Khartoum, et Leopoldville

R. P. LEJAY and J. DURAND
Laboratoire National de Radioélectricité, Bagneux/Seine

Abstract—Results for the *F2* region are compared for the eclipse of 25 February 1952. At Bangui and Ibadan, those stations nearest to the magnetic equator, there is clear evidence for the appearance during the latter half of the eclipse of a new *F2* stratification below the "old *F2*" layer. Although less marked, an analogous phenomenon is detected at Khartoum, but at Leopoldville (magnetic dip = 33·5° S) it is absent. Rapid increase in height of the "old *F2*" layer, without change of its critical frequency, also seems to accompany the appearance of the "new *F2*" layer.

Possible causes of this effect of proximity to the magnetic equator are considered.

1. INTRODUCTION

NOUS avons comparé les résultats des sondages ionospériques effectués lors de l'eclipse du 25 février 1952 dans les stations ci-après, particulièrement en ce qui concerne la région *F*.

Station	Coord. géographi.	Incl. magnét	Heures T.U. de l'éclipse			Heures locales du maximum
			Commenct.	Totalité	Fin	
Bangui	4° 21′ N, 18° 35′ E	12° S	env. 0700	totale à env. 0820	env. 0950	env. 0830
Ibadan	7° 22′ N, 4° E	2·5° S	env. 0650	éclipse partielle 75%	env. 0920	env. 0830
Khartoum	15° 35′ N, 32° 35′ E	13° N	env. 0750	totale à env. 0915	env. 1040	env. 1015
Leopoldville	4° 22′ S, 15° 15′ E	33·5° S	env. 0640	éclipse partielle 75%	env. 0920	env. 0810

Les fig. 1, 2, 3a, et 4 donnent les courbes de variation de la f_0F2 pour ces quatre stations durant l'éclipse.

A titre de comparaison, la fig. 3b indique la variation de la f_0F2 à Huancayo (12° S, 75° 20′ W), situé sur l'équateur géomagnétique (Incl. Mag. voisine de zéro), au cours de l'éclipse partielle du 25 janvier 1944 (WELLS et SHAPLEY, 1946).

2. APPARITION D'UNE NOUVELLE COUCHE *F2*

Il est remarquable de constater que les sondages de Bangui, de Ibadan, et de Huancayo, c'est-à-dire ceux des stations le plus voisines de la ligne d'inclinaison magnétique nulle, accusent toutes nettement l'apparition d'une nouvelle couche

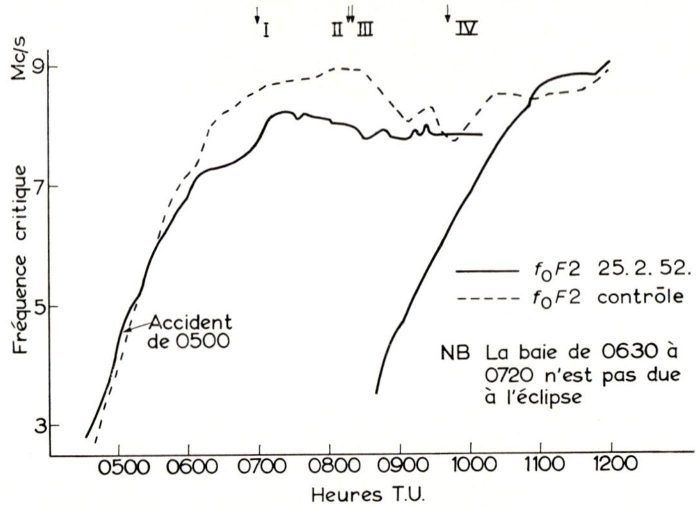

Observations de l'éclipse à Bangui, 25 Février 1952

Fig. 1.

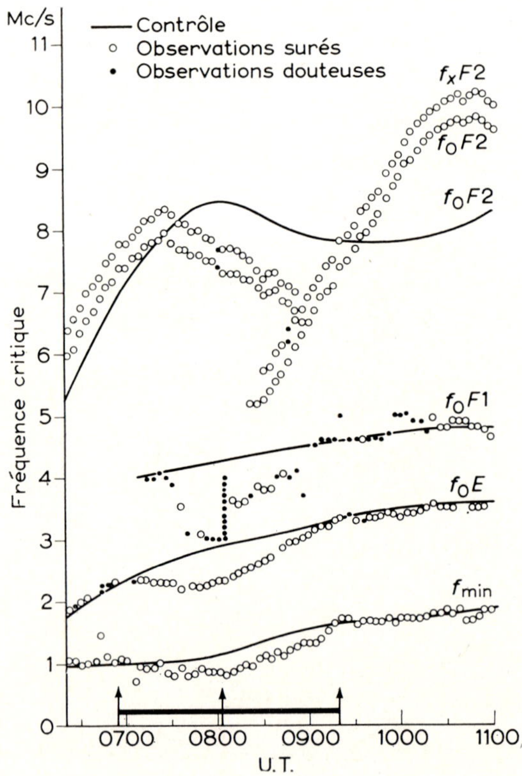

Fig. 2. Observations de l'eclipse à IBADAN Niger 25 fev., 1952.

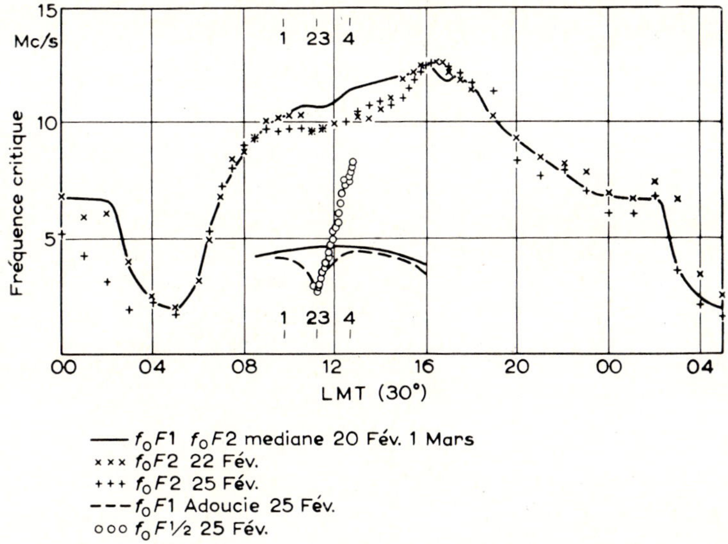

Fig. 3a Observations de l'éclipse à KHARTOUM le 25 fev., 1952.

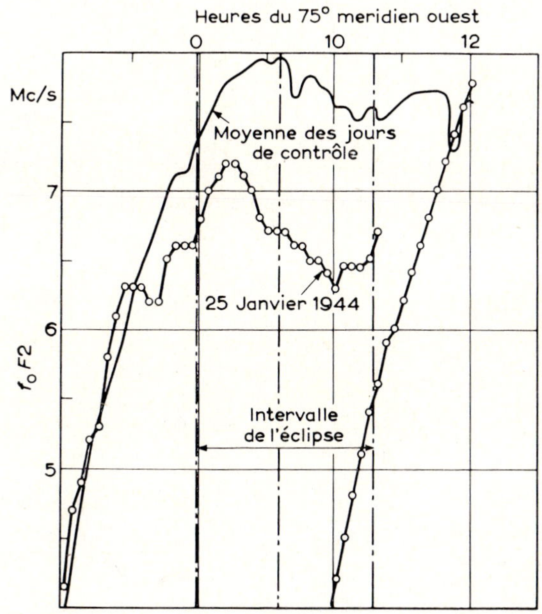

Fig. 3b Observations de l'éclipse à HUANCAYO le 25 janv., 1944.

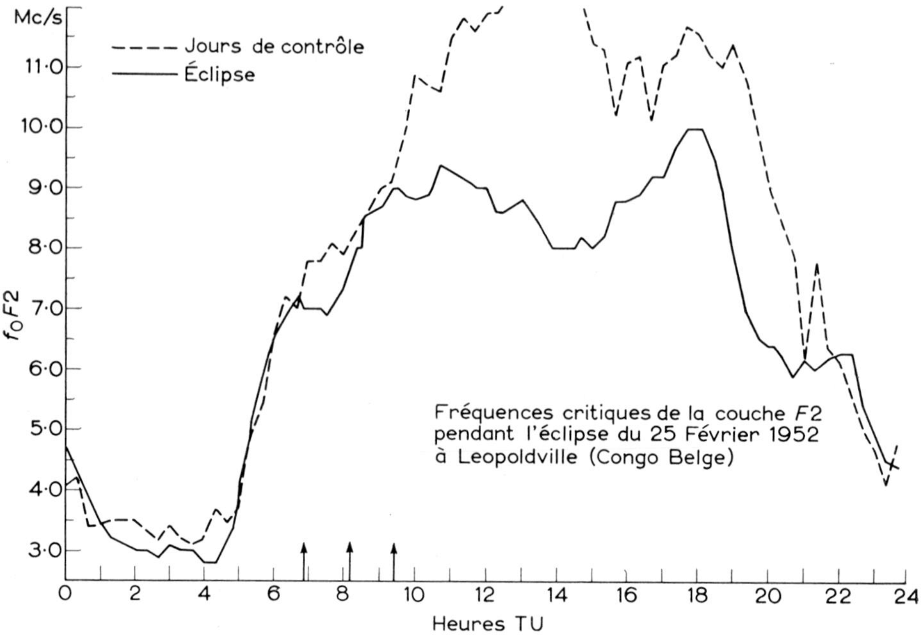

Fig. 4

F2 au-dessous de l'ancien *F2*, soit après le 2e contact si l'éclipse est totale, soit vers la fin de l'éclipse si celle-ci est partielle. Cette nouvelle couche *F2* coexiste durant un certain temps avec l'ancien *F2* pour s'y substituer finalement. Son évolution est approximativement celle qui a lieu au lever du Soleil.

A Khartoum (fig. 3a) un phénomène analogue se produit mais la nouvelle stratification qui apparaît dans la région *F* et qui a été désignée par C. M. MINNIS (1955) par $F1\frac{1}{2}$ est moins marquée, surtout si l'on se réfère aux enregistrements des sondages (fig. 7 et 8).

Au contraire à Leopoldville (inclinaison mag. = 33·5° S), à part une légère baie entre 0650 et 0900 (voir fig. 4), probablement attribuable à l'éclipse, on ne voit pas d'effet particulier de l'éclipse sur l'évolution de la f_0F2; la baisse des fréquences enregistrée le 25 février 1952 coincidant avec l'agitation magnétique, on ne peut lui donner une valeur significative, comme divers auteurs l'ont d'ailleurs fait remarquer pour d'autres stations.

Aucune nouvelle couche *F2* ou stratification ne s'y manifeste nettement.

La comparaison des enregistrements de Bangui et de Khartoum où l'éclipse a été totale montre qu'à Khartoum, l'ancien et le nouvel *F2* finissent par se fondre l'un dans l'autre, tandis qu'à Bangui les fréquences critiques du nouvel *F2* sont toutes très nettes, à tangente verticale, et l'ancien *F2* dont la fréquence critique reste stationnaire, est finalement occulté par le nouvel *F2* situé en dessous et qui subit une croissance rapide.

D'après C. M. MINNIS (1955), la différence de comportement de *F2* entre les enregistrements de la station d'Ibadan (où les phénomènes ont été les mêmes qu'à Bangui) et ceux de Khartoum, proviendrait de la différence dans les heures locales

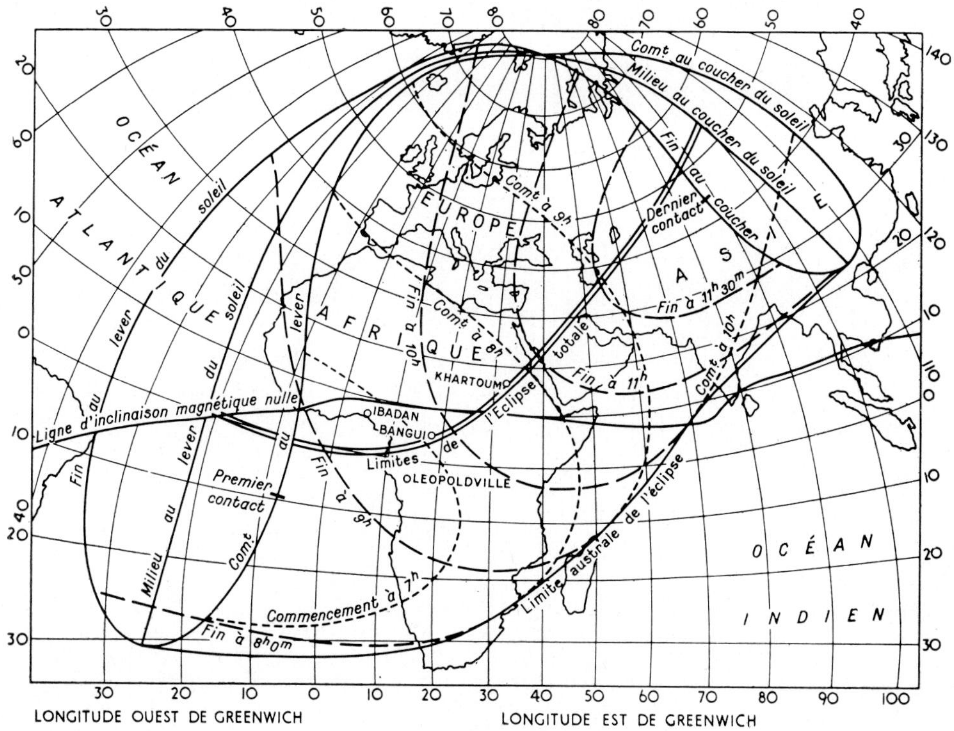
OCÉAN
ATLANTIQUE
EUROPE
AFRIQUE
ASIE
OCÉAN
INDIEN
KHARTOUM
IBADAN
BANGUI
LEOPOLDVILLE
Ligne d'inclinaison magnétique nulle
Comt au coucher du soleil
Milieu au coucher du soleil
Fin au coucher
Dernier contact
Premier contact
Comt à 9h
Comt à 8h
Comt à 10h
Fin à 10h
Fin à 11h
Fin à 11h30m
Fin à 9h
Fin à 8h0m
Commencement à 7h
Limites de l'Eclipse totale
Limite australe de l'éclipse
LONGITUDE OUEST DE GREENWICH
LONGITUDE EST DE GREENWICH

Fig. 5

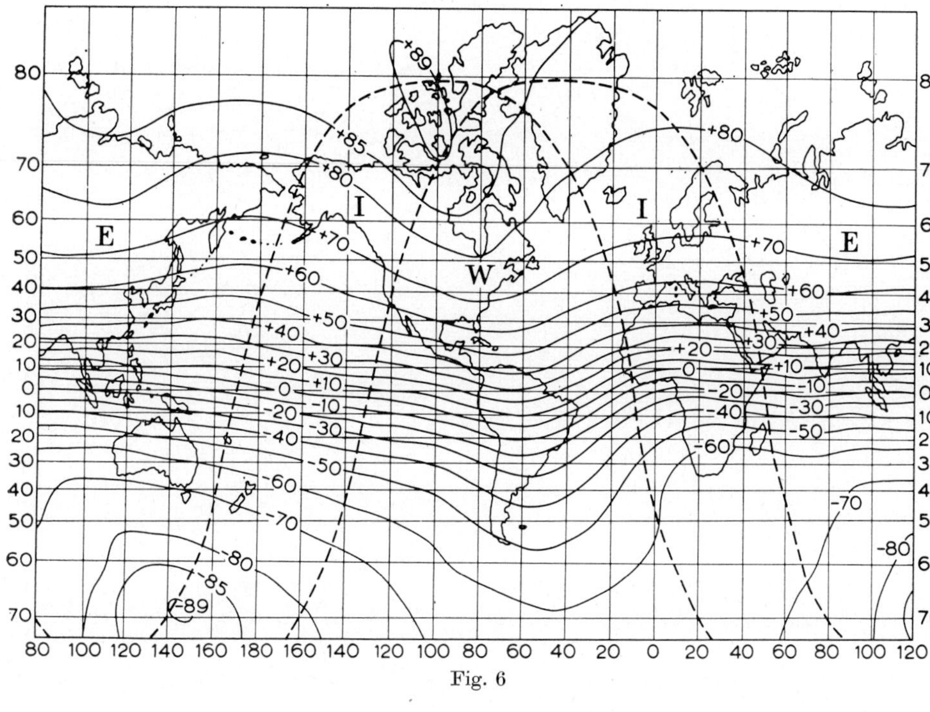
E
I
W
I
E
+89
+85
+80
+70
+60
+50
+40
+30
+20
+10
0
-10
-20
-30
-40
-50
-60
-70
-80
-85
-89

Fig. 6

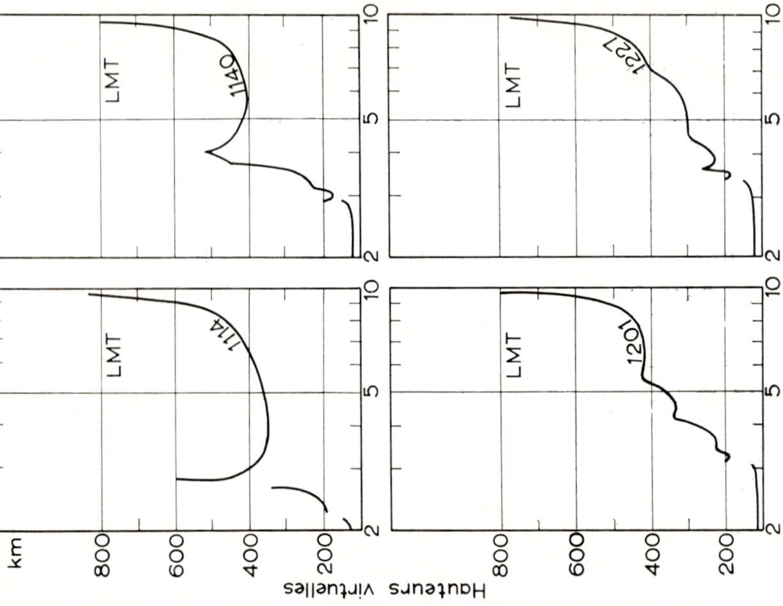

Fig. 8. Observations ionosphériques durant l'éclipse du 25.2.1952 à KHARTOUM après la totalité.

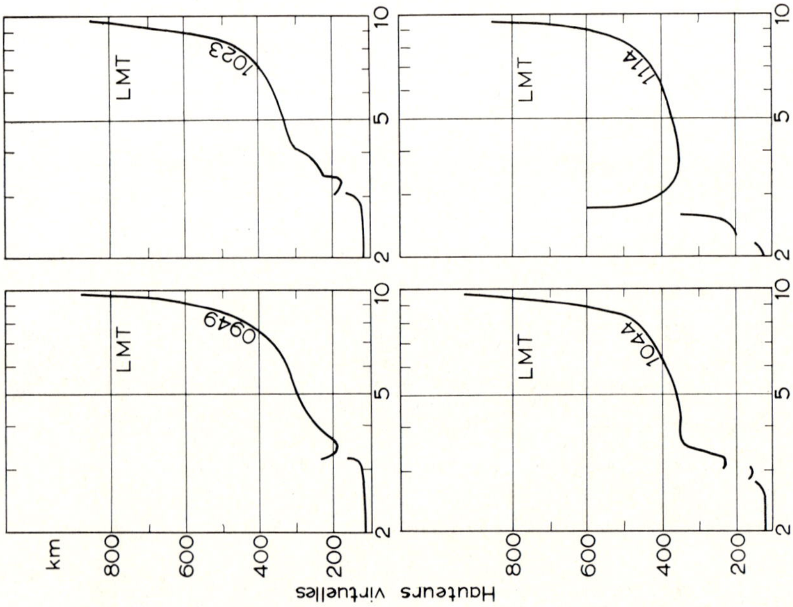

Fig. 7. Observations ionosphériques durant l'éclipse du 25.2.1952 à KHARTOUM avant la totalité.

de l'éclipse aux deux endroits. En effet, à Ibadan comme à Bangui, la couche *F2* (éclipse à 0830 locales env.) n'avait pas terminé sa formation au moment de l'éclipse, tandis qu'à Khartoum elle était déjà stable à ce moment (maximum de l'éclipse à 1015 locales). Ce raisonnement se trouve malheureusement contredit par le comportement de la région *F* à Leopoldville où l'heure locale du maximum de l'éclipse a été pratiquement la même qu'à Ibadan (0810) et où, bien que la grandeur de l'éclipse ait été pratiquement la même qu'à Ibadan (75%) aucune stratification n'est apparue.

Il semble donc bien que l'apparition du nouvel *F2* ou d'une stratification nette de *F2* soit plutôt en rapport étroit avec la proximité de l'équateur magnétique.

3. L'élévation rapide de l'"ancien *F2*".

L'examen des enregistrements de sondage de Bangui (fig. 9 et 10), comme ceux de Khartoum, où l'éclipse a été totale, révèle en plus de l'apparition d'un nouvel

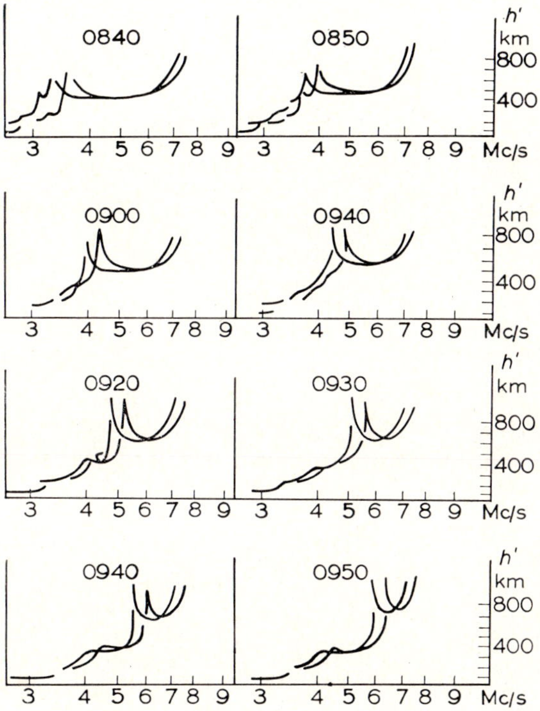

Fig. 9. Eclipse du 25.2.52 BANGUI (A.E.F).

F2 ou d'une stratification inférieure de *F2*, une rapide élévation de l'ancien *F2* après le 2e contact.

La hauteur virtuelle du maximum d'ionisation de l'ancien *F2* à Bangui (h'_m) est en effet passée de 400 km à 0840 à 800 km à 1020, accusant une vitesse de translation verticale virtuelle de l'ordre de 65 mètres par seconde.

D'autre part, durant cette ascension rapide, la fréquence critique de l'ancien

F2 est demeurée invariable (env. 7, 8 Mc/s), tandis que celle du nouvel *F2*, situé en dessous ne cessait de croître et de s'approcher de celle de l'ancien *F2* pour finir par la dépasser, le nouvel *F2* occultant alors complètement la couche supérieure.

Nous avons mesuré, en utilisant la méthode de RATCLIFFE (1951), la demi-épaisseur du nouvel *F2* durant cette évolution ainsi que le retard apporté par la traversée de cette couche aux hauteurs virtuelles de l'ancien *F2* sur les enregistrements de sondage. Nous en avons ainsi déduit l'évolution de la hauteur *réelle* (h_m) du maximum d'ionisation de l'ancien *F2*.

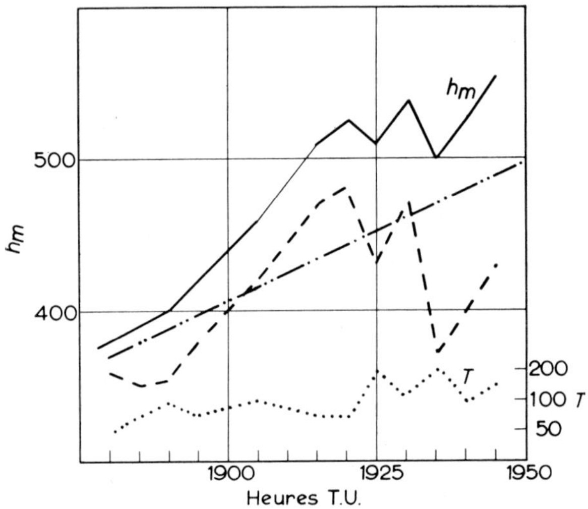

Fig. 10 Variation de h_m en fonction de l'heure (ancien *F2*). Éclipse du 25 fevrier 1951 à BANGUI.

Les résultats sont indiqués dans le graphique de la fig. 10 où:

T est la demi-épaisseur en km du nouvel *F2*,

Δh_m est le retard provoqué par la traversée de la nouvelle couche,

h_m est la hauteur réelle en km du maximum d'ionisation de l'ancien *F2*.

La courbe en trait plein indique les valeurs de h_m d'après la méthode de RATCLIFFE (1951), qui ne tient compte que du retard imputable à la moitié inférieure de la couche parabolique traversée (nouvel *F2*).

La courbe en trait tireté indique les valeurs de h_m obtenues en doublant ce retard, c'est-à-dire en supposant le nouvel *F2* parabolique entièrement distinct de l'ancien.

L'évolution moyenne générale de la hauteur réelle h_m du maximum d'ionisation de l'ancien *F2* est indiqué par la droite en trait mixte dont la pente correspond à une vitesse verticale d'environ 100 km en 55 minutes, soit de 30 metres/sec environ.

Enfin la courbe inférieure en pointillés représente l'évolution de la demi-épaisseur T du nouvel $F2$, qui croît de 50 km environ à 0840 environ 150 km à 0950.

A partir de 0950, la valeur de h_m cesse d'être mesurable, tandis que h_m continue à croître jusqu'à la disparition de la couche.

4. Conclusions

Nous sommes donc en présence de deux phénomènes remarquables:

(a) apparition d'un nouvel $F2$ d'autant plus distinct que le lieu du sondage est plus voisin de l'équateur magnétique;

(b) élévation rapide de l'ancien $F2$, sans changement de sa fréquence critique.

Ce phémomène semble également se produire uniquement dans les régions voisines de l'équateur magnétique.

Ainsi, par exemple, l'élévation de l'ancien $F2$ a été aussi très rapide à Huancayo lors de l'éclipse du 25-1-1944.

Au contraire, à Leopoldville aucune élévation sensible de $F2$ après le maximum de l'éclipse du 25-2-1952, n'a été constatée.

L'explication de ce comportement n'est pas facile. Il semble bien toutefois que la théorie du "drift" due à D. F. Martyn appliquée à l'équateur magnétique, où le "drift" vertical a sa valeur maximum, puisse être invoquée.

Toutefois, la valeur de 30 mètres/sec trouvée ci-dessus pour la vitesse de transport vertical de l'ancien $F2$ apparaît trop élevée. En effet rappelons que M. Hirono et H. Maeda ont calculé cette vitesse pour Huancayo et trouvé qu'elle devrait passer au milieu de la journée locale par le maximum de 12 mètres/sec seulement.

Il est possible également d'envisager une augmentation anormale du coefficient de recombinaison à la base de l'ancien $F2$, due à la présence du nouvel $F2$ inférieur, ec qui entraînerait une élévation purement apparente du niveau d'ionisation maximum h_m de l'ancien $F2$.

Bibliographie

Wells, H. W., et Shapley, A. H. 1946 *Terrest. Mag. Atmos. Elect.*, **51,** 401.

Minnis, C. M. 1955 *J. Atmosph. Terr. Phys.*, **6,** 105.

Ratcliffe, J. A. 1951 *J. Geophys. Res.*, **56,** 463.

Piggott, W. R. 1952 *Commission mixte de l'ionosphère. Compte-rendu de la 3ème réunion, Canberra*, p. 44.

Estrabaud, S. 1954 L'ionosphère à Bangui en février 1952 et pendant l'éclipse du 25 février 1952. *Note préliminaire No. 174, du Laboratoire National de Radioélectricité.*

Lepechinsky, D., et Pillet, G. 1952 Contribution à l'étude de l'influence du magnétisme terrestre sur l'ionisation de la couche; *F2. Note préliminaire No. 139, du Laboratoire National de Radioélectricité.*

Solar Eclipses and the Ionosphere (Special Suppl. Vol. 6 to J. Atmosph. Terr. Phys.), Pergamon Press, London, 1956

Movements in the *F2* Layer of the Ionosphere during some Solar Eclipses

J. O. THOMAS AND A. R. ROBBINS
Cavendish Laboratory, Cambridge

Abstract—Five different sets of ionosphere soundings made during eclipses have been analysed in an attempt to see whether vertical transport of electrons plays an important part in the *F2* layer during an eclipse. The assumption was made that the rates of production and loss of electrons were those tentatively suggested by RATCLIFFE *et al.* (1956). It has been found that the five sets of observations lead to fairly similar conclusions. Vertical movements of importance seem to have occurred during the eclipses. Their possible causes are considered.

1. INTRODUCTION

THE ordinary daily and seasonal variations of the *F2* layer of the ionosphere are different from what would be expected on the theory of a "simple" layer, and are not well understood. It is therefore not surprising that its behaviour during an eclipse is also difficult to explain. The behaviour on an ordinary day is almost certainly much influenced by vertical transport of electrons; indeed, this transport may well be the determining factor (MARTYN, 1955). It might therefore be expected that vertical transport might play an important and decisive part in determining the layer's response to an eclipse. It is the purpose of this paper to estimate from experimental results the magnitude of that transport during eclipses.

2. THE CONTINUITY EQUATION FOR THE *F2* LAYER

The equation of continuity of the electron density N at a fixed height may be written

$$\mathrm{d}N/\mathrm{d}t = I(t) \,.\, f(\chi,h) - \beta(h) \,.\, N - M(h,t) \tag{1}$$

where

$$M(h,t) = \mathrm{d}(Nw)/\mathrm{d}h \tag{2}$$

and $I(t)$ represents the intensity of the ionizing radiation outside the earth's atmosphere (during an eclipse it is a function of time, t); $f(\chi,h)$ represents the way in which the rate of production of electrons depends on the solar zenith angle χ and the height h in the ionosphere; $\beta(h)$ represents an attachment coefficient which varies with height (reasons will be given shortly for selecting attachment rather than recombination for the process of electron loss); w represents the upward transport velocity of electrons. The effects of horizontal drifts are neglected in equation (1).

By analysing the $h'(f)$ records made during an eclipse, it is possible to determine N and $\mathrm{d}N/\mathrm{d}t$ at a series of heights and times (e.g., SAVITT, 1950). If, then, reasonable assumptions could be made about $I(t)$, $f(\chi, h)$, and $\beta(h)$, it would be possible to compute the magnitude of $M(h, t)$ at different times during the eclipse. It is easy to make reasonable assumptions about $I(t)$; the other two terms are more difficult.

A recent examination of the $h'(f)$ curves obtained at several places, seasons, and epochs of the solar cycle (SCHMERLING and THOMAS, 1956) has led to suggestions about the quantities $f(\chi, h)$ and $\beta(h)$ on ionospherically quiet days (RATCLIFFE *et al.*, 1956). It has been shown that the experimental results are consistent with the idea, first suggested by BRADBURY (1938), that the *F2* layer is formed by the ionizing process which produces the *F1* layer, and that the electrons are lost by an attachment-like process with the loss-coefficient β changing with height in the way illustrated in Fig. 1, which is reproduced from RATCLIFFE *et al.* (1956).

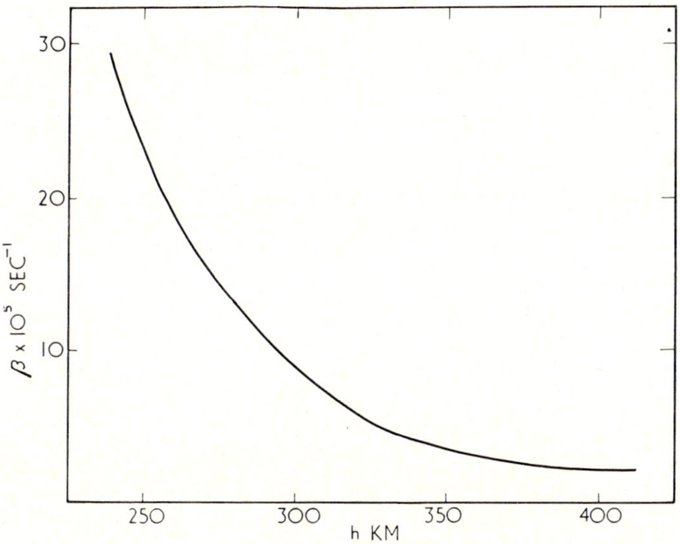

Fig. 1. To illustrate the variation of the loss coefficient β, with height (from RATCLIFFE *et al.* (1956)).

We have taken

$$f(\chi, h) = \exp\{1 - z - e^{-z} \sec \chi\}$$

with

$$z = \{h\ (km) - 180\}/45$$

and I, when there is no eclipse, is given by

$$I = 280\,\{1 + 1{\cdot}4 \times 10^{-2}\,\bar{R}\}\ \text{cm}^{-3}\,\text{sec}^{-1} \tag{3}$$

where $\bar{R}$ is the monthly average relative Zürich sunspot number.

It was emphasized by RATCLIFFE *et al.* that, although these hypotheses are consistent with all the experimental results they examined, they cannot be shown to give a unique description of the *F2* layer. It seems nevertheless to be worth while to insert them into equation (1) in order to see what movements would have to be postulated to explain the eclipse results on the basis of this ionosphere model.

In order to illustrate how different the observed *F2* layer results are from what might be expected on simple theory, data appropriate to the eclipse of 30 June 1954 observed at Cambridge were inserted into equation (1) with the transport term omitted, and periodic solutions were obtained, using the EDSAC digital computer of the University of Cambridge Mathematical Laboratory. The observed and calculated eclipse and control-day curves for 280 km are shown in Fig. 2.

The curves have been normalized to N_0, the noon value of N. The differences between the observed curves and those calculated are at once obvious.

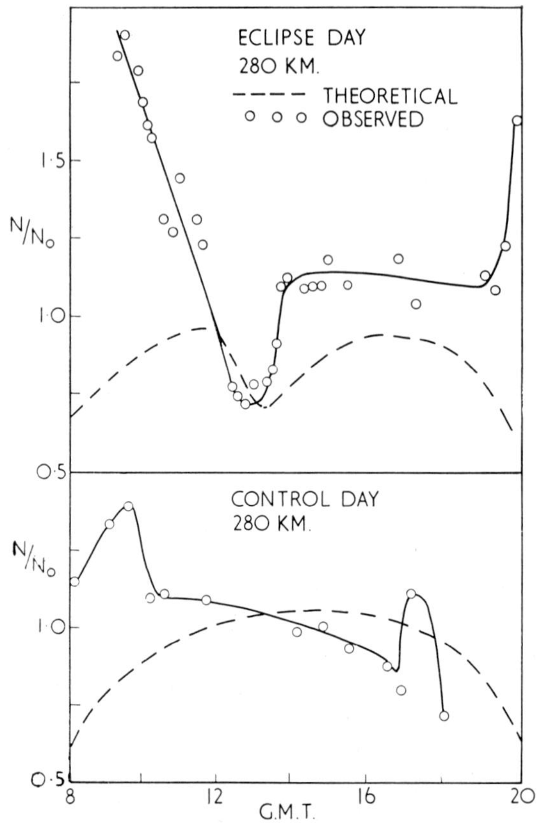

Fig. 2. Periodic solutions of equation (1) (with the transport term omitted) for the control and eclipse days at Cambridge plotted as curves of N/N_0 against time for 280 km (N_0 is the noon value of N).

3. RESULTS

Details of the eclipses studied are given in Table 1.

The magnitudes of N and dN/dt required in the calculations were obtained, for the eclipse of 1954, by KELSO's (1952) method, applied to $h'(f)$ curves obtained at Cambridge. For the eclipse of 1940 they were calculated by the same method from the $h'(f, t)$ information contained in the paper by PIERCE (1948). For the eclipse of 1952 observed at Khartoum they were taken from Fig. 11 in the paper by MINNIS (1955), and the corresponding information deduced from observations made at Ibadan was given by PIGGOTT (1955). Values for the 1947 eclipse were taken from the paper by SAVITT (1950).

The magnitude of $I(t)$ was calculated from equation (3) with the additional

Table 1. Details of the eclipses analysed

Site of observation	Date	Geographic latitude	Geographic longitude	Magnetic latitude	Geo-magnetic latitude	Dip (deg)	Time and duration of eclipse L.M.T. (250 km)	% of sun's area eclipsed at max.	Reference
Cambridge	30.6.54	52° 13′ N	00° 42′ E	51° 03′ N	54° 49′ N	+68	1113–1233–1351	79	
Bocayuva	20.5.47	17° 30′ S	43° 00′ W	06° 04′ S	07° 12′ S	−12	0813–0932–1056	100	Savitt, J. (1950)
Khartoum	25.2.52	15° 35′ N	32° 35′ E	07° 06′ N	13° 00′ N	+14	0944–1112–1241	97	Minnis, C. M. (1955)
Queenstown	1.10.40	31° 54′ S	26° 52′ E	45° 42′ S	32° 24′ S	−64	1454–1605–1706	100	Pierce, J. A. (1948)
Ibadan	25.2.52	07° 22′ N	03° 58′ E	03° 00′ S	10° 36′ N	−06	0652–0800–0918	71	Piggott, W. R. (1955)

assumption that the intensity of the ionizing radiation was proportional to the area of the visible sun.

It is desirable that the magnitudes of $M(h, t)$ (equation 2) deduced from observations made during eclipses should be compared with corresponding magnitudes calculated for control days. It may be possible at a later date to compare the eclipse results obtained at Cambridge with the results for the corresponding "mean quiet $F2$ layer" deduced in the way discussed in the paper by SCHMERLING and THOMAS (1956). Here, however, the control data for the 1954 eclipse correspond to the results obtained on one of the control days (4 July). For comparison with the results from Khartoum, the control data given by MINNIS (1955) were used, and for Bocayuva those of SAVITT (1950). No control data were available for the other two places.

The magnitudes of $M(h, t)$ (equation 2), calculated from equation (1), are shown, for the occasions of eclipse, by a continuous line in Fig. 3, and for the control days by a broken line. The differences ($\Delta M = M_E - M_C$) between the values on the eclipse day (M_E) and on the control day (M_C) are shown in Fig. 4.

4. DISCUSSION OF RESULTS

If the $F2$ layer is formed in the way suggested by RATCLIFFE *et al.* (1956), then Figs. 3 and 4 indicate what movements or extra movements must occur during eclipses. There has not yet been time to deduce the magnitudes of the vertical transport velocity by the use of equation (2), but some interesting points emerge, even from a consideration of the magnitude of $M \{= d(Nw)/dh\}$.

Fig. 4 shows that the changes of ΔM which accompanied the eclipse were somewhat similar on the different occasions. ΔM appears to be positive during the first part of the eclipse and negative during the second. The same type of variation of M seems also to have occurred during the other two eclipses, included in Fig. 3, for which no control-day information was available. The similarity of the changes under the different conditions encourages us to believe that the movements deduced in this way have some real significance.

The maximum change of M produced during the eclipse at Cambridge has been compared with some results obtained by SCHMERLING and THOMAS (1956), who deduced the magnitudes of $N(h, t)$ for the average of the five international magnetically quietest days in June 1953 from $h'(f)$ observations made at Slough. The magnitude of $M(h, t)$ was deduced for this month. The results showed that the greatest change of M during the eclipse period at Cambridge was about 60 per cent of the greatest change occurring on the (mean) control day at Slough. This seems to imply that the changes of vertical velocity during the two hours of an eclipse may be of the same order as the changes which occur over twenty-four hours on a normal day.

No attempt will be made here to present a theory of these movements. It has sometimes been suggested that important changes of temperature might occur in the $F2$ layer during an eclipse, and that these might cause movements. It seems doubtful, however, whether the thermal relaxation time of this part of the atmosphere could be small enough for these changes to occur (see BATES, 1951). A more likely explanation relates the extra movements in the $F2$ layer to the effect of the

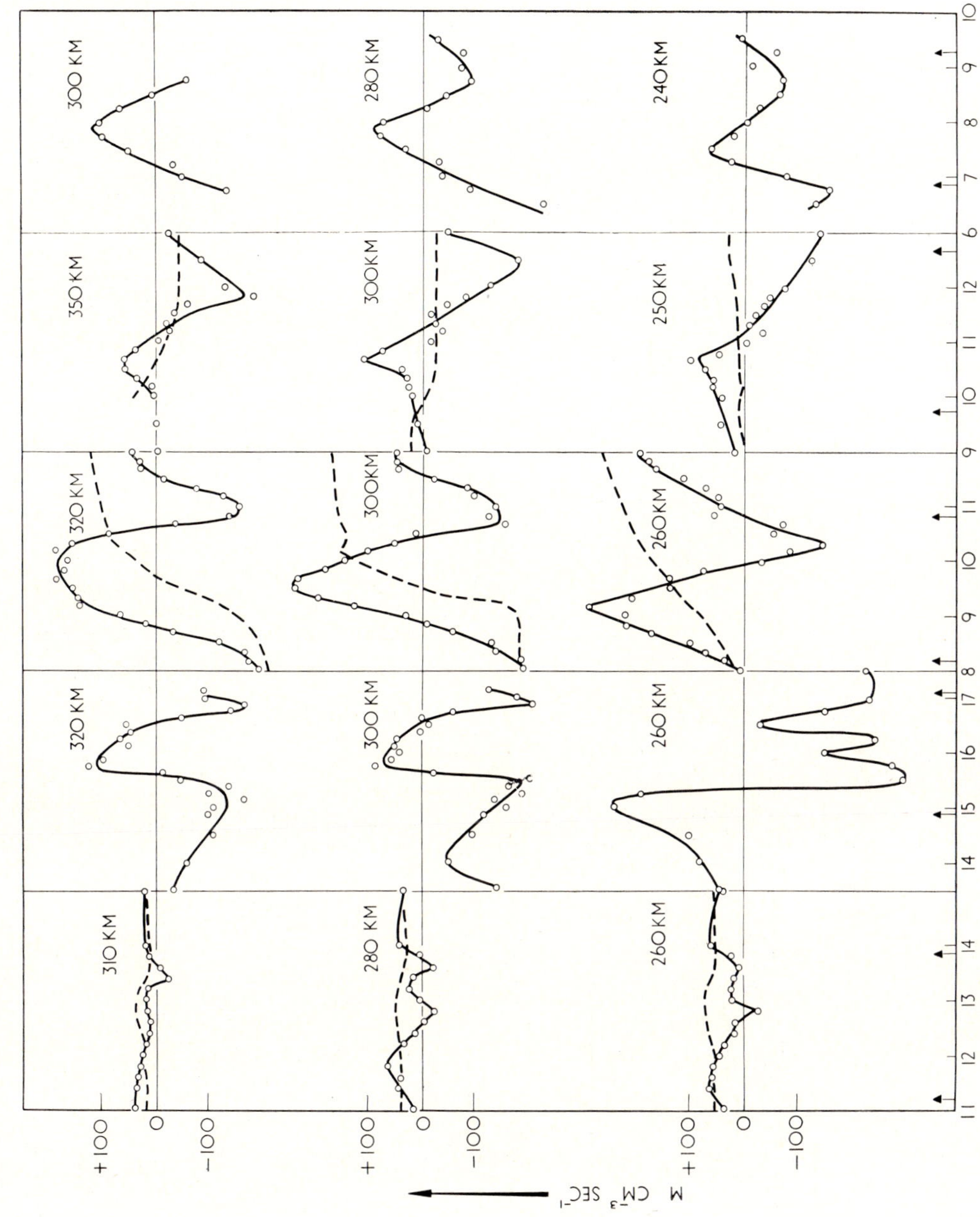

Fig. 3. To illustrate how the vertical transport term $M[= d(Nw)/dh]$ of equations (1) and (2) varies with time at the five places, (a) when there is an eclipse (continuous line) and (b) when there is no eclipse (broken line).

eclipse in lower parts of the ionosphere (PIERCE, 1948). For example, the eclipse might decrease the conductivity of that part of the ionosphere in which the geomagnetic S_q currents flow, and, as a result, an electrostatic polarization might

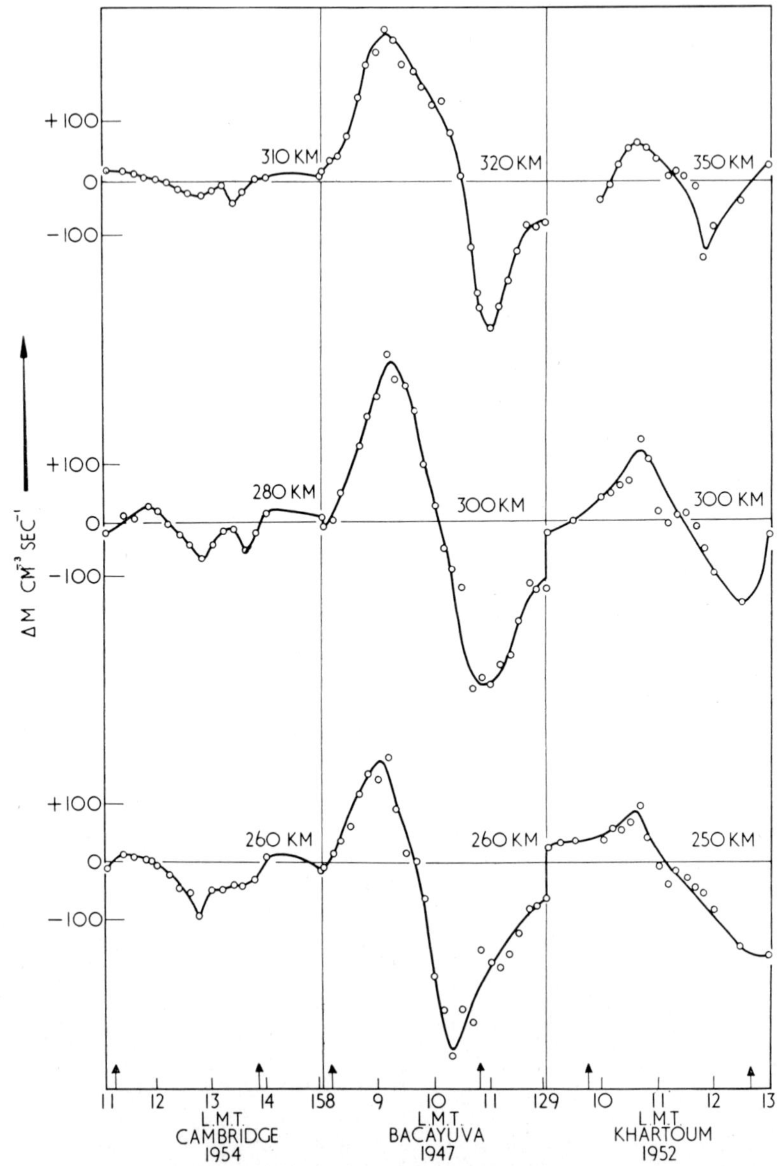

Fig. 4. To illustrate the difference ΔM $(= M_E - M_C)$ between values of M during eclipses (M_E) and on control days (M_C).

be built up in this region which would produce an electric field in the $F2$ layer. This field, crossed with the earth's magnetic field, would cause that layer to move. The details of an interaction of this kind have not been worked out.

It might prove that the electric field produced by the polarization could spread out so as to influence the *F2* layer outside the area of the visible eclipse. It would be valuable in future eclipses to have observations made at a series of places across the line of the eclipse and both inside and outside the belt of totality.

Acknowledgements

We are indebted to Mr. W. R. Piggott for making his deductions about the electron density at Ibadan available to us, and to Dr. M. Spencer for assistance both in the setting up of the apparatus for obtaining the $h'(f)$ records at Cambridge and for making some of the observations used in the analysis.

The theoretical $N(t)$ curves were computed on the University of Cambridge Mathematical Laboratory EDSAC. We are grateful to the Director of the Mathematical Laboratory and to Dr. K. G. Budden who carried out the computations.

This work was partly financed by a grant from the Department of Scientific and Industrial Research.

References

Bates, D. R. 1951 *Proc. Phys. Soc.*, **64**, 805.
Bradbury, N. E. 1938 *Terr. Mag. Atmos. Elect.*, **43**, 55.
Kelso, J. M. 1952 *J. Geophys. Res.*, **57**, 357.
Martyn, D. F. 1955 *Report Camb. Ionosph. Conf.* (Phys. Soc. Lond.), p. 254.
Minnis, C. M. 1955 *J. Atmosph. Terr. Phys.*, **6**, 91.
Pierce, J. A. 1948 *Proc. I.R.E.*, **36**, 8.
Piggott, W. R. 1955 private communication.
Ratcliffe, J. A., Schmerling, E. R. Setty, C. S. G. K., and Thomas, J. O. 1956 *Phil. Trans.*, **A248**, 621.
Savitt, J. 1950 *J. Geophys. Res.*, **55**, 385.
Schmerling, E. R., and Thomas, J. O. 1956 *Phil. Trans.*, **A248**, 609.

Solar Eclipses and the Ionosphere (Special Suppl. Vol. 6 to J. Atmosph. Terr. Phys.), Pergamon Press, London, 1956

Absorption Measurements for the Eclipse of 30 June 1954

K. Rawer

Service de Prévision Ionosphérique Militaire, Freiburg

Abstract—Results are presented of field strength and virtual height measurements made at Freiburg on frequencies of 2·9 and 1·95 Mc/s during the partial eclipse of 30 June 1954. The different behaviours observed on the two frequencies are explained in terms of the relative contributions of deviative and non-deviative absorptions. It is found possible to isolate these two types. Some observations on 1·95 Mc/s at oblique incidence are also described.

The first eclipse absorption measurements of which I know were made by Denisse, Gallet, and Seligmann (1947). They could estimate an effective recombination value from observations of only one frequency. During the totality they found almost night-time conditions.

For the eclipse of 30 June 1954 similar observations on several frequencies were made by Bibl and Delobeau (1955) near Bergen (Norway) and also at Freiburg (Germany). The essential features of these observations are rather similar, and I shall only present here the results for Freiburg.

We used two frequencies, 2·9 and 1·95 Mc/s. Fig. 1 gives the observed echo

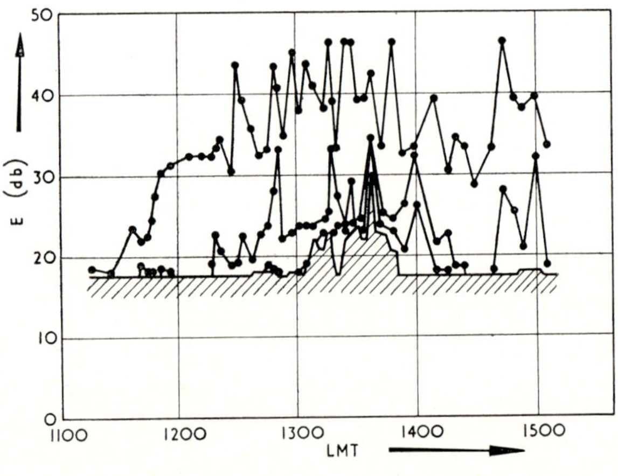

Fig. 1. Field strength on 2·9 Mc/s.

amplitudes for 2·9 Mc/s. There was nearly complete absorption at the beginning, then the first echo appeared and increased rapidly in size. Later the second and even the third echo also appeared with the increased screening of the sun. This seems to be a very regular eclipse result. It is seen that the noise level (atmospherics and jamming transmissions) also increased when the partial eclipse was near its maximum; this is also due to the change in propagation conditions.

The result for the other frequency is given in Fig. 2. There is no visible eclipse influence. One would say that the eclipse strongly influenced the field on 2·9 Mc/s, but not at all on 1·95 Mc/s.

Fig. 3 gives the virtual height variation and shows that there was a decrease of h' on 2·9 Mc/s, but on 1·95 Mc/s there was an increase to a first maximum, then a minimum, and then a second maximum roughly in a position symmetrical to the first relative to the maximum of the eclipse.

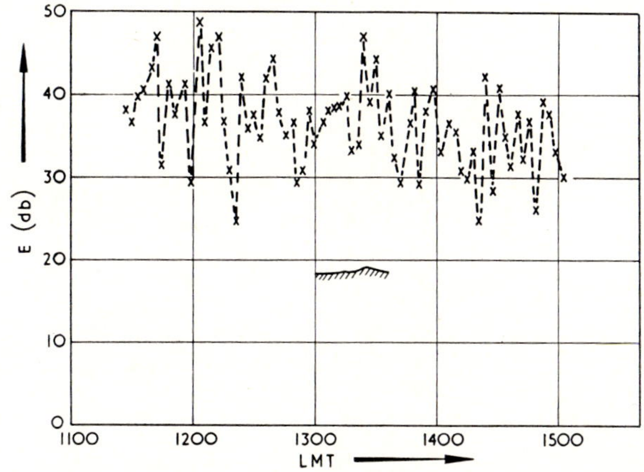

Fig. 2. Field strength on 1·95 Mc/s.

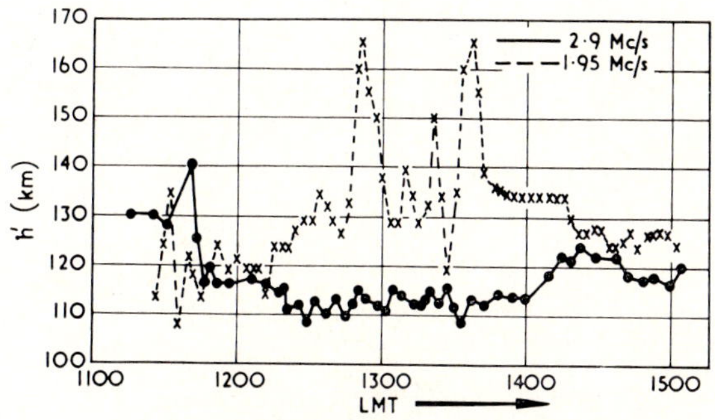

Fig. 3. Virtual heights on two frequencies.

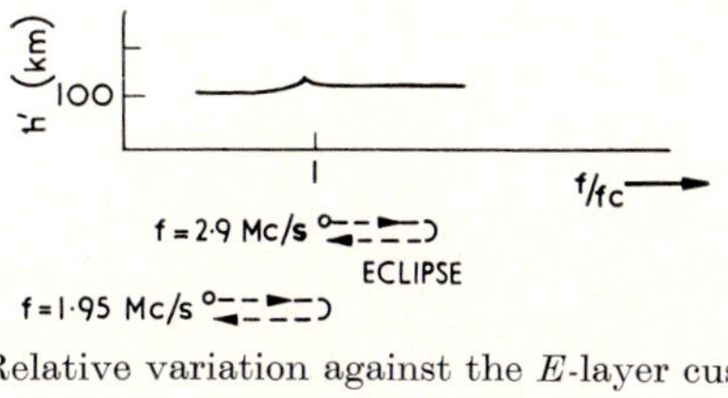

Fig. 4. Relative variation against the E-layer cusp frequency.

Fig. 4 indicates some ionograms with the frequencies of observation. We see that 2·9 Mc/s was first near the cusp frequency and was reflected later by the Es layer. But 1·95 Mc/s passed twice across the cusp. Now, there is a deviative absorption in the E layer which is maximum at the cusp frequency. Thus we have for 2·9 Mc/s in addition to the non-deviative absorption, an important deviative one at the beginning of the eclipse which disappeared with the screening of the sun. Thus the observed effect on this frequency is greater than would be the case with D region (non-deviative) absorption alone. On the other hand for 1·95 Mc/s there was a compensation of the variation of deviative and non-deviative absorption; this explains that no effect was directly visible on this frequency.

The deviative absorption has been calculated with a suitable layer model. In this way (BIBL and RAWER, 1951) the non-deviative term was isolated and the absorption constant B was obtained (RAWER, 1951). Fig. 5 shows that the values

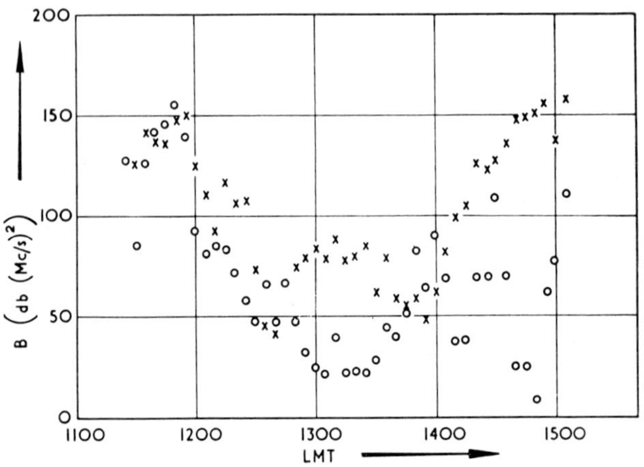

Fig. 5. Parameter B of non-deviative absorption.

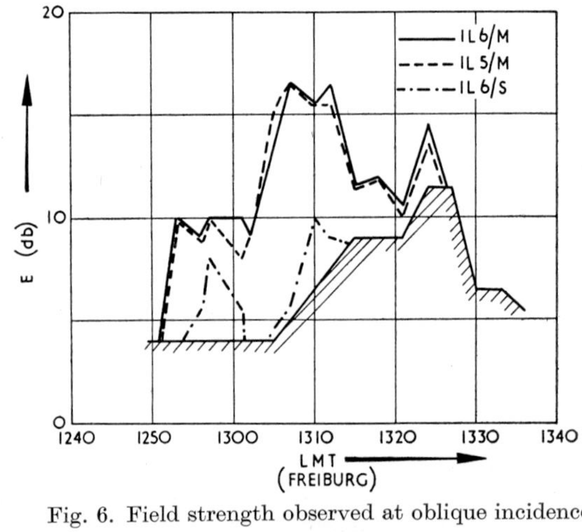

Fig. 6. Field strength observed at oblique incidence.

are now consistent (with the dispersion of values usual for all instantaneous absorption work).

We represented these values according to the expression

$$B \simeq (g \cdot \cos \chi)^n$$

(g is a screening factor, i.e., $g \cdot I_0$ is the effective solar radiation). We found a median value of 1·1 for the exponent n, which is fairly consistent with the results obtained from the daily variation of absorption.

Some observations have also been made at oblique incidence on 1·95 Mc/s. We used the Loran transmitters north of the British Isles. Fig. 6 shows that an

E echo appeared (at a distance of 1500 and 1750 km) during the middle part of the eclipse. The time delay corresponds roughly to the geometry (difference in eclipse time between Freiburg and the reflexion point).

The following conclusions may be stated from these observations:

(1) Eclipse absorption measurements are useful, but they must be done on *more than one frequency.*

(2) *Selective* (non-deviative) *absorption* in the *E* layer must be taken into account.

(3) The main difficulties are not different from those found in normal absorption work. The way to overcome them is to obtain a *large number* of eclipse absorption values.

References

BIBL, K., and DELOBEAU, F. 1955 *Z. Geophys.* (in press).

BIBL, K., and RAWER, K. 1951 *J. Atmosph. Terr. Phys.*, **2**, 51.

DENISSE, F., GALLET, R., and SELIGMANN, P. 1947 *Compt. Rend.*, **225**, 1196.

DENISSE, F., and BOSSON, F. 1947 *Relations entre les Phénomènes Solaires et Géophysiques*, Lyon, p. 117.

RAWER, K. 1951 *J. Atmosph. Terr. Phys.*, **2**, 38.

Solar Eclipses and the Ionosphere (Special Suppl. Vol. 6 to J. Atmosph. Terr. Phys.), Pergamon Press, London, 1956

Absorption Measurements during an Eclipse

W. R. Piggott

Radio Research Station, Slough

Abstract—The measurement and interpretation of the changes in intensity of radio waves reflected from the ionosphere during an eclipse present many difficulties. Some methods of overcoming these are discussed and applied to observations made at Ibadan, Nigeria, during the eclipse of 25 February 1952. It is found that the eclipse perturbation of the absorbing region implies a distribution of ionizing radiation over the sun's disk closely similar to those found for the *E* and *F1* layers during the same eclipse.

Introduction

Although measurements of field-intensity changes during an eclipse provided some of the first evidence of eclipse perturbations of the ionosphere (Eccles, 1912, 1919; Appleton and Barnett, 1925, 1927), they have seldom given information which can be used to indicate the ionization changes in the lower ionosphere qualitatively. The perturbations observed are usually due to the superposition of the effects of changes in the absorption of the lower ionosphere and changes in the mode of propagation or conditions of reflection.

There are two main methods by which simple data about the lower ionosphere might be obtained:

(a) to keep the mode and loss during reflection constant, and thus measure the changes in the lower ionosphere alone;

(b) to arrange an experiment so that changes in the reflection attenuations can be balanced out.

The first method involves in practice making measurements at oblique incidence, using an *E*-layer reflection, the range being chosen so that the deviative losses in the reflecting layer are negligibly small, e.g., greater than 500 km, and the frequency or frequencies being selected to give measurable reflections obeying these conditions throughout the eclipse. The main difficulties encountered in this type of measurement are: establishing the absolute value of the total absorption at any time, finding a suitable frequency, and errors due to the layer curvatures or tilts which often accompany an eclipse. It is, however, a valuable technique, particularly when the eclipse is partial and the band of frequencies usable is correspondingly wide. Pulse measurements on several frequencies are most desirable, since unexpected stratification of the *E* layer is particularly liable to occur during eclipses and may seriously alter the changes in field strength with time. This phenomenon was found in the absorption measurements taken at vertical incidence at Slough on frequencies between 1·6 Mc/s and 3·4 Mc/s during the eclipse of 9 July 1945, and was prominent on the oblique-incidence pulse transmissions at 5·1 Mc/s (equivalent frequency about 1·45 Mc/s for the *E* reflection) between Inverness and Slough, 710 km, during the eclipse of 30 June 1954. It was also present at Loth, Scotland, during the former eclipse, and undoubtedly modified the eclipse perturbation found there (Smith-Rose, 1946). There seems little point in carrying out detailed analysis when this phenomenon is severe.

In the present paper we are mainly concerned with the second possibility and, in particular, with the exploitation of combined absorption measurements, made on a few selected frequencies, and observations using conventional *h'f* recordings from an automatic $h'f$ recorder. Neither technique is, of course, new; for example, $h'f$ recordings were used to estimate eclipse changes in absorption as far back as 1936 (MAEDA and ISAGAWA, 1936). Our results are, however, of interest in that they appear to enable us to measure rapid changes in the absorption during an eclipse with sufficient accuracy to indicate the main features of the distribution of ionizing radiation over the sun's disk. This cannot be done using the manual absorption results alone (PIGGOTT, 1953a).

2. The Deduction of Absorption Changes During the Eclipse

While it is not difficult to measure the slower changes in absorption during an eclipse, using a limited number of frequencies and standard techniques (PIGGOTT, 1953b), the interpretation of the changes found presents considerable difficulties. The absorption measured on a given frequency is the sum of the absorptions in different parts of the trajectory, and each component will, in general, vary independently. Thus the absorption on a frequency reflected from the E layer may be split into the non-deviative absorption in the lower part of the E and in the D layers, where the refractive index, μ, is approximately unity, and a deviative absorption incurred during the process of reflection. During an eclipse the former will decrease, whereas the latter will tend to increase. If the frequency had been reflected from the F region, on the other hand, both the non-deviative and the deviative absorption in the E layer would decrease during the eclipse, but an additional deviative loss during reflection in the F region would be incurred, dependent on the behaviour of the reflecting layer. Usually observations on frequencies reflected from the E layer underestimate the changes of non-deviative absorption whereas those from the F tend to overestimate them. Ideally, therefore, we should attempt to obtain measurements for both cases, selecting the frequencies so as to obtain a balanced picture. In practice, the rapid changes in layer structures make this very difficult unless a large number of independent recorders are available. Thus it was necessary to make most of the manual observations on frequencies reflected from the most stable reflecting layer, E. Even then quick changes in absorption cannot be separated uniquely from changes due to curvature of the reflecting layer (focus fading), and it is necessary to minimize the latter by taking averages over about fifteen minutes.

No unique or satisfactory method of resolving the interpretation difficulty due to deviative absorption has yet been evolved. We shall, therefore, attempt to estimate the probable value of the non-deviative absorption for the D and E layers for deviative absorption absent, using the simplest possible conversion rules.

The absorption $\int Kds$, where K is the absorption coefficient per unit path s in a non-deviating absorbing structure for the ordinary wave of frequency f Mc/s (large compared with the gyrofrequency f_L Mc/s about the longitudinal component of the earth's magnetic field), may be written:

$$\int Kds = \frac{A_1(t)}{(f + f_L)^2} \tag{1}$$

where $A_1(t)$ is an absorption index numerically equal to the absorption, at time t, for an effective frequency $(f + f_L)$ of 1 Mc/s.

The corresponding deviative absorption depends on the critical frequency f_0 of the deviating layer and a function dependent, *inter alia*, on the shape of the ionization and collisional frequency variations with height

$$\int K ds = B\left(\frac{f}{f_0}, t\right). \tag{2}$$

If these are known, it is possible to calculate the non-deviative equivalent absorption index $A_2(t)$ for the reflecting structure, e.g., by putting $\frac{A_2(t)}{(f+f_L)^2} = B\left(\frac{f}{f_0}, t\right)$, when $\frac{f}{f_0} \gg 1$.

Thus we should like to be able to estimate $A_1(t)$, $A_2(t)$, or at least $A_1(t) + A_2(t)$, but can only measure an apparent reflection coefficient ρ representing losses due to $A_1(t)$ and $B\left(\frac{f}{f_0}, t\right)$ for a signal reflected once in the ionosphere, i.e.,

$$-\log \rho = \frac{A_1(t)}{(f+f_L)^2} \quad B + \left(\frac{f}{f_0}, t\right). \tag{3}$$

In general, the form of B may change with time and can only be guessed rather crudely.

The manual absorption measurements at Ibadan have been made at frequencies between 2 Mc/s and 4 Mc/s, using a D.S.I.R. absorption-measuring equipment (PIGGOTT, 1955) on the eclipse day and eight control days. Technical and staffing difficulties restricted the number of hours of observations possible and, apart from the eclipse day, when measurements were maintained from sunrise to sunset, observations on any one day are restricted to the morning, afternoon, or night period. The number of days with afternoon or night observations is insufficient to establish a normal variation of absorption during this period.

On two of the eight control days the mean mid-day absorption was equal to that on the eclipse day. In view of the limited number of observations available, we have multiplied the observed absorption throughout the days by the factors given in Table 1, thus making the mean noon absorptions equal. This gives us the normalized absorption variations for each frequency, examples of which are shown in Figs. 1 and 2.

Table 1. Normalizing factors for standard noon absorption

Date	February							March
	23	24	25	26	27	28	29	1
Factor	1	1	1	1·2	0·95	1·15	1·2	1·15

The mean curves shown in these figures give our adopted values for the normal variation of absorption on each frequency.

We now consider the possibility of deducing absorption changes from the variations of the lowest usable frequencies on which echoes can be observed, using the $h'f$ recorder.

The lowest frequency on which the xth mode of reflection can be observed, $f_{\min}(x)$, occurs when the spatial attenuation and ionospheric absorption reduce

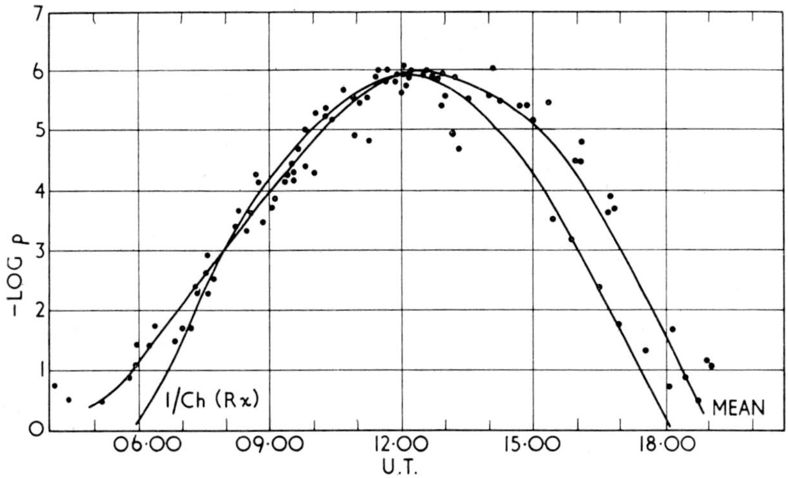

Fig. 1.

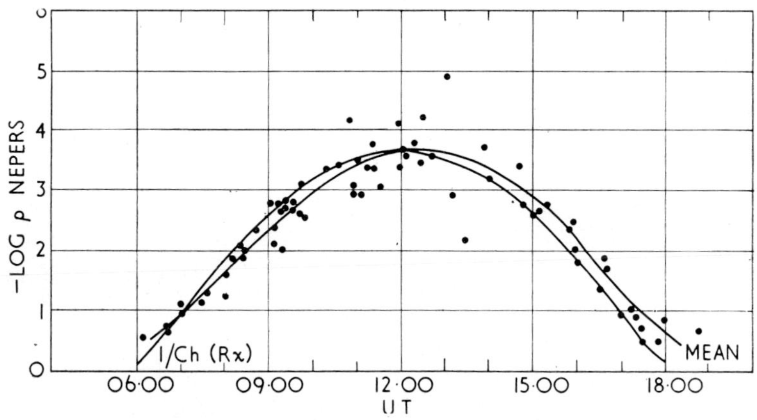

Fig. 2.

the reflected signal to the minimum intensity which can be recorded. The spatial attenuation, as usual, may be measured by the equivalent path traversed $P'(x)$ compared with a standard path length P'_0, and the minimum signal intensity by a sensitivity factor $-\log \rho_c$. Then if the xth mode involves n passages from ground to ionosphere

$$n(-\log \rho) = -\log \rho_c + \log P'_0/P'(x) \tag{4}$$

All three terms can vary with frequency and time, thus altering the relation between $-\log \rho$ and $f_{\min}(x)$.

The form of $-\log \rho_c$ is mainly determined by the design of the recorder and, in particular, by the characteristics of the transmitting aerials. However, it may also vary with the intensity of noise and interference picked up by the receiving aerials, which change diurnally and seasonally. On the average, $-\log \rho_c$ decreases when the frequency decreases. Omitting the absorption bands associated with the critical frequencies, $-\log \rho$ decreases when the frequency increases. Thus the

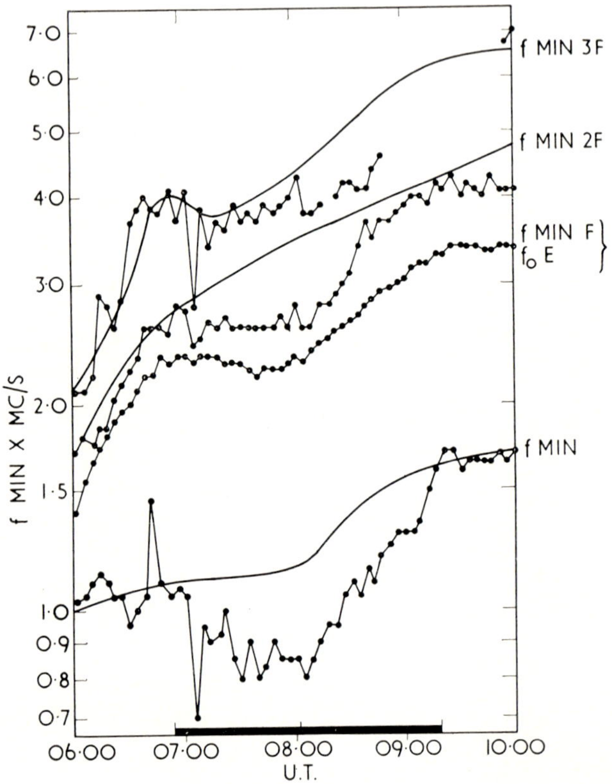

Fig. 3.

changes in $f_{\min}(x)$ with absorption correspond to a more rapid apparent change of absorption with frequency than that actually present. A D.S.I.R. automatic $h'f$ recorder (CLARKE and SHEARMAN, 1953) was available at Ibadan.

Turning now to our particular problem, we can measure (see Fig. 3) $f_{\min}$, $f_{\min}F$, $f_{\min}2F$, and $f_{\min}3F$ on the majority of records. Of these, $f_{\min}F$ is usually determined by the absorption band associated with f_0E or by screening due to Es, and is not a measure of mean absorption; $f_{\min}3F$ is mainly determined by group retardation and deviative attenuation losses in the F layer during the day, and can, therefore, only be used at night. However, $f_{\min}$ and $f_{\min}2F$ are usable, and are found, on the average, to fluctuate together at corresponding times on different days.

At Ibadan, $f_{\min}$ usually varies between 0·6 Mc/s and 2·0 Mc/s; $f_{\min}2F$ between

1·0 Mc/s and 7·0 Mc/s. Also, one is above and one below f_0E, thus enabling the effect of the deviative absorption to be partly balanced out.

The average variation of $-\log \rho$ as a function of frequency, deduced from our normalized diurnal variations, is shown in Fig. 4, the scales being proportional to $\log (f + f_L)$ and $\log (-\log \rho)$ respectively. Clearly the total absorption varies approximately inversely with frequency during the eclipse time of day. Comparing $f_{\min}2F$ with absorption measurements during the control period by the same method, we find

$$-\log \rho \propto (f_{\min}2F + f_L)^{-n} \tag{5}$$

where $n = 1{\cdot}4 \pm 0{\cdot}2$, the correlation coefficient being 0·87 for twenty-five sets of

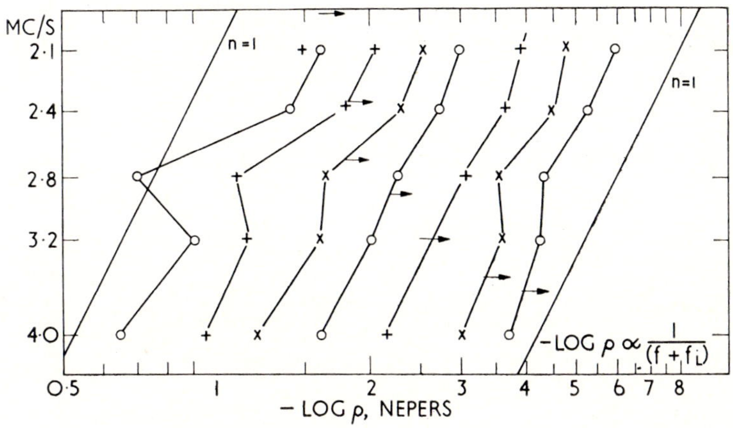

Fig. 4.

comparisons. It is interesting that, if the absorption is measured on frequencies above f_0E only, thus making the deviative loss similar for both terms, $n = 1{\cdot}5$ and the correlation coefficient rises to 0·96.

Comparing the changes of $f_{\min}$ and $f_{\min}2F$ on different occasions, we conclude that equation (5) is also true for $f_{\min}$, using the same value of n. Thus we may combine our observations of absorption changes during the eclipse with those deduced using $f_{\min}$, $f_{\min}2F$, and equation (5). We use running means of $f_{\min}$, $f_{\min}2F$ for three successive records, marking the observations doubtful $\odot$ when one observation is greatly inconsistent with the remaining two. We also include the averages with these observations omitted.

The absorption measurements are divided into those on frequencies and at times for which deviative losses will decrease the apparent eclipse perturbation ($\wedge$) and those for which they will increase it ($\vee$). These are deduced from the $h'f$ patterns obtained at the same time with the aid of equation (3). Thus if f/f_0 is nearer unity for the eclipse observation than for observations at the same time on the control days, or if $\partial h'/\partial f$ is greater at f for the eclipse observation, the apparent eclipse perturbation is probably decreased by additional deviative absorption. The converse is also true. Perturbations corrected for deviative losses should lie between these limits. We should also note that the original records suggest that the $f_{\min}2F$ observations between 0830 and 0850 underestimate the eclipse perturbation at this time owing, mainly, to changes in apparent height. We then find the apparent absorption perturbation relative to a normal day shown in Fig. 5.

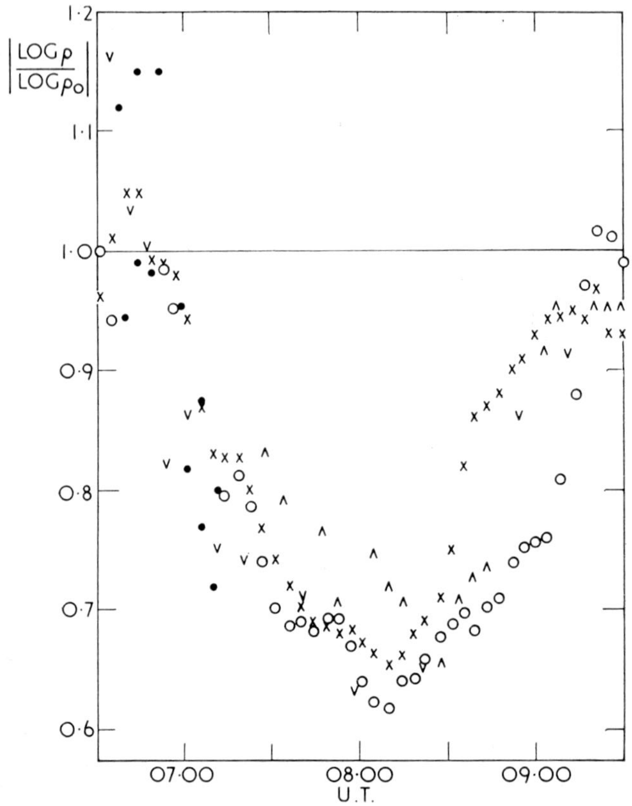

Fig. 5.

3. THE APPARENT RATE OF ION PRODUCTION IN THE ABSORBING LAYERS

Considering Fig. 5 as a whole, we adopt the mean of the changes of absorption deduced from $f_{\min}\,2F$ and $f_{\min}$ as a convenient probable representation of the absorption changes during the eclipse, noting that considerable errors may be present at certain times and that the type of analysis usually used for the E-layer ion production (MINNIS, 1955; PIGGOTT, 1956) is not applicable. We must also remember that the absorption perturbation represents a change in the product $N\,.\,\nu$ integrated throughout the absorbing region, where ν is the collisional frequency.

Thus we can only draw conclusions about the changes in $q_0(t)$, assuming a known change in ν and proportional changes in N at different heights. We assume ν is not altered by the eclipse.

Comparing the rates of ion production at corresponding times on the eclipse day, $q(t)$, and the normal days, q, and the ionization densities produced N, N_0, we find

$$\frac{q(t)}{q} = \frac{N^2 + \dfrac{1}{\alpha}\dfrac{\partial N}{\partial t}}{N_0^2 + \dfrac{1}{\alpha}\dfrac{\partial N_0}{\partial t}}. \tag{6}$$

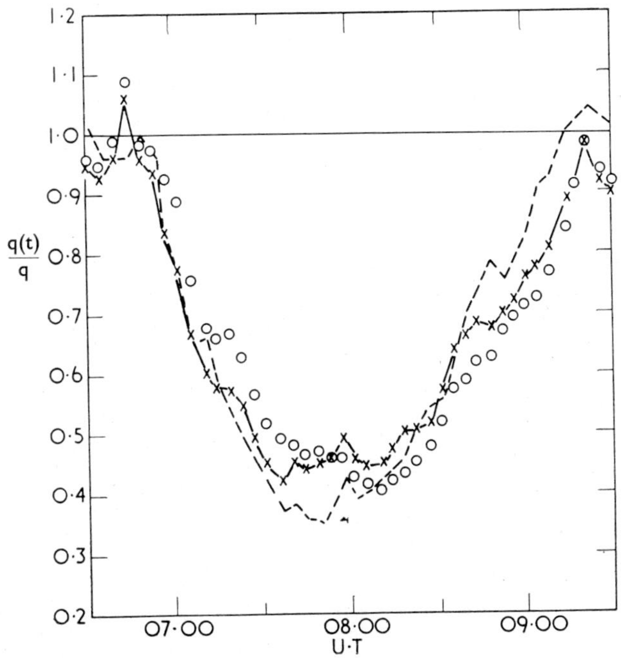

Fig. 6.

Putting $1/\alpha N = T$, $1/\alpha N_0 = T_0$ where T, T_0 are relaxation times corresponding to N, N_0

$$\frac{q(t)}{q} = \left(\frac{N}{N_0}\right)^2 \left\{1 + \frac{T}{N}\frac{\partial N}{\partial t}\right\} \bigg/ \left\{1 + \frac{T_0}{N_0}\frac{\partial N_0}{\partial t}\right\} = K(t)\left(\frac{N}{N_0}\right)^2. \tag{7}$$

Fortunately the relaxation times of the absorbing and E layers are almost equal, and the variations of $(\partial N/\partial t)/N$ are comparable for both layers. Thus we may use the calculated (PIGGOTT, 1956) values of $K(t)$ for the E layer, assuming a reasonable (MINNIS, 1955; PIGGOTT, 1956) value for α_E ($\alpha = 1{\cdot}3 \times 10^{-8}$) to indicate the probable magnitude of the correction for the absorbing layer. In Fig. 6 we show $q(t)/q$ when $K(t) = 1$ (points ○); when $K(t)$ is calculated, using E data, (points ×); and, for comparison, $q(t)/q$ for E.

The most striking feature of the latter two curves is the remarkable way in which the changes in q for the E and absorbing layers fluctuate together.

We must conclude that the radiation causing the two phenomena varied over the sun's disk in a very similar manner; it is, in fact, questionable whether the apparent differences between the E layer and absorbing-layer variations of q are real. If we assume that the absorbing layer q variations are reliable, the pattern as a whole indicates considerably more radiation from the edge of the sun than was found for the E layer, about one-third of the radiation coming from the disk, the bright areas, and the limb respectively.

4. The Measurement of Absorption During Eclipses

The measurements discussed above were originally planned as two independent experiments:

(a) the measurement of absorption changes during an eclipse deduced from manual measurements,

(b) the measurement of absorption changes during an eclipse deduced from automatic $h'f$ records.

Our experience with these experiments suggests that it would be advantageous to use the manual technique to fix the absorption changes at certain epochs in the eclipse, leaving the detailed variation to be determined from the recorder records. This would greatly improve the accuracy of estimating the percentage change in absorption during the eclipse, since the continuity demanded by experiment (a) above is no longer needed. We recommend this combined technique to the attention of those proposing to make absorption measurements in future eclipses. It is clear that the changes in $f_{\min}(x)$ above cannot give the magnitude of the eclipse perturbation of absorption unless the law connecting $f_{\min}(x)$ and absorption during the eclipse can be determined with reasonable accuracy. In our example the noise-level at the crucial frequencies did not change appreciably during the eclipse, so that the calibrations for normal days are reasonably consistent with those deduced from the absorption measurements made during the eclipse. Greater changes in noise-level appear to have occurred during the other eclipses mentioned above.

Acknowledgements

The sites and temporary buildings used for these measurements were provided by the University College, Ibadan, Nigeria, whose helpful co-operation is gratefully acknowledged. The $h'f$ equipment was operated mainly by Dr. R. Wright and Mr. R. A. Brown of the Physics Department of the College, who also helped considerably in making the manual absorption measurements. Without this assistance no measurements could have been made.

The work described above was carried out as part of the programme of the Radio Research Board. This paper is published by permission of the Director of Radio Research of the Department of Scientific and Industrial Research.

References

Appleton, E. V., and Barnett, M. A. F. 1925 *Proc. Camb. Phil. Soc.*, **22**, 672.
1927 *D.S.I.R. Special Report No.* 7 (H.M.S.O.).
Clarke, C., and Shearman, E. D. R. 1953 *Wireless Engr.*, **30**, 211.
Eccles, W. H. 1912 *Electrician*, **69**, 109.
1919 *Nature*, **104**, 323.

MAEDA, K., and ISAGAWA, Y.	1936	*Rep. Radio Res. Japan*, **6**, 91.
MINNIS, C. M.	1955	*J. Atmosph. Terr. Phys.*, **6**, 91.
PIGGOTT, W. R.	1953a	*Proceedings of Third Meeting of the Mixed Commission on the Ionosphere (U.R.S.I.), Canberra*, p. 42.
	1953b	*J. Inst. Elec. Engrs.*, Part III, **100**, 61.
	1955	*Wireless Engr.*, **32**, 164.
	1956	(In course of publication).
SMITH-ROSE, R. L.	1946	*J. Brit. Instn. Radio Engrs.*, **6**, 82.

Solar Eclipses and the Ionosphere (Special Suppl. Vol. 6 to J. Atmosph. Terr. Phys.), Pergamon Press, London, 1956

Some Observations of the Lower Ionosphere during the Solar Eclipse of 30 June 1954

K. Weekes
Cavendish Laboratory, Cambridge

Abstract—Amplitude and phase measurements were made on long and medium waves during the eclipse of 30 June 1954 at vertical and oblique incidence. The midpoints of the oblique trajectories corresponded in general to 75–90 per cent obscurity. Changes in reflection height were usually observed, and considerable increases in the reflection coefficient for the larger obscurations.

1. Introduction

Deductions can be made about the effects of an eclipse on the lower ionosphere by observing the behaviour of long and very long waves reflected at vertical incidence and of long and medium waves reflected at oblique incidence. An extensive series of observations of this kind was made during the eclipse of 30 June 1954, and this paper gives an account of the results.

The results are summarized in Section 5. The theory of the reflection of long waves from the ionosphere is not yet sufficiently advanced to allow detailed conclusions to be drawn from the observations.

The observations included the study of 16-kc/s waves reflected at near-vertical incidence and at oblique incidence, and the study of waves of frequencies between 70 kc/s and 250 kc/s reflected at various obliquities. In addition, experiments were made at vertical incidence, using radio waves of frequency 719 and 1420 kc/s. Most of the observations were made at Cambridge, England, and the midpoints of the trajectories were not in the zone of totality, but corresponded to obscured areas from 75 to 90 per cent.

As most of the observations were made using continuous wave-senders, special aerial arrangements were needed if the sky-wave was to be separated from the ground-wave. Such arrangements were possible at moderate obliquities, but for very oblique incidence it became impracticable to separate the two waves, which were arriving at very nearly the same angle of incidence; at these greater obliquities the total signal (ground-wave + sky-wave) was observed, and interpreted in the light of previous experience of the variations during the day and night.

After reflection at the ionosphere, the sky-wave is elliptically polarized and has, in general, a component vector $\boldsymbol{E}_\parallel$ in the plane of incidence, and a component $\boldsymbol{E}_\perp$ perpendicular to the plane of incidence; these two components are often referred to as the normal and abnormal components of the sky-wave. We call the ratio of the quantities to the incident vector $\boldsymbol{E}_0$, the *reflection coefficient* ${}_\parallel R_\parallel$ $(= |\boldsymbol{E}_\parallel|/|\boldsymbol{E}_0|)$, and the *conversion coefficient* ${}_\parallel R_\perp (= |\boldsymbol{E}_\perp|/|\boldsymbol{E}_0|)$. The phases of the components $\boldsymbol{E}_\parallel$ and $\boldsymbol{E}_\perp$ relative to the ground-wave give information about the polarization of the sky-wave, and their changes may be interpreted as changes of phase height of the reflecting layer.

2. Observations on Sender GBR

Observations were made at eight receiving sites of the signals from the P.O. sender GBR (16 kc/s) situated at Rugby.

(a) *Observations at Cambridge* (90 km, maximum obscuration 76 per cent)—Two methods of observation were used at Cambridge. In one method the abnormal component of the sky-wave was selected with a suitable loop-aerial, and its phase and amplitude relative to the ground-wave both recorded automatically (Straker, 1955) and visually. In the other method the signal strength in a loop-aerial oscillating about the position of minimum signal was recorded. From the values

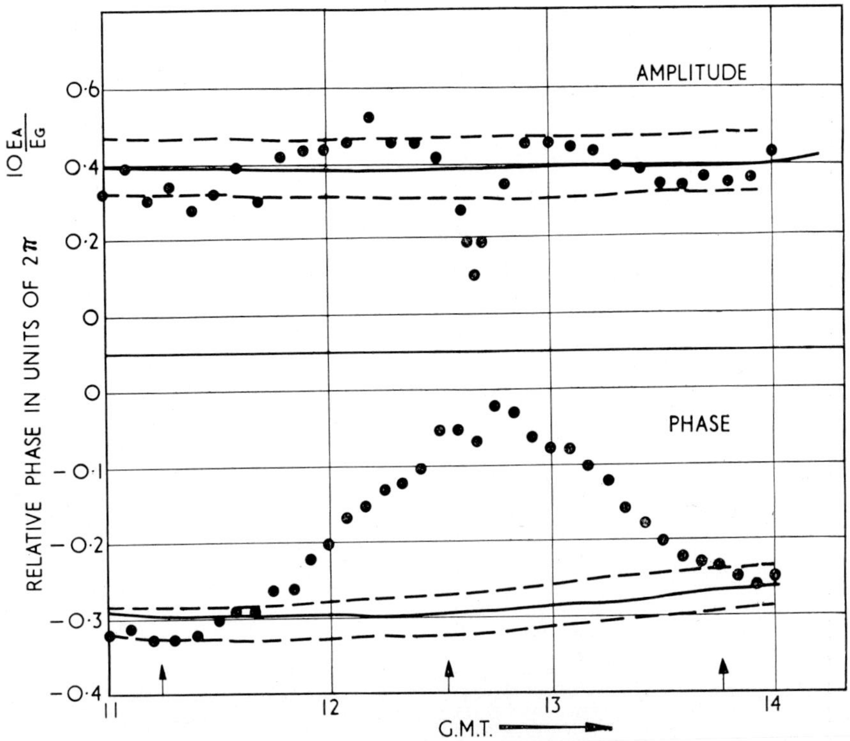

Fig. 1. Variation of amplitude and phase of the abnormal component of the downcoming wave on 16 kc/s during the eclipse compared with a normal day. The dotted lines give the scatter of normal days.

of signal at the turning points and the minimum value it is possible to determine the ellipse traced out on the ground by the electric vector of the wave. The characteristics of this ellipse depend on the normal and abnormal components of the sky-wave. The results of both methods of observation are in good agreement. In Fig. 1 the phase and amplitude of the abnormal component of the sky-wave relative to the ground-wave are plotted during the period of the eclipse, and for the same times of day on a normal day. The effect of the eclipse was rather complex. In the first half of the eclipse up to the time of maximum obscuration, the amplitude increased and the phase difference increased (corresponding to a rise in the height of reflection). The increase in amplitude was scarcely outside the normal day-to-day scatter, but seemed to be systematic, and hence was probably significant. The total amplitude-increase was about 1·5/1, and the phase path increased by 0·28λ. Shortly after maximum obscuration the phase height and amplitude appeared to be decreasing again, but 6 min after maximum the amplitude rather suddenly decreased to about one-quarter of its normal midday value, and the phase increased again suddenly. The phase and amplitude then returned to normal relatively slowly by the end of the eclipse.

Since the phase before the eclipse was slightly less than normal, and at the end was about average, it was difficult to give an exact value for the change at maximum, but reasonable limits seem to be $0{\cdot}25\lambda$–$0{\cdot}30\lambda$. Such a change of phase would suggest a change in height of reflection of 2·8–3·3 km.

(b) *Observations at oblique incidence*—At four receiving sites up to 600 km from the sender, the observations were made with the oscillating loop equipment which allows the ground ellipse to be determined. The bearing-swing and the value of the minor axis of this ellipse give information about the abnormal component of the sky-wave. The normal component of the sky-wave is effective only in altering the loop maximum signal. At this time of year the wave is almost linearly polarized, with no abnormal component, at distances greater than 300 km. Thus the ground ellipse is virtually a straight line at the two more distant sites.

The detailed analysis of the ground ellipse involves a knowledge of the site errors at each place. These have not been evaluated accurately at present, so that the results for the abnormal component quoted here are based upon a tentative estimate of the site errors, and may be revised as a result of accurate values of the site errors. The ratio of the loop maximum signal to the ground-wave at the four sites has been plotted in Fig. 2, together with the normal variation for the same period of the day.

These changes may be compared with what would be expected for a rise in the height of reflection for the Hollingsworth standing wave-pattern observed on a normal day (WEEKES, 1950). When the reflection height rises, the pattern moves out from the sender; the effects at these sites are exactly what would be expected qualitatively from such a change.

In Table 1 the results are summarized. The values of Δh are estimated from

TABLE 1.

Place	Distance (km)	% totality	Δh (km)	$E_{\parallel}/E_0$	$(E/E_0)_{\max}$	Δt
Giggleswick	200	78	3 ± 1	0·2	1·23	6 min
Newcastle	300	80	4 ± 1	0·2	1·00	2 min*
Edinburgh	400	82	—	0·35	2·00	0
Inverness	600	83	—	0·5	0·5	5 min

* Estimated from the changes in abnormal component.

the behaviour of the abnormal component when it is observable; the value of $E_{\parallel}/E_0$ is taken from the standing wave-pattern; Δt is the time-delay of the estimated maximum effect in the reflected wave from the time of maximum obscuration.

The changes of loop maximum signal at Edinburgh and Inverness may be compared with what would be expected in terms of the phase change associated with a change of height of 3–4 km, as is suggested by the experiments at steeper incidence. On this assumption, the amplitude of the sky-wave would have to be about double its normal value at the maximum obscuration.

Observations of the signal from GBR observed on a vertical aerial were also very kindly made for us by Dr. F. LIED at Kjeller and Tromso and by Professor

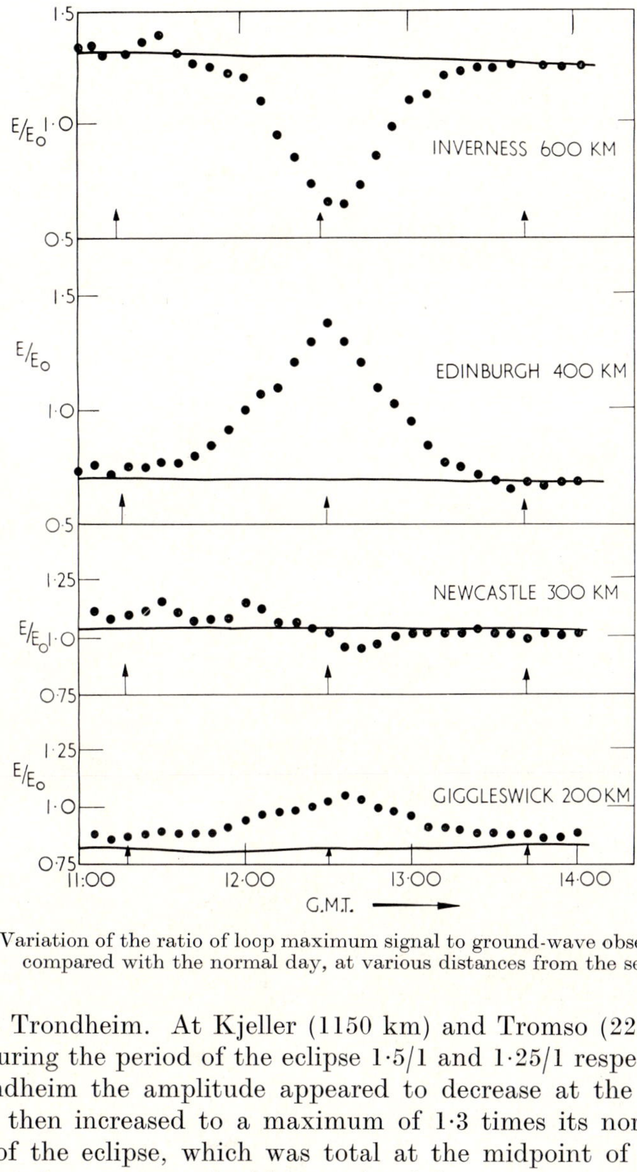

Fig. 2. Variation of the ratio of loop maximum signal to ground-wave observed on 16 kc/s compared with the normal day, at various distances from the sender.

JENSSEN at Trondheim. At Kjeller (1150 km) and Tromso (2200 km) the signal increased during the period of the eclipse 1·5/1 and 1·25/1 respectively.

At Trondheim the amplitude appeared to decrease at the beginning of the eclipse and then increased to a maximum of 1·3 times its normal value at the maximum of the eclipse, which was total at the midpoint of the path. At all places the maximum occurred within 2 min of the maximum obscuration.

These results also were qualitatively what would be expected from the Hollingsworth Interference Pattern observed at great distances (BUDDEN, 1953). Detailed calculations are too involved to be worthwhile at the present stage, but the increase in signal seemed larger than would be predicted, so that there were indications of an improved reflection at the great distances.

The interpretation of these results obtained from such long radio wavelengths requires the use of a full wave treatment of the kind recently initiated by BUDDEN (1955). The increase of reflection coefficient at oblique incidence may be explicable in terms of a change in the ionization gradient in the layer. The curious behaviour of the amplitude at steep incidence (90 km) may possibly involve partial penetration of a layer, and destructive interference between the waves reflected from two levels.

3. OBSERVATIONS ON HIGHER FREQUENCIES

Transmissions from a number of senders with frequencies between 70 kc/s and 250 kc/s were observed at Cambridge and, in addition, those from Droitwich (200 kc/s) and Kalundborg (245 kc/s) were observed for us at Trondheim by Professor JENSSEN. All the paths were very oblique and the signal-strength was observed, using a vertical aerial. From the interference effects between sky-wave and ground-wave it is possible to make estimates of the change in reflection height and the amplitude of the sky-wave.

Changes in Reflection Heights

Limits could be set to the change Δh in reflection height in four cases only. These changes of height which are listed in Table 2 are consistent with those deduced from the 16 kc/s observed: $\Delta h = 3$–4 km.

Changes in Reflection Coefficient

Table 2 shows how the reflection coefficient observed at the maximum of the eclipse compared with the normal day- and night-reflection coefficients, and also lists the time-delay Δt between the maximum amplitude and the time of maximum obscuration of the sun.

TABLE 2

	Distance (km)	% Totality	$\|R\|$ Day	$\|R\|$ Night	$\|R\|$ Eclipse	Δt	Δh Min (km)	Δh Max (km)
Droitwich—Trondheim (200 kc/s)	1460	93	0·025	0·25	0·25	−2 min	2·3	4·5
Motala—Cambridge (191 kc/s)	1170	91	0·016	0·25	0·08	−2 min	2·3	4·5
71 kc/s—Cambridge	940	87	0·06	0·70	>0·17	0	—	4·5
85·4 kc/s—Cambridge	850	85	0·02	0·50	>0·05	0	—	4·5
Kalundborg—Cambridge (245 kc/s)	820	85	0·0055	0·20	<0·005	—	—	—
Kalundborg—Trondheim	860	100	0·0055	0·20	>0·08	3 min	—	—

The most remarkable result indicated in this table is the behaviour of the waves from Kalundborg, which showed no detectable eclipse effect at Cambridge, but a very marked effect at Trondheim; detailed examination of the curve (Fig. 3) observed at Trondheim shows that there was no detectable effect until 90 per cent of the solar surface was eclipsed. The signal did not rise above the normal daytime fluctuations until the eclipse reached 94 per cent; the return to normal was somewhat delayed, and occurred when the eclipsed area had fallen to 80 per cent.

Some confirmation of this surprising result is found in the observation of the abnormal component of the sky-wave from Kalundborg at a distance of 180 km taken by Dr. LAUTER (1955). He found that, whereas the normal daytime value

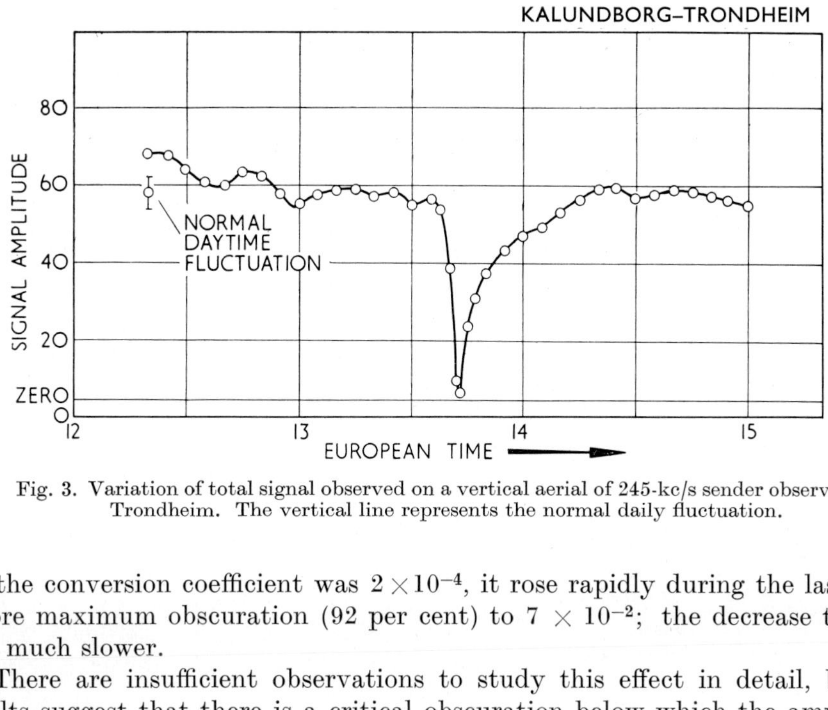

Fig. 3. Variation of total signal observed on a vertical aerial of 245-kc/s sender observed in Trondheim. The vertical line represents the normal daily fluctuation.

for the conversion coefficient was 2×10^{-4}, it rose rapidly during the last 10 min before maximum obscuration (92 per cent) to 7×10^{-2}; the decrease to normal was much slower.

There are insufficient observations to study this effect in detail, but these results suggest that there is a critical obscuration below which the amplitude of the sky-wave is not appreciably enhanced.* This may provide an explanation for the rather contradictory results that have been obtained in previous eclipses for the propagation of waves of various frequencies at oblique incidence.

A number of transmission paths showed no detectable eclipse effect. Consideration of the normal daily values of the reflection coefficient at different frequencies, and the behaviour of the stations which did give positive effects, shows that the fact that these transmissions exhibited no effect was to be expected.

4. Observations at Vertical Incidence

Mr. J. MAWDSLEY made observations throughout the eclipse at vertical incidence, using radio frequencies of 719 and 1420 kc/s. He used the frequency

* This sudden rise in amplitude may be related to the very sharp rise in reflection coefficient observed in the frequencies in the evening when the sun sets.

modulation equipment described by GNANALINGAM (1954), which is capable of detecting very weak echoes. These waves are normally reflected in the lower part of the E region during the summer day.

The results are summarized in Table 3.

TABLE 3.

f (kc/s)	Height in km		Absorption (nepers)	
	Normal	Eclipse max	Normal	Eclipse max
719	96·7 ± 1·0	98·1	6·4 ± 0·45	4·5
1420	101·5 ± 0·7	102·9	5·0 ± 0·35	3·5

The change in equivalent height was only just outside the normal day-to-day scatter, but showed as a consistent change developing during the eclipse and returning to normal after the end of the eclipse. Thus the existence of a change of height seems established, although the actual amount may be somewhat uncertain.

The changes in absorption were certainly significant, and it appears from these results that $\Delta \log \rho \propto \log \rho$.

The actual change seems to be consistent with the hypothesis that it is due to the removal of a slab of ionization from the lower part of the D layer which reflects the very long waves.

5. SUMMARY OF RESULTS

In this section we may conveniently collect the results described above.

16 *kc/s* (at 90 km from the sender—maximum obscuration 76 per cent)—Reflection height increased 3 ± 0·3 km during the eclipse. The conversion coefficient increased by a factor 1·5 to maximum obscuration and showed a sudden drop to one-quarter its normal value 6 min after maximum obscuration.

At distances up to 600 km (see Table 1) the reflection height changed 3·5 ± 1 km, and the reflection coefficient was about doubled at distances beyond 400 km. The increase of reflection coefficient seems necessary also to explain results obtained at 1150–2200 km from the sender.

70–250 *kc/s at oblique incidence* (see Table 2)—There were considerable increases in reflection coefficients at obscuration exceeding 85 per cent. In some cases the reflection coefficient reached its night-time value. 245 kc/s observed at the same distance showed no appreciable increase in reflection coefficient at 85 per cent obscuration, but a very large increase when the obscuration exceeded 90 per cent. In cases where it could be estimated, there appeared to be a change in reflection height of 3 ± 1 km.

719 *and* 1420 *kc/s at vertical incidence* (see Table 3)—A small increase in height of reflection was observed, and a considerable reduction in absorption, which was greater on the lower frequency.

Acknowledgements

The work on which this note is based was carried out with the help of a number of my colleagues. Mr. G. McK. Allcock took records at Cambridge of the signals from Motala, Kalundborg, and other senders; Mr. J. S. Belrose was responsible for the records of the Decca Navigator senders; and Mr. E. W. Taylor was responsible for the operation of the oscillating-loop records at the four sites in England. I am particularly indebted to Dr. F. Lied and Professor Jenssen for the recordings which they very kindly took in Norway. The 16-kc/s work was carried out as part of the work on very long waves supported by a contract from the D.S.I.R.

References

Budden, K. G.	1953	*Phil. Mag.*, **44**, 504.
	1955	*Proc. Roy. Soc.*, **227**, 516.
		Phil. Trans., **A248**, 45.
Gnanalingam, S.	1954	*Proc. Inst. Electr. Engrs.*, **101**, III, 72, 243.
Sprenger, K., and Lauter, E. A.	1955	*Gerlands Beit. z. Geophys.*, **64**, 284.
Straker, T. W.	1955	*Proc. Inst. Electr. Engrs.*, **102c**, 122.
Weekes, K.	1950	*Proc. Inst. Electr. Engrs.*, **97**, III, 100.

Solar Eclipses and the Ionosphere (Special Suppl. Vol. 6 to J. Atmosph. Terr. Phys.), Pergamon Press, London, 1956

Measurements on Medium and Low Frequencies and of Atmospheric Noise during the Solar Eclipse of 30 June 1954

E. A. LAUTER
Meteorologisches Observatorium, Kühlungsborn

Abstract—Eclipse effects on the intensity of sky-waves received on various frequencies at distances varying from 180 to 700 km are described. The dependence of the duration of the effect (delayed start and premature ending relative to the eclipse) on frequency and distance can be interpreted in terms of a simple *D* region of double structure.

Atmospheric-noise measurements on 8 to 69 kc/s indicate an increase up to the eclipse maximum (depending on frequency and distance) and a steep decrease afterwards.

SOLAR eclipse observations of the lower ionosphere have been made at Kühlungsborn (Germany), using the sky-wave of c.w.-transmitters at six frequencies between 185 and 1223 kc/s over distances from 180 to 700 km. The magnitude of the eclipse was 90–100 per cent for the different propagation paths. The downcoming wave was received on a loop aerial turned to eliminate the ground-wave and to receive only the horizontally-polarized component of the sky-wave. The intensity of the sky-wave relative to the ground-wave was taken to compute the conversion coefficient in the usual manner. On account of the strong ionospheric absorption during summer days, we observe normally a signal ratio of $1 : 10^{-3}$ between night and noon values. Therefore for all frequencies the expected clear effects were obtained. By drawing the eclipse anomaly relative to the normal field strength in decibels, we can describe the properties as follows:

1. There is a remarkable lag in the beginning of the ionospheric effect at all frequencies, and the effect disappears well before the end of the eclipse. The duration of the effect increases with the frequency and decreases with the distance of the transmitter.
2. The time-delay of the maximum of the ionospheric effect is about 4 min. This is in agreement with earlier eclipse observations of the *D* region for short waves.
3. The maximum effect on the field strength of the sky-wave is about 40 db, consistent with calculations based upon the daily and seasonal variations interpreted by a simple *D*-region model.

The systematic variation of the duration of the effect at medium and low frequencies is possibly caused by the double structure of the *D* region. At the beginning of the eclipse there might be a compensation between the decreasing absorption of the lower region and the decreasing reflection coefficient of the upper layer. Therefore an increase of the downcoming waves is only perceptible as long as the selective absorption in the upper *D* layer disappears and the reflections are changing to the lower *E* layer. Then the attenuation decreases rapidly and field strength of the sky-wave is steeply increased. Shorter wavelengths will penetrate to the *E* layer earlier than longer wavelengths. Also the *E* reflections occur earliest at a steep angle of incidence, i.e., at the shortest transmitter distance. In the same manner the re-establishment of the *D* region causes the disappearance of the anomaly well before the end of the eclipse. These considerations seem to confirm the idea of a double structure of the D region stated by other observations at very low and high frequencies. Only during the maximum of the eclipse are the reflections from the E layer nearly consistent with a simple attenuation model of the D layer.

Further observations were made on very-low-frequency atmospheric noise at 8, 14, 27, and 69 kc/s. The measured characteristics were the mean-square value and the average number of noise-pulses per minute (above 50 μV/m). The most remarkable results deviating from the normal daily variations of atmospheric noise are as follows:

1. There is a positive anomaly up to the maximum of the eclipse at all frequencies and for all characteristics of atmospheric noise.
2. The magnitude of the positive anomaly increases with increasing frequency.
3. The positive anomaly is relatively stronger in the records of the mean-square value than in the number of noise-pulses.
4. The maximum of the positive anomaly seems to occur earlier at the lower frequencies.
5. After the maximum of the eclipse there is a steep decrease below the normal values at all frequencies.

A distinct interpretation of these results is somewhat complicated because the propagation paths of the atmospherics are quite differently affected by the solar eclipse, but the general behaviour can be understood. The increase of atmospheric noise in the first phase, due to the decreasing ionospheric absorption, is to be expected to be strongest on the shorter propagation paths in the vicinity of the zone of total eclipse. Therefore the mean-square value is more affected than the number of pulses, which contains a greater number of discharges in far distances. In the same manner, the dependence on frequency can be explained, because the percentage of noise-pulses from distant sources increases with decreasing frequency. The negative anomaly starting just after the maximum phase and lasting for several hours seems to be due to meteorological effects of the eclipse, such as the disappearance of thunderstorms by cooling in the lower atmosphere in the eclipse region.

Solar Eclipses and the Ionosphere (Special Suppl. Vol. 6 to J. Atmosph. Terr. Phys.), Pergamon Press, London, 1956

The Influence of the Solar Eclipse of 30 June 1954 on Low-frequency Atmospherics

A. EHMERT and K. REVELLIO

Max-Planck-Institut für Physik der Stratosphäre, Weissenau bei Ravensburg

Abstract—Records of atmospherics on 13 kc/s obtained at Weissenau during the solar eclipse are compared with other European records. For small atmospherics from distant centres, the variation shows a distinct peak during the eclipse. This peak is absent for stronger atmospherics of the same frequency and also for higher frequencies from nearer discharge centres. Narrow-sector d.f. observations at Payerne and Zürich, together with meteorological information, enable the source of these atmospherics to be located near the Azov Sea and the Caucasus at distances from 2000 to 2500 km. Best transmission occurred when the sun was more than 75 per cent obscured at both reflection points for the doubly reflected waves.

THE first report of an effect on atmospherics during an eclipse was that of LUGEON (1932) who reported an increase of the atmospherics recorded in Poland during the solar eclipse of 31 August 1932. The central line was at that time moving from north to south across the North American continent. The same recording at Tromsö did not show this effect.

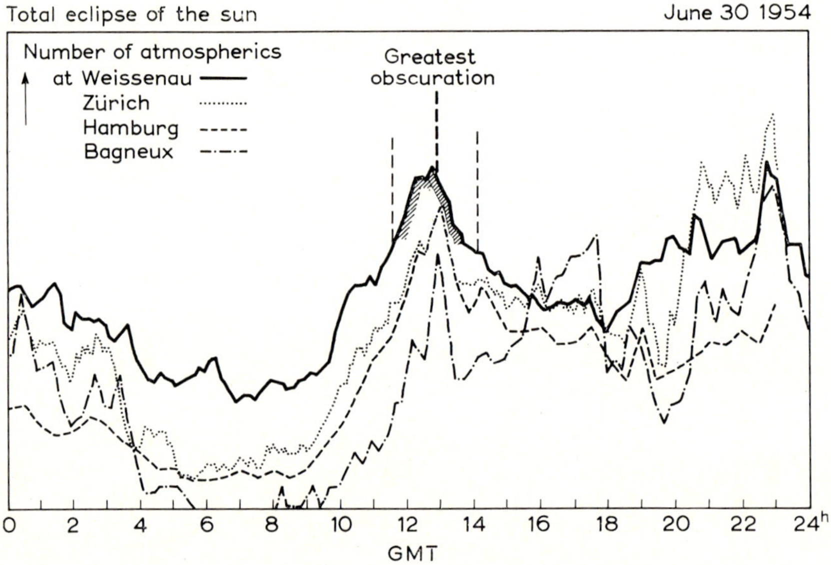

Fig. 1.

For the solar eclipse of 30 June 1954 the recordings of the European stations show a very distinct influence in the form of an increase of atmospherics.

The upper curve of Fig. 1 shows our results obtained at Weissenau, where we record, with a high resolution, the atmospherics which are received by a receiver tuned to 13 kc/s. The dotted curve is the recording by the Zürich atmo-radiograph* which operates with a receiver tuned to 27 kc/s (LUGEON, 1935). In Fig. 1 we also show the variation of the mean voltage of a broad-band receiver at

* We are most grateful to Prof. LUGEON for kindly placing this curve at our disposal.

Hamburg* on a range of 3 to 50 kc/s and a record obtained at Bagneux with the radiocinemagraph of BUREAU† for 25 kc/s.

Each curve shows a striking peak about the time of local maximum obscuration of the sun, the base of this peak having a width of about two hours. Fig. 2 shows

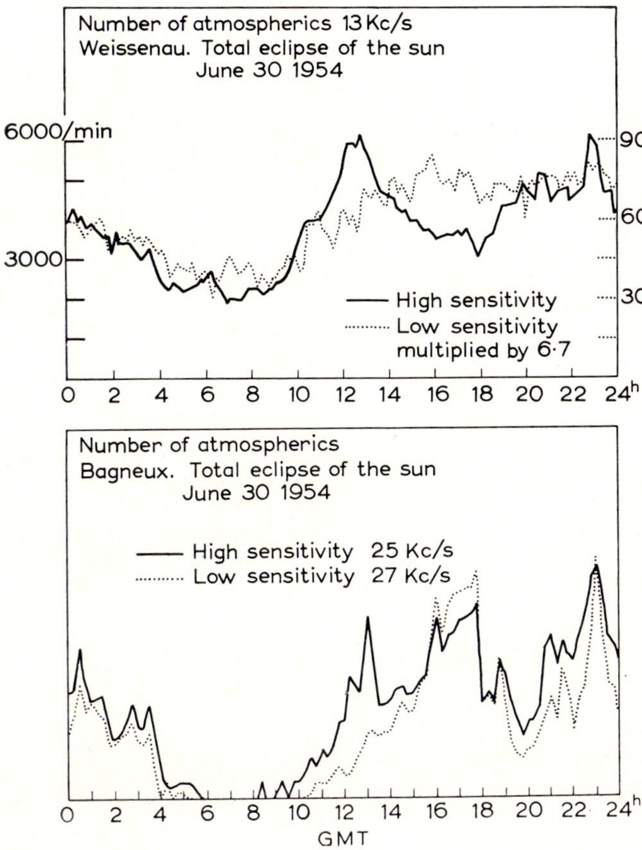

Fig. 2.

the same recordings at Weissenau and Bagneux (full lines), together with curves for the same frequencies and recording systems, but with a threshold-value nearly three times as great. The scales for these curves are different. These records were responding to atmospherics from nearer discharges, and hence permit an estimation of what is arriving from greater distances. The similarity of the increase of both Bagneux curves permits a more precise determination of the influence of the eclipse than the other recordings, because here the beginning of the eclipse occurred in an earlier period of the daily increase. This increase, however, varies daily with regard to time and extent, as is shown by some examples for Weissenau in Fig. 3.

At the outset it does not seem unreasonable to suppose that, due to cooling by

* *Ionosphären-Bericht der Arbeitsgemeinschaft Ionosphäre Jehrg.* **6**, Nr. 19, with kind permission of Prof. SCHULZE, Hamburg.

† We thank Director Dr. BUREAU very much for kindly placing this recording at our disposal.

the solar eclipse, some of the sources diminished their activity, and therefore the curve of the 30 June in Fig. 3 should be compared with the curve of 4 or 5 July. The recordings at Kühlungsborn* show that there was such a cooling influence of a limited extent. These recordings show that, after the eclipse for all frequencies (recordings on 8, 27, 40, and 69 kc/s), there was a decrease of intensity of about

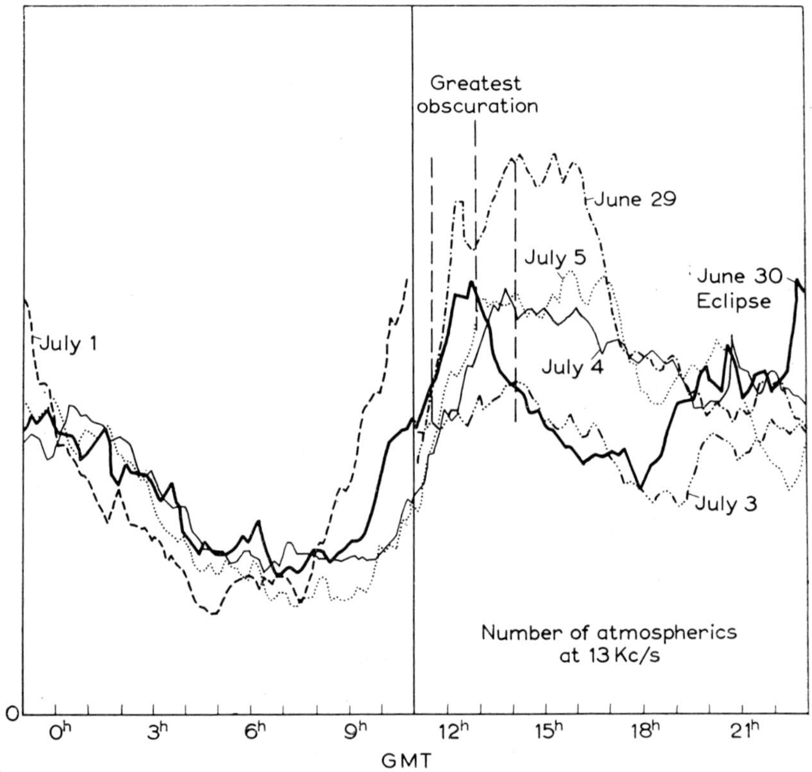

Fig. 3.

20 per cent, whereas at 69 kc/s there is a peak with an increase of up to 40 per cent during the eclipse. According to this comparison with higher frequencies on one side, and with the curves for greater source-distance and longer waves on the other, we can estimate that the hatched part in Fig. 1 is to be attributed to the improved conditions of ionospheric transmission.

In the narrow-sector direction-findings at Zürich, Payerne,† and Bagneux,‡ the above-mentioned peak manifests itself as an accumulation of recording points in a narrow section of a time/direction map. Fig. 4 shows a part of such recordings at Payerne with the instrument of LUGEON and NOBILE. The sectors of these point-clouds are given in Fig. 5. The hatched sector corresponds to the

* Information kindly supplied by Dr. LAUTER of Observatory Kühlungsborn.
† See footnote *, p. 126.
‡ See footnote †, p. 127.

point-clusters during the eclipse and locates (with little precision, however) the region of the Azov Sea and Caucasus (2000- to 2500-km distance from Weissenau) as the discharge-centre position. The meteorological reports on thunderstorms

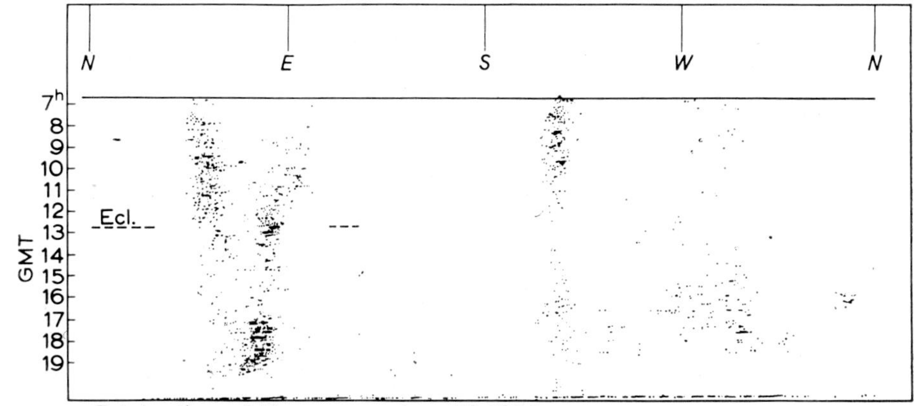

Total eclipse of the sun June 30 1954
Part of the registration of the narrow-sector recorder (Lugeon-Nobile) at Payerne (Switzerland)

Fig. 4.

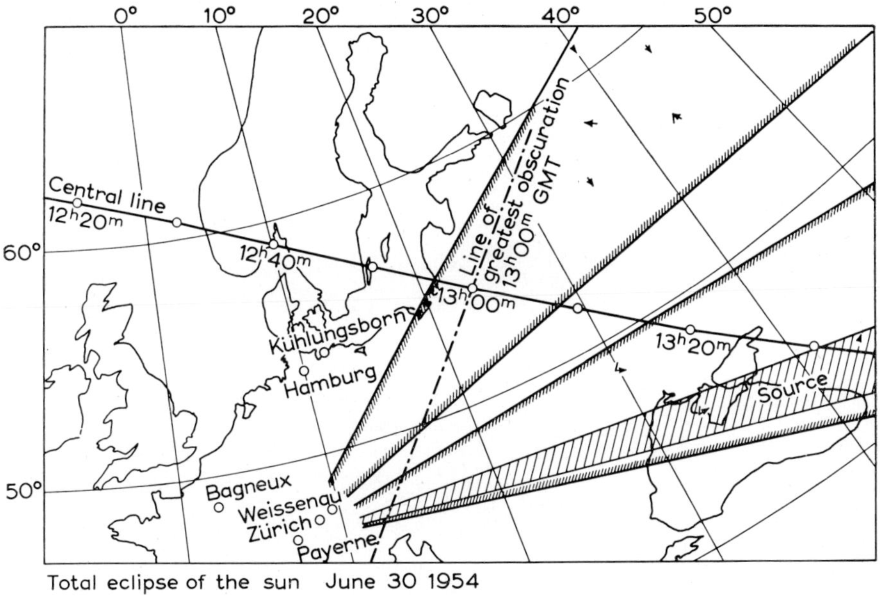

Total eclipse of the sun June 30 1954

Fig. 5.

and summer lightnings confirm this location. The fronts drawn as full lines in Fig. 6 are for 30 June 1954 0000 G.M.T., the pointed lines twenty-four hours later. They also confirm the position of the discharge-centre mentioned above.

The central line of the eclipse is given in Fig. 5 together with the Greenwich time of totality at certain places. At the location of the effective sources the total eclipse took place between 1315 and 1330 G.M.T. At this time the maximum of the curves in Fig. 1 had already been exceeded.

Fig. 7 shows for Weissenau, Hamburg, and Bagneux the obscuration of the sun along the transmission path between the source and the recording station for

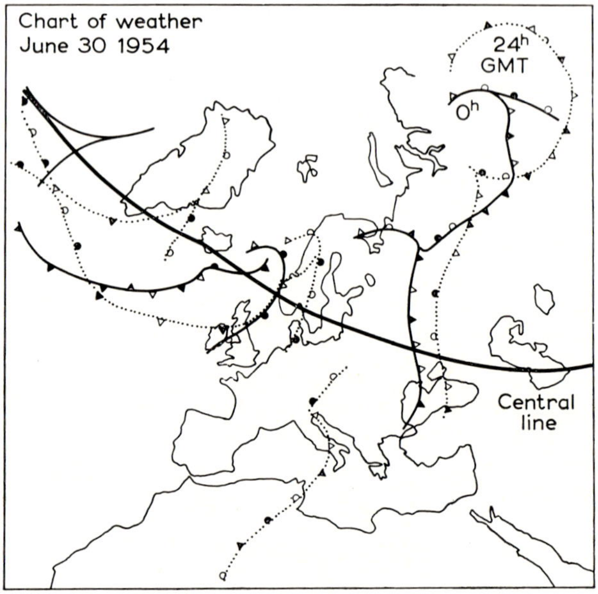

Fig. 6.

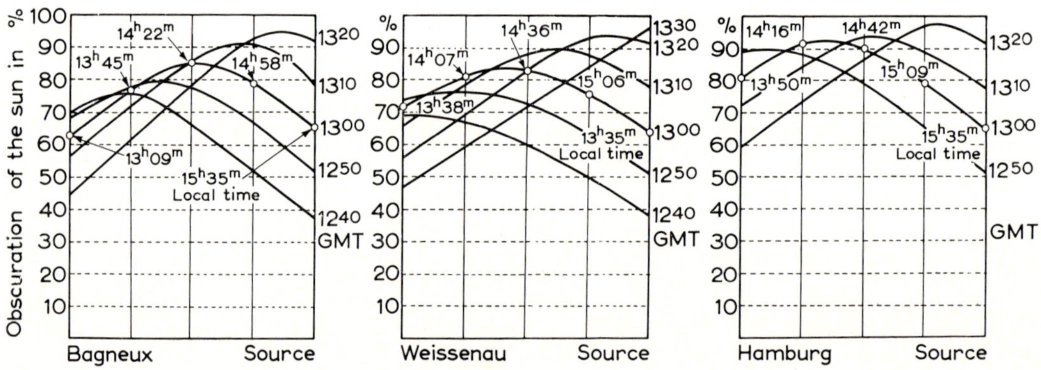

Fig. 7.

different Greenwich times. Also, for some points, the local times are given which correspond to the G.M.T. given in the margin for the individual curves. These curves were not calculated with astronomical precision, but by an approximate procedure of sufficient accuracy.

At the above-mentioned stations the beginning of the effect was accordingly determined by the extent of obscuration at the reflection points near the source, whilst the obscuration at the reflection points near the recording station was the determining factor for the fading of the effect. In the vicinity of the reflection points the atmospherics undergo the greatest damping.

It is striking that the source recorded farther north in Fig. 5 does not show the above-mentioned effect, although a part of these atmospherics may come from the region beyond the central line of the eclipse. There is not yet sufficient material available to clarify these conditions, because the decrease of near sources in the region North Bavaria–Bohemia also takes a part. This renders the analysis more difficult.

References

LUGEON, J. 1932 *Compt. Rend.*, **195**, 817.
1935 *Sur la Nécessitée d'une Station Polaire Permanente d'Observations radiométéorologiques pour les Services de Prévision du temps.* Warsaw.

Solar Eclipses and the Ionosphere (Special Suppl. Vol. 6 to J. Atmosph. Terr. Phys.), Pergamon Press, London, 1956

Influence de l'Eclipse du 30 juin 1954 sur la Propagation des Atmosphèriques sur la Fréquence de 27 kc/s

R. Bost
Laboratoire National de Radioélectricité, Bagneux/Seine

Abstract—A brief account is given of recordings of atmospherics on several frequencies during the eclipse of 30 June 1954. Quite pronounced enhancement was observed for the weaker atmospherics on 27 kc/s, but no eclipse effects were observed at frequencies in the range 4–12 kc/s.

La journée du 30 juin 1954 est caractérisée par la présence d'un foyer proche de Bagneux qui domine fortement sur tous les enregistrements. Cet orage donne des niveaux d'après-midi très élevés entre 1500 et 1900 h T.U.

Les goniomètres à secteur étroit enregistrent ce foyer dans les directions allant de l'ouest vers le sud sur l'enregistrement de Bagneux, et de l'est vers le sud-est sur celui de Brest.

Si nous faisons abstraction de l'effet de cet orage, nous voyons que les enregistrements $N/8$ et $S'1$ (11 km) (Fig. 1) font apparaître un renforcement entre 1200 et 1330 (T.U.). L'enregistrement $B/1$ (11 km) n'a pas mis ce phénomène en évidence car sa sensibilité (seuil de 220 maxwells par mètre (max/m) n'est pas suffisante pour capter les atmosphériques ayant été renforcés par l'eclipse. La comparaison de ces deux courbes, $N/8$ (seuil 80 max/m) et $B/1$ (seuil 220 max/m) permet donc de dire que les atmosphériques ayant été renforcés au cours de l'éclipse après avoir subi le trajet ionosphérique ont des *intensités inférieures à* 200 *max/m.*

La courbe du champ moyen $S'1$ montre que le champ des atmosphériques reçus s'est accru de 7 microvolts par mètre pendant la durée de l'éclipse (entre 1300 et 1400 /T.U.).

En ce qui concerne les autres longueurs d'onde: les enregistrements $H/1$ accordés sur 1·6 km et $G/1$ sur 5 km ne sont pas sensibles au renforcement, la portée de ces instruments étant assez limitée. L'enregistrement $P/5$ sur 24 km a aussi un seuil trop peu sensible (430 max/m) pour enregistrer le phénomène.

Les enregistrements $X/1$ sur 35 km; $Y/1$ sur 57 km et $Z/1$ sur 70 km qui ont un seuil de 1200 max/m ne l'ont pas enregistré non plus. L'enregistrement du champ moyen $U'1$ sur 57 km qui est très sensible n'a pourtant pas de renforcement; ceci semblerait prouver que la propagation de ces ondes n'a pas été affectée soit parce qu'elles n'empruntent pas, dans l'ionosphère, les mêmes chemins que les ondes de 11 km, soit parce que leur absorption est très rapide.

L'examen des goniogrammes nous montre:

d'une part, le foyer proche dont nous avons parlé au début;

d'autre part, un foyer dont l'azimut reste constant dans la direction est à Bagneux. C'est ce foyer, situé dans la direction du cône d'ombre de l'éclipse (Fig. 2), qui se renforce pendant les heures où l'on observe un accroissement de niveau sur les en registrements des niveaux *moyens* non seulement de Bagneux, mais aussi de Hambourg, Weissenau et de Zurich.* Ce renforcement de la source est est nettement

* Ce phénomène nous a été également signalé par correspondance par le Prof. Ehmert de Weissenau et le Dr. Lugeon de Zurich.

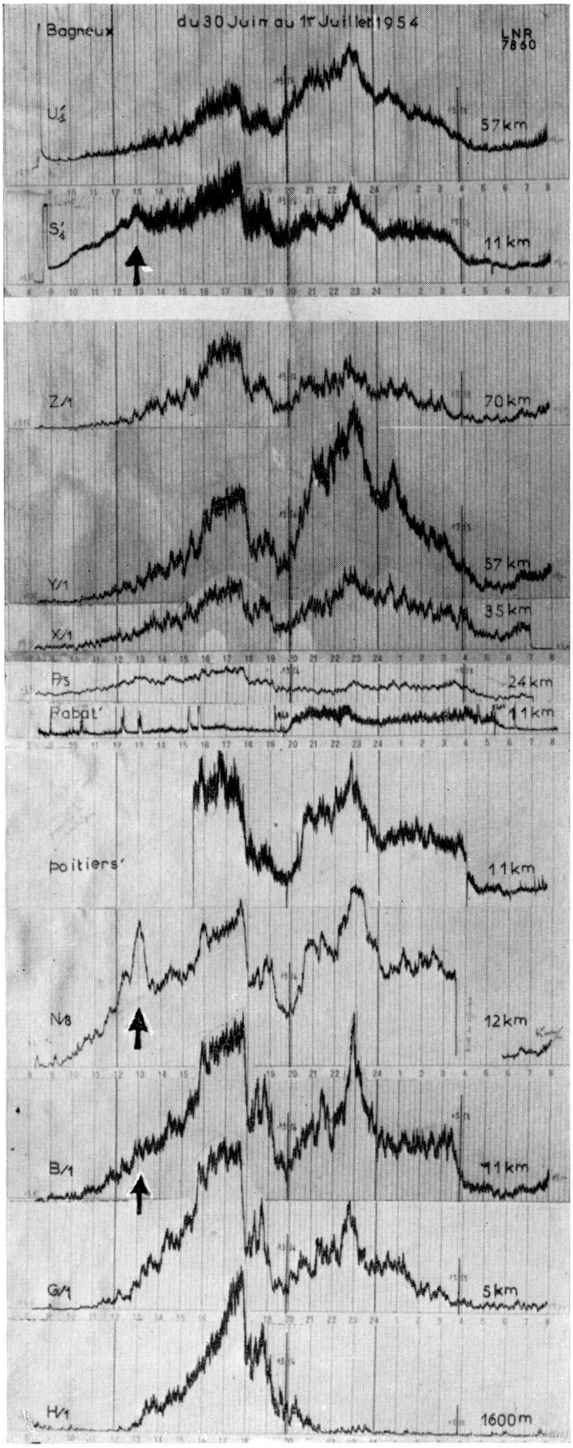

Bagneux
du 30 Juin au 1er Juillet 1954
LNR 7860
57km
11km
70km
57km
35km
24km
Rabat'
11km
Poitiers'
11km
12km
11km
5km
1600m

Fig. 1.

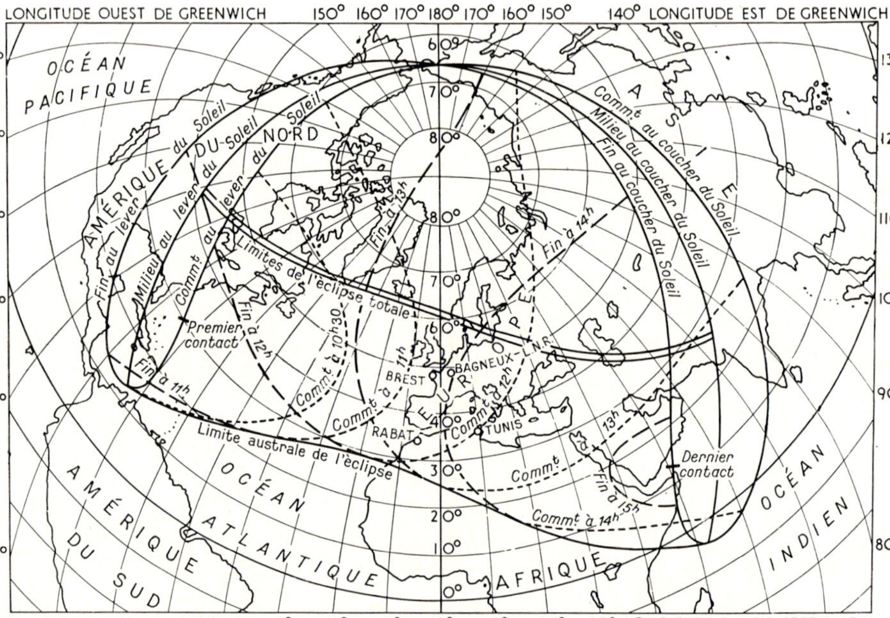
LONGITUDE OUEST DE GREENWICH
150° 160° 170° 180° 170° 160° 150°
140° LONGITUDE EST DE GREENWICH
OCÉAN PACIFIQUE
AMÉRIQUE DU NORD
Fin au lever du Soleil
Milieu au lever du Soleil
Comm.t au lever du Soleil
Limites de l'éclipse totale
Premier contact
Fin à 11h
Fin à 12h
Fin à 13h
Fin à 14h
Fin à 15h
Comm.t à 10h30
Comm.t à 11h
Comm.t à 12h
Comm.t à 13h
Comm.t à 14h
BREST
BAGNEUX-L.N.R.
RABAT
TUNIS
EUROPE
ASIE
Comm.t au coucher du Soleil
Milieu au coucher du Soleil
Fin au coucher du Soleil
Dernier contact
Limite australe de l'éclipse
AMÉRIQUE DU SUD
OCÉAN ATLANTIQUE
AFRIQUE
OCÉAN INDIEN
LONGITUDE OUEST DE GREENWICH 30° 20° 10° 0° 10° 20° 30° LONGITUDE EST DE GREENWICH

Fig. 2.

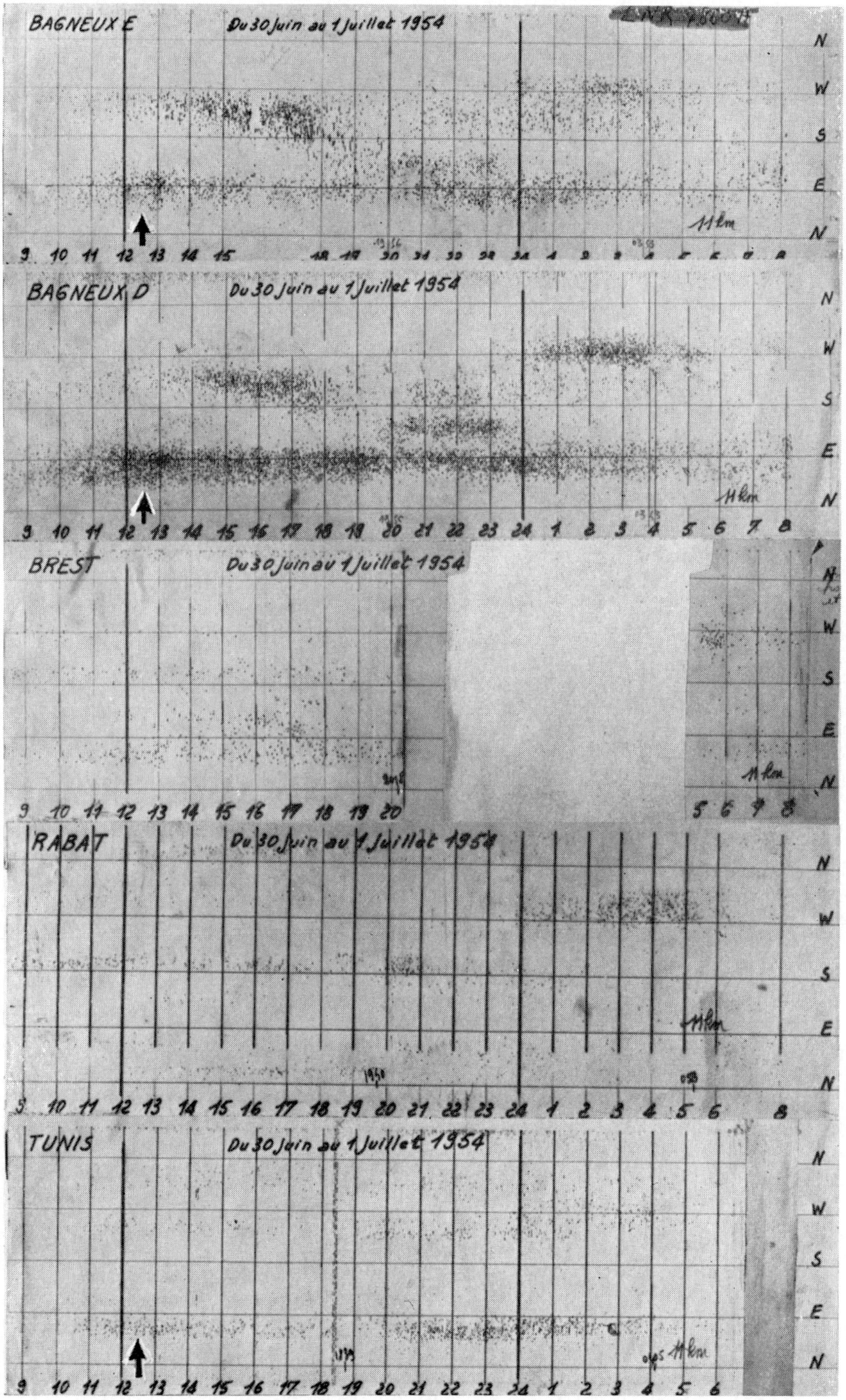
BAGNEUX E
Du 30 juin au 1 Juillet 1954
BAGNEUX D
Du 30 juin au 1 Juillet 1954
BREST
Du 30 juin au 1 Juillet 1954
RABAT
Du 30 juin au 1 Juillet 1954
TUNIS
Du 30 juin au 1 Juillet 1954
N
W
S
E
N
11 km
9 10 11 12 13 14 15 16 17 18 19 20 21 22 23 24 1 2 3 4 5 6 7 8

Fig. 3.

visible sur les goniogrammes Bagneux E, Bagneux D à 1300. A Brest et à Rabat les goniomètres enregistrent aussi les foyers d'est, mais ils ne sont pas réglés à une sensibilité suffisante pour faire apparaître le renforcement.

Notons enfin, que les enregistrements à Bagneux du champ d'émetteurs de fréquences ëtalons ont accusés, de 1315 à 1400 T.U. une propagation anormale qui a permis à l'émission de WWV sur 15 Mc/s d'etre enregistrée avec un champ de 2 microvolts par mètre. Aucun effet n'a cependant été observé sur l'onde de 10 Mc/s de WWV.

Les Figs. 1 et 3 sont la reproduction des principaux enregistrements effectués à Bagneux et dans les stations annexes. Notons enfin que le goniogramme de Tunis qui accuse aussi le renforcement des sources d'est au moment de l'éclipse, indique des directions légèrement inexactes du fait d'une imperfection matérielle.

Solar Eclipses and the Ionosphere (Special Suppl. Vol. 6 to J. Atmosph. Terr. Phys.), Pergamon Press, London, 1956

Some Measurements of the Signal Strengths of Radio Waves reflected from the Ionosphere during the Solar Eclipses of 30 June 1954 and 20 June 1955

R. G. Rastogi, R. M. Sheriff, and N. G. Nanda
Physical Research Laboratory, Ahmedabad

Abstract—The signal strengths of radio waves from B.B.C. transmissions on 15·07 Mc/s were measured at Ahmedabad during the 1954 eclipse, with control observations on a few preceding and succeeding days. The signals, which could have been either by two-hop of three-hop reflections, showed well-marked increases in intensity during the eclipse. The increase of signal strength is considered to be due to a decrease of attenuation in the *F1* layer.

Similar measurements were carried out for the 1955 eclipse at Trivandrum, on the radio transmission from Colombo on 4·87 Mc/s directed skyward, and at Ahmedabad on 7·19 Mc/s. Here again the signals showed well-marked increases of strength during the eclipse. In both cases the signal intensity returned to its normal value well before the end of the eclipse. The reflections observed at Trivandrum could have been from *F1*, if there was an *F1* layer. Near the geomagnetic equator, however, *F1* is not usually observable at this hour of the day. Probably the reflection took place from *F2*, with decrease of attenuation mainly in *F1*. The reflections observed at Ahmedabad should have been from *E*, and the decrease of attenuation during the eclipse should have taken place below *E*.

Measurements made at Ahmedabad (23° 02′ N, 72° 38′ E) during the two solar eclipses on 30 June 1954 and 20 June 1955, and at Trivandrum in the 1955 eclipse, showed significant increases of signal strength during the period of the eclipse. The present note gives a summary of the observations.

In the eclipse of 30 June 1954 the track of totality lay between London and Ahmedabad (Fig. 1). Rastogi and Sheriff (1955) measured the strength of B.B.C. transmission on 15·07 Mc/s at Ahmedabad with a calibrated communication receiver between 26 June and 4 July from 1530 to 2130 I.S.T. (1000 to 1600 G.M.T.) on each day. The average signal strength on control days and the signal strength on the eclipse day are plotted in Fig. 2. The increase in signal strength between 1820 and 1930 I.S.T. on the eclipse day is very pronounced.

The distance between London and Ahmedabad is 6800 km. Considering the time of the day and the normal critical frequencies of the ionospheric layers at the places of reflection, the principal mode of propagation between London and Ahmedabad should have been by two-hop reflection from *F2*, although the possibility of a three-hop reflection is not altogether excluded. The two-hop reflection points are *G* and *H* and the three-hop points *A*, *B*, and *C* in Fig. 1. The times at which the maximum phase of the eclipse was reached at these different points are indicated in Fig. 2. The maximum signal strength was recorded at 1845 I.S.T. (1315 G.M.T.), with a value nearly five and a half times the control-day average at that time.

A probable explanation for this increase of signal strength is that, while the critical frequencies of the *F2* layer remain comparatively unaffected during the eclipse, there is a decrease of electron density and attenuation in the lower layers during the eclipse. The maximum usable frequency via *F1* on normal days would be close to 15 Mc/s; there would thus be a large attenuation of waves reflected from *F2* on normal days. The decrease in electron density in *F1* which is known to occur during eclipses would mean less attenuation of the signals reflected from *F2*.

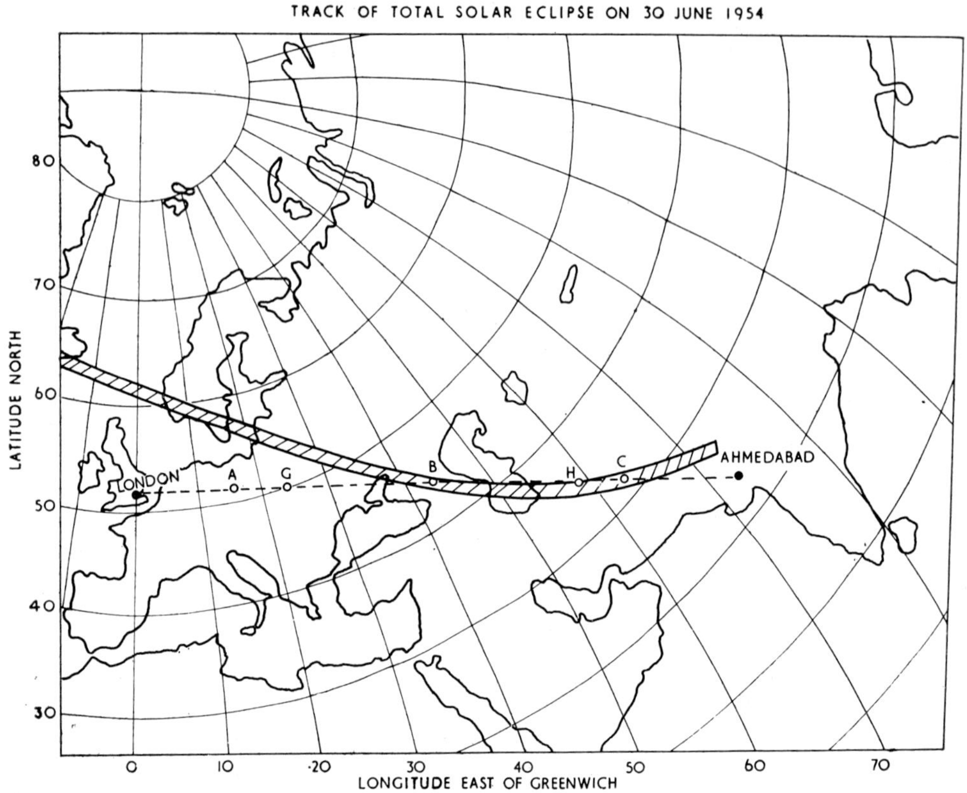
TRACK OF TOTAL SOLAR ECLIPSE ON 30 JUNE 1954
LATITUDE NORTH
80
70
60
50
40
30
LONDON
A
G
B
H
C
AHMEDABAD
0
10
20
30
40
50
60
70
LONGITUDE EAST OF GREENWICH

Fig. 1.

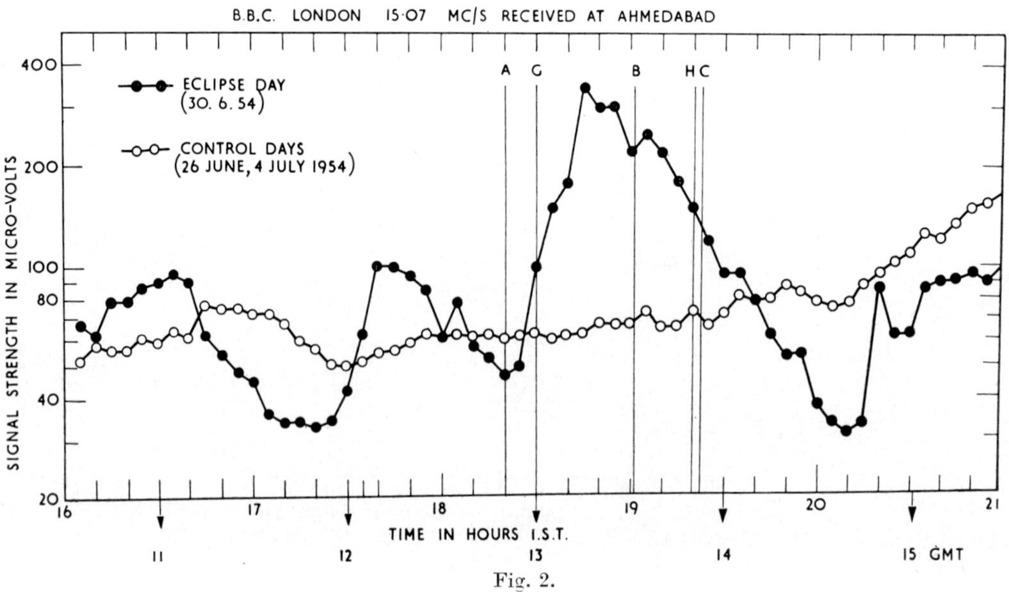
B.B.C. LONDON 15·07 MC/S RECEIVED AT AHMEDABAD
ECLIPSE DAY (30. 6. 54)
CONTROL DAYS (26 JUNE, 4 JULY 1954)
A G B H C
SIGNAL STRENGTH IN MICRO-VOLTS
400
200
100
80
40
20
16
17
18
19
20
21
TIME IN HOURS I.S.T.
11
12
13
14
15 GMT

Fig. 2.

This is considered to be the main reason for the increase in signal strength during the eclipse, though the changes in electron density in E would also produce a similar but less-marked effect.

The track of the total eclipse on 20 June 1955 passed over Ceylon, and the magnitude of the eclipse at Trivandrum was 0·97 (Fig. 3). The signal strength of

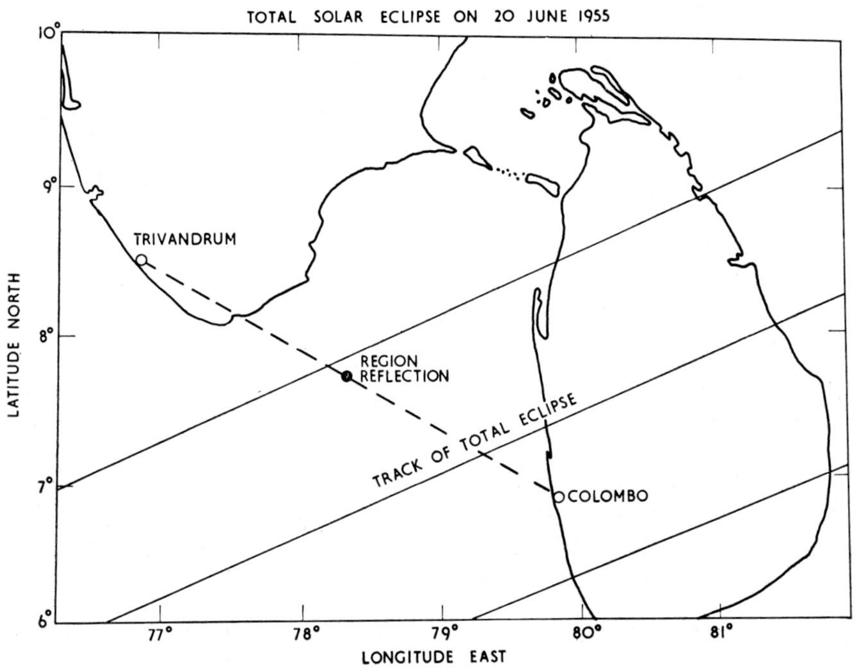

Fig. 3.

Colombo transmissions on 4·87 Mc/s was measured at Trivandrum (8·5° N, 77·0° E) between 0600 and 1030 I.S.T. (0030 to 0500 G.M.T.) on each day from 17 to 24 June. A modified communication receiver was used to receive the signals. The automatic control was removed from the R.F. and I.F. stages, and the voltage at the second detector was measured with a bridge-type d.c. amplifier. The meter readings of the d.c. amplifier were calibrated daily against signal voltages fed to the receiver from a signal generator. The average of the meter readings was recorded every 30 sec on all control days and every 15 sec on the eclipse day. The readings were converted into signal input voltages, and these were averaged for successive five-minute periods. Fig. 4 shows the signal strengths on the eclipse day and the averages of the strengths on the control days. The ranges of values observed on control days are also shown by thin lines. The perturbation during the eclipse is seen even better in Fig. 5, in which the percentage differences of signal strength on the eclipse day from the average values at the same time on control days are plotted.

The distance between Colombo and Trivandrum is 420 km. Since the transmissions from Colombo were directed upwards, the signal strengths measured

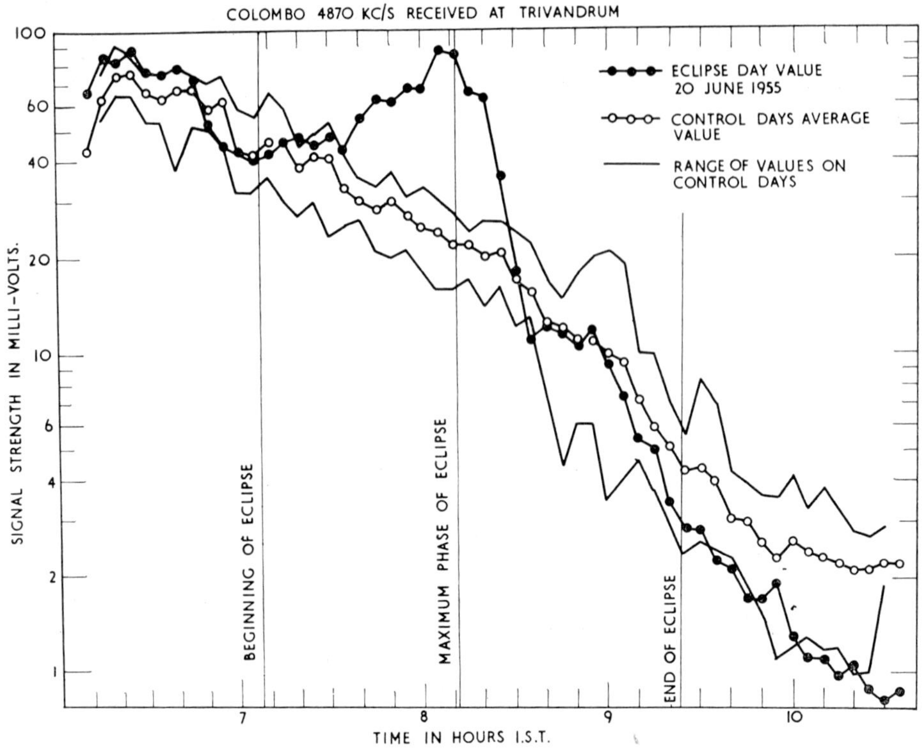
COLOMBO 4870 KC/S RECEIVED AT TRIVANDRUM
ECLIPSE DAY VALUE 20 JUNE 1955
CONTROL DAYS AVERAGE VALUE
RANGE OF VALUES ON CONTROL DAYS
SIGNAL STRENGTH IN MILLI-VOLTS.
BEGINNING OF ECLIPSE
MAXIMUM PHASE OF ECLIPSE
END OF ECLIPSE
TIME IN HOURS I.S.T.

Fig. 4.

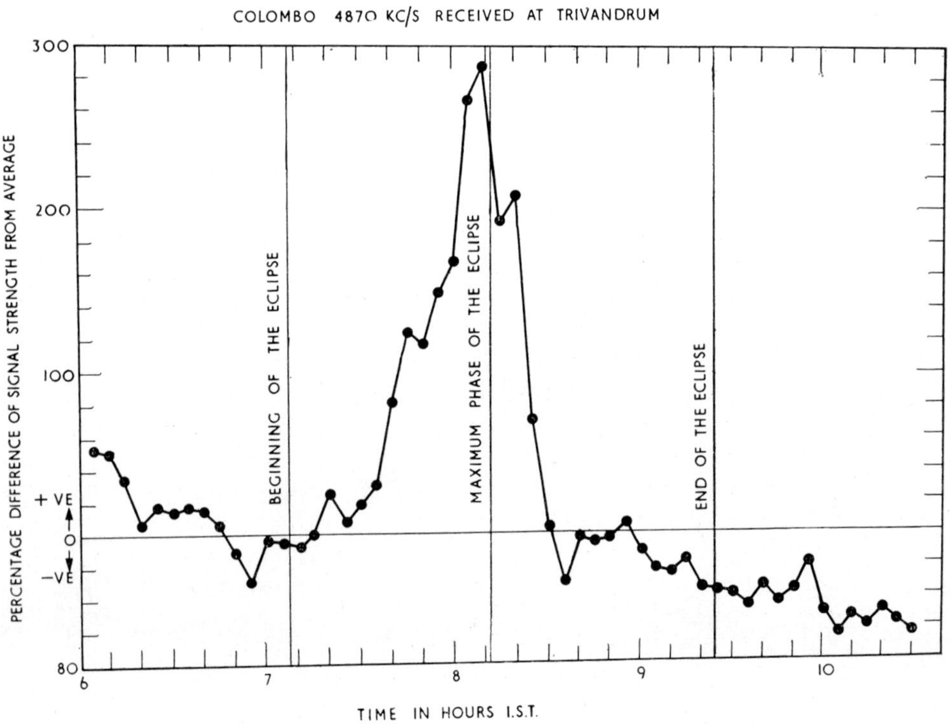
COLOMBO 4870 KC/S RECEIVED AT TRIVANDRUM
PERCENTAGE DIFFERENCE OF SIGNAL STRENGTH FROM AVERAGE
+VE
−VE
BEGINNING OF THE ECLIPSE
MAXIMUM PHASE OF THE ECLIPSE
END OF THE ECLIPSE
TIME IN HOURS I.S.T.

Fig. 5.

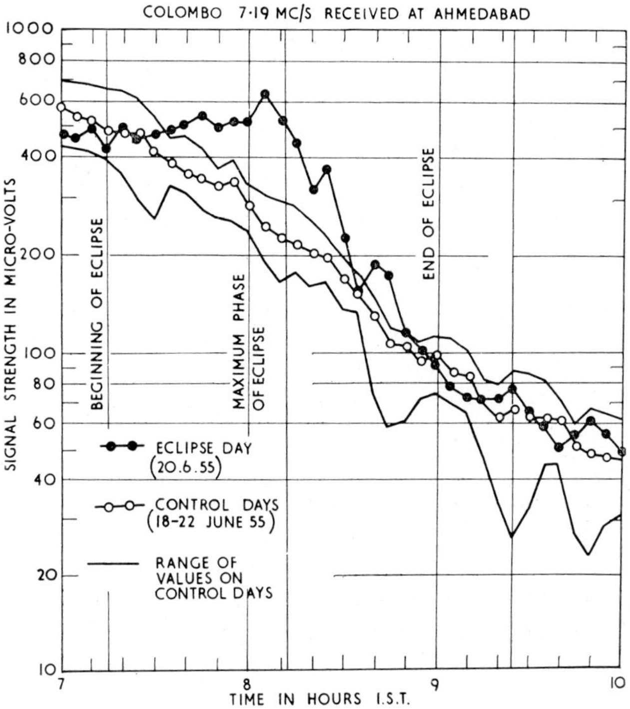
COLOMBO 7·19 MC/S RECEIVED AT AHMEDABAD
SIGNAL STRENGTH IN MICRO-VOLTS
1000
800
600
400
200
100
80
60
40
20
10
BEGINNING OF ECLIPSE
MAXIMUM PHASE OF ECLIPSE
END OF ECLIPSE
ECLIPSE DAY (20.6.55)
CONTROL DAYS (18-22 JUNE 55)
RANGE OF VALUES ON CONTROL DAYS
7
8
9
10
TIME IN HOURS I.S.T.

Fig. 6.

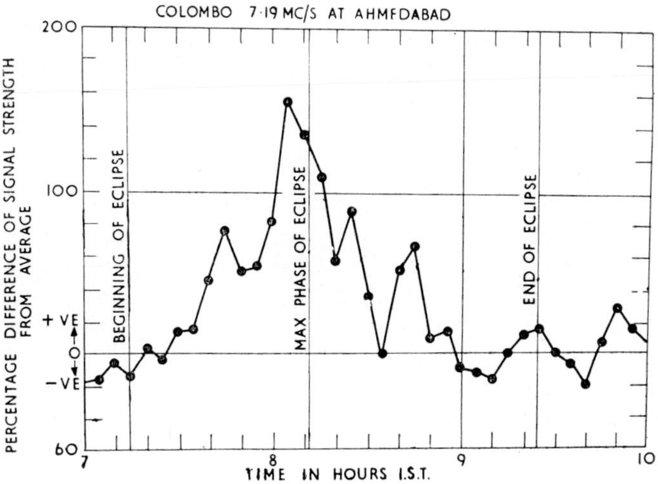
COLOMBO 7·19 MC/S AT AHMEDABAD
PERCENTAGE DIFFERENCE OF SIGNAL STRENGTH FROM AVERAGE
200
100
+VE
0
−VE
60
BEGINNING OF ECLIPSE
MAX PHASE OF ECLIPSE
END OF ECLIPSE
7
8
9
10
TIME IN HOURS I.S.T.

Fig. 7.

at Trivandrum may be assumed to be mainly due to singly-reflected sky-waves reflected from *F2*.

It will be noticed from Figs. 4 and 5 that the signal strength began to increase shortly after the beginning of the eclipse, and that it reached its maximum at about the maximum phase of the eclipse. It then dropped fairly rapidly and attained the normal control-days' value well before the end of the eclipse. The peak signal strength was about four times the strength on control days. This pronounced increase can be attributed to a decrease of electron density and of attenuation mainly in *F1* and to a lesser extent in *E*.

Similar measurements were taken at Ahmedabad on 7·19 Mc/s transmissions. The magnitudes of the eclipse at Ahmedabad and at the midpoint of the transmission path were 0·47 and 0·77 respectively. The distance between Colombo and Ahmedabad is 1950 km. Considering that the equivalent vertical frequencies for one-hop transmission by reflection from a height of 110 km would be 1·33 Mc/s, and for two-hop transmission 1·84 Mc/s, it may be taken that the reflection was from the *E* layer by either of these modes. Figs. 6 and 7 show the comparative signal strengths on the eclipse day and on the control days.

In this case also, the signal strength began to increase shortly after the beginning of the eclipse and the maximum strength of a signal was observed at about the time of maximum phase. Curiously enough, here too the signal strength returned to its normal value well before the end of the eclipse.

It is not possible to attach any significance to the small difference between the times of the maximum signal strength and of the maximum phase of the eclipse.

Acknowledgements—The authors are grateful to Professor K. R. RAMANATHAN for directing the work and providing all the necessary facilities. The Director of Radio Ceylon kindly arranged an earlier start of the morning transmissions on 4·87 Mc/s for a few days before and after the eclipse. Thanks are also due to Dr. C. S. VENKATESWARAN, Principal of the Trivandrum College, for providing facilities for observation at the Trivandrum Observatory, and to Mrs. MALATI RASTOGI for help in taking the observations at Trivandrum.

REFERENCE

RASTOGI, R. G., and SHERIFF, R. M. 1955 *J. Sci. Indust. Res.* (India), **14A,** 159.

In addition to the above papers the following reports on eclipse measurements were received and distributed at the symposium. Only abstracts are given here.

Ionospheric Observations during the Solar Eclipse of 30 June 1954 in Sweden

(Royal Board of Swedish Telecommunications)

IONOSPHERIC soundings were made at Luleå (65° 36′ N., 22° 07′ E) every 10 min during the eclipse, and every half-hour on the six control days. Results are shown in Figs. 1 and 2 overleaf, giving critical frequencies and virtual heights for the *E*, *F1*, *F2*, and *Es* layers. The magnitude of the eclipse at Luleå was 81 per cent. The first contact was at 1135 G.M.T. and the last contact at 1348 G.M.T. Maximum occurred at 1243 G.M.T., and the decrease in electronic density from estimated undisturbed values was: for the *E* layer 58 per cent, for the *F1* layer 42 per cent, and for the *F2* layer 36 per cent.

Relative field-strength and reflection-height measurements on 1·5, 2·0, and 2·5 Mc/s were also made at Värnamo (57° 12′ N, 14° 08′ E), where the maximum eclipse was 100 per cent.

A long-wave transmitter at Karlsborg (58° 29′ N, 14° 28′ E), ordinarily used for fixed traffic, was equipped for pulse transmission on 115·65 kc/s during the eclipse, and the signals from this transmitter were received on a special broadband longwave pulse receiver at Värnamo. Measurements were carried out during the day and the night. Owing to the high interference level and the low value of the reflection coefficient during the day, no echoes could then be observed. During the eclipse, echoes could be observed, but only for a few minutes around totality. A very rapid apparent change of reflection height was observed at totality. A similar but slow change in reflection height was observed during the evening hours.

The Influence of the Eclipse of 30 June 1954 on Radiocommunication Circuits of the Netherlands

(Netherlands P.T.T.)

FIELD-strength measurements on a number of wavelengths for several radio circuits between the Netherlands, Scandinavia, Japan, and Indonesia show a decrease in field strength in some, but in others no definite change. A study of field-strength records for several other circuits, some well outside the actual eclipse zone, shows a remarkable tendency for a worldwide eclipse effect in the ionosphere. It seems possible that ionospheric movements especially within the *F2* layer may result in eclipse effects spreading outside the eclipse zone, and it is suggested that, in future eclipse observations, attention should be given to this possibility.

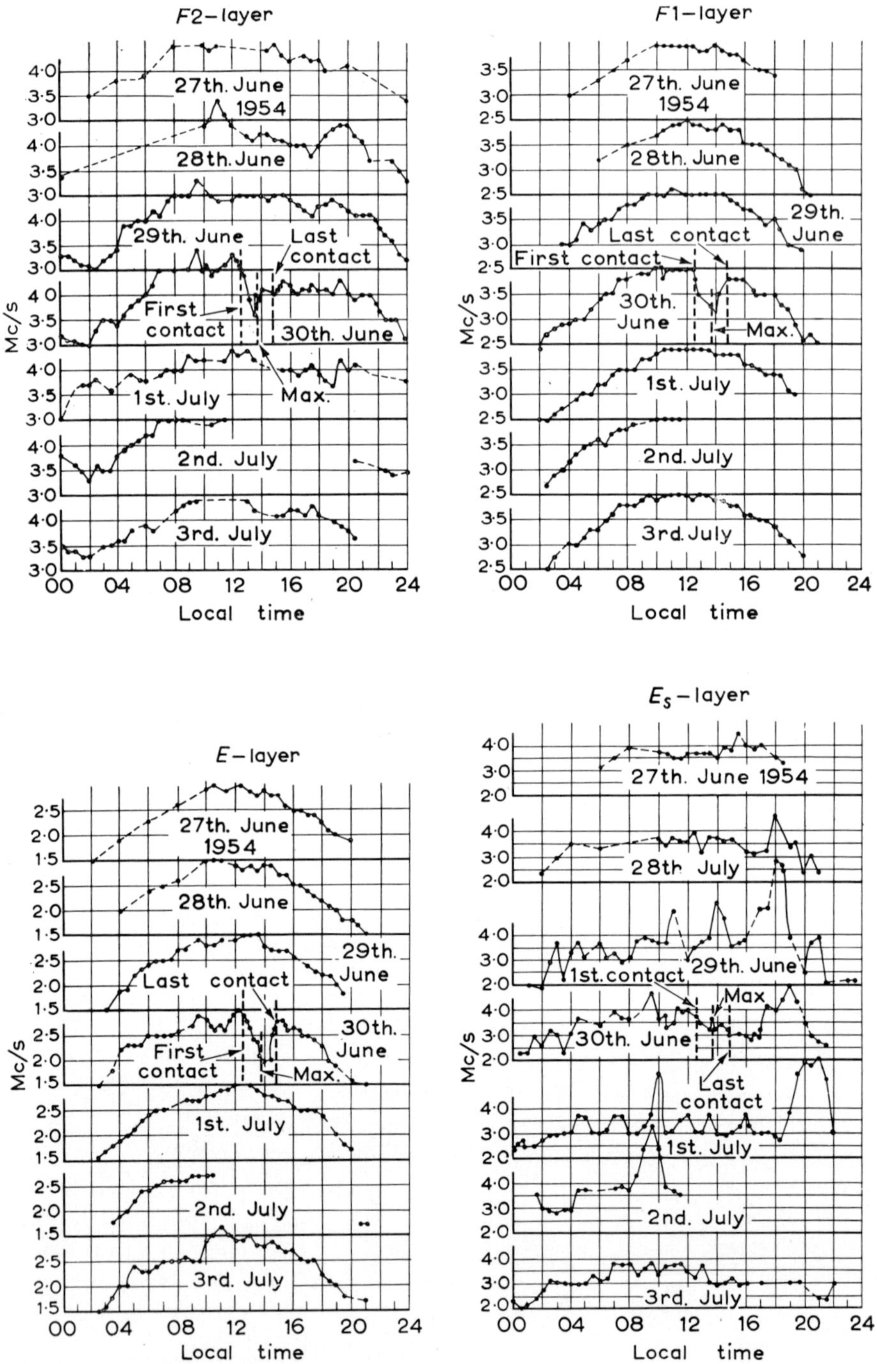
F2-layer
27th. June 1954
28th. June
29th. June
30th. June
1st. July
2nd. July
3rd. July
First contact
Last contact
Max.
Mc/s
Local time
F1-layer
27th. June 1954
28th. June
29th. June
30th. June
1st. July
2nd. July
3rd. July
First contact
Last contact
Max.
Mc/s
Local time
E-layer
27th. June 1954
28th. June
29th. June
30th. June
1st. July
2nd. July
3rd. July
First contact
Last contact
Max.
Mc/s
Local time
E_s-layer
27th. June 1954
28th July
29th. June
30th. June
1st. July
2nd. July
3rd. July
1st. contact
Last contact
Max
Mc/s
Local time

Fig. 1.

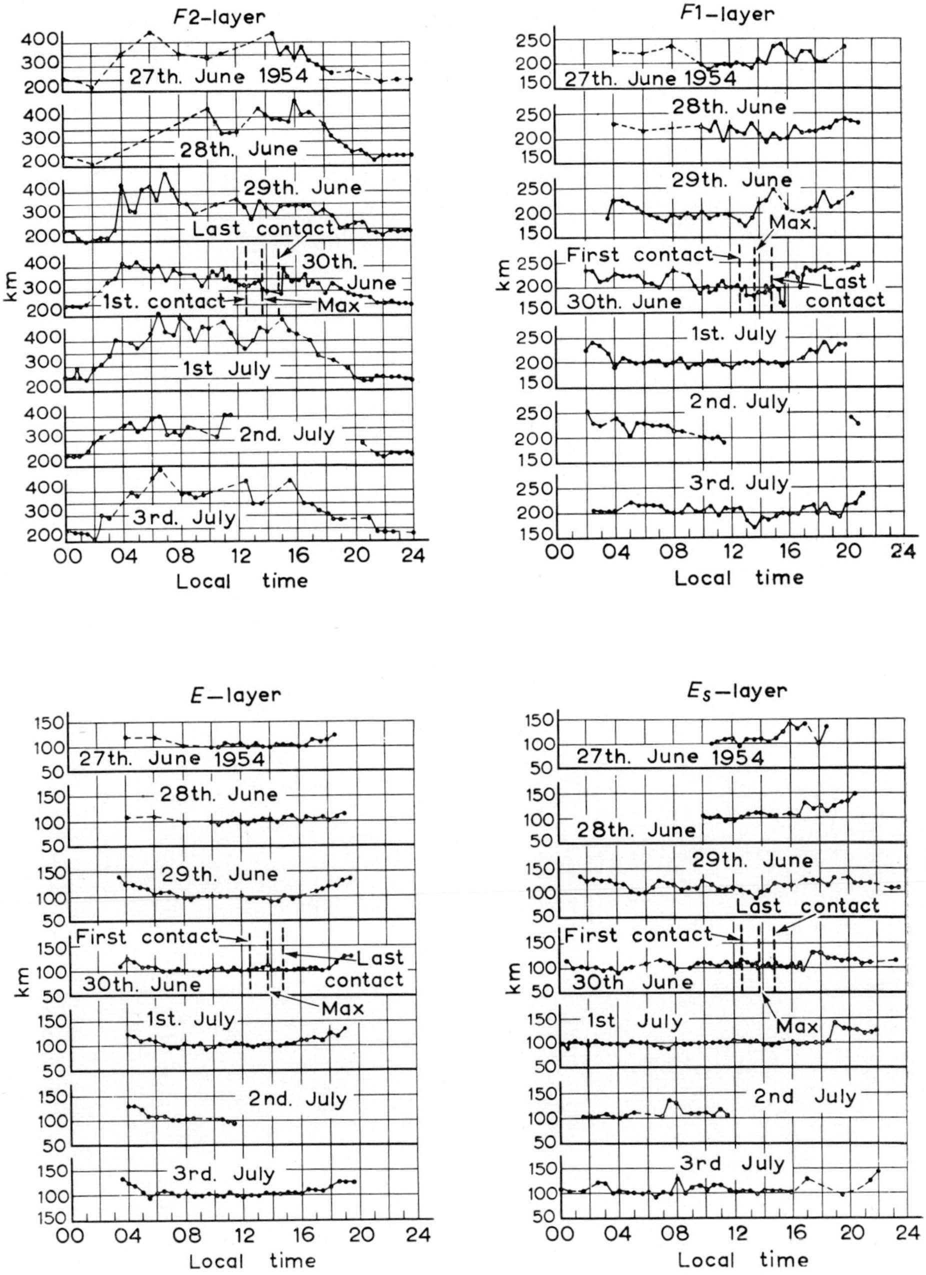
F2-layer
F1-layer
E-layer
E_s-layer
27th. June 1954
28th. June
29th. June
30th. June
1st. July
2nd. July
3rd. July
First contact
1st. contact
Max
Last contact
km
00 04 08 12 16 20 24
Local time

Fig. 2.

Report on Signal-Intensity Measurements during the Solar Eclipse of 30 June 1954

(Canadian Broadcasting Corporation)

THE report summarizes some measurements of relative field-strength, made in Canada and in Europe, of transmissions on 11·7 and 15 Mc/s during the eclipse period. Generally all transmissions in the 15-Mc/s band showed a decrease in signal strength during the eclipse period. The B.B.C. transmission on 11·7 Mc/s observed in Canada was not affected by the eclipse, whereas some reduction in strength of the C.B.C. transmission on the same frequency was observed by the B.B.C.

DISCUSSION OF PAPERS IN SECTION III

Chairman: Mr. J. A. Ratcliffe

The $F1\frac{1}{2}$ Layer

Prof. Ramanathan It seems to be generally assumed that the $F2$ layer is formed out of the $F1$ layer. I have some doubts about it, since at sunrise at Ahmedabad the $F2$ layer is formed first and at a much greater height than the $F1$ layer.

Mr. Ratcliffe We have made a study of the variation of electron density with height at stations distributed over the earth, and also a detailed consideration of what happens at sunrise. The facts can be made to fit in with the Bradbury hypothesis.

Dr. Aono At sunrise we always observe the $F1\frac{1}{2}$ layer at all Japanese stations.

Mr. Ratcliffe What about the $F1\frac{1}{2}$ layer which appears during eclipses? In my survey paper I suggested that it was only observed at localities where the magnetic dip is low.

Prof. Nagata At the eclipse of 1955 we observed it at Yamagawa, but not at our other stations. The eclipse magnitude at Yamagawa was 0·4. The most northerly Japanese station is Wakkanai (magnetic dip = 59°N) and the most southerly is Yamagawa (magnetic dip = 44°N). We usually get the $F1\frac{1}{2}$ layer appearing at all eclipses.

Mr. Shapley I think we should be quite clear about what we mean by "$F1\frac{1}{2}$." It is well known that transitory stratification of the $F1$ and $F2$ layers often occurs, but I think that by $F1\frac{1}{2}$ we mean a layer occurring in the second half of an eclipse in which the critical frequency increases with time.

Mr. Minnis There is a difficulty about what we actually observe. The peak in the retardation on the $h'f$ record may be very obvious, but on an (N, h) curve it may be no more than a point of inflexion.

Mr. Shapley My general impression is that $F1\frac{1}{2}$ does not occur at any station where the eclipse magnitude is less than about 0·5.

Mr. Brown A paper on the 1955 eclipse at Baguio (magnetic dip = 19° N) has been submitted to this symposium. The eclipse occurred near noon and was of magnitude 0·96. The authors refer to considerable stratification of the F layer on both the eclipse day and the control days, and report the formation on the eclipse day of a distinct intermediate layer between the $F1$ and $F2$ layers after the maximum phase of the eclipse.

Dr. Rawer At the 1952 eclipse (magnitude 0·6) at Djibouti (magnetic dip = 5° N) the $F1\frac{1}{2}$ layer was not observed.

Mr. Delobeau The $F1\frac{1}{2}$ layer did not appear at Dakar (magnetic dip = 26° N) during the eclipse of 1951.

Mr. Piggott For the 1952 eclipse (magnitude 0·75) at Ibadan (magnetic dip = 5° S) we saw a ripple similar to a typical $F1\frac{1}{2}$ pattern starting in the latter part of the eclipse, but its development was fundamentally different from that of an $F1\frac{1}{2}$ layer. This ripple developed smoothly into a regular, new $F2$ layer which lasted for some hours after the eclipse, and was only abnormal in being much denser and lower than the normal $F2$ layer at these times of day. This was quite different from the behaviour of the $F1\frac{1}{2}$ ripple at Khartoum, which was essentially a transitory phenomenon observed between the classical $F1$ and $F2$ layers.

Father LEJAY Nothing of this kind appeared during the 1952 eclipse (magnitude 0·75) at Leopoldville (magnetic dip = 34° S). The eclipse occurred in the early morning (maximum phase about 0800).

Mr. MINNIS For the 1955 eclipse (magnitude 0·45) at Singapore (magnetic dip = 16° S) the $F1\frac{1}{2}$ layer was just detectable. This eclipse occurred near noon.

Mr. LEPECHINSKY At Bangui (magnetic dip = 12° S) for the 1952 eclipse (which was total) there was a definite stratification and an $F1\frac{1}{2}$ layer.

Mr. RATCLIFFE In the paper by Professor NAKATA I noticed that stratification was observed *before* and after totality.

Mr. SHAPLEY It is seen coming down on the $h'f$ record 5 to 10 min before totality. There is a change of layer-shape with the formation of more and more of a cusp a few minutes before totality.

Mr. MINNIS At Khartoum it was certainly observed immediately after totality, and very doubtfully some 5 min before.

Mr. SHAPLEY We have observed it with a high-speed recorder (4 sweeps per minute), and its development is very systematic and smooth.

Mr. RATCLIFFE The phenomenon seems to be a fairly common occurrence in Japan, and I think it would be very useful if some Japanese $h'f$ records could be made available for study by other workers. [Dr. AONO and Prof. NAGATA agreed to arrange this—Ed.].

Dr. BEYNON One wonders whether the fact that the focus of the S_q current system passes over the latitude of Japan may not have some bearing on this phenomenon. It is true that the current system is located in region E but there may well be some influence in the F layer too. Mr. G. M. BROWN and I have found strong ionospheric evidence for the S_q current system in region E, and studies of f_0F1 which we have made give evidence for believing that such influence may extend at least to region $F1$.

Prof. NAGATA The geomagnetic latitude of Japan is low (about 25°), whereas the dip is large (about 50°).

Mr. VAN BERKEL It seems to me that the places where this $F1\frac{1}{2}$ layer is observed during an eclipse are those at which stratification of the F layer is a fairly frequent and normal occurrence.

Mr. SHAPLEY I have studied many hours of 4-min periods shown on motion pictures and noticed many complex transitory stratifications. However, most of these are short-lived and almost always indicate *decreasing* critical frequencies with time. I do not think I have ever observed a phenomenon like this, with *increasing* critical frequencies.

Dr. RATCLIFFE It would appear that there are places near the geomagnetic equator (such as Ibadan, where a detailed analysis has been carried out by SKINNER and WRIGHT) at which on ordinary days there are some subdivisions of the F layer which have some considerable degree of permanence.

Dr. WRIGHT The stratifications at Ibadan to which we referred were continuous even over a period of 4 to 5 hours.

Mr. DELOBEAU This is also true for Dakar.

Dr. BECKER Stratification is observed at Lindau more often at sunspot-minimum than at sunspot-maximum. Is this the case elsewhere?

Mr. PIGGOTT At Slough there is certainly little stratification at sunspot-maximum, but at Singapore it is definitely observed at sunspot-maximum and -minimum.

Mr. RATCLIFFE Possibly, then, the occurrence of the $F1\frac{1}{2}$ layer at an eclipse depends on the epoch of the sunspot cycle, but there is, as yet, insufficient data to examine this point.

Layer Movements during an Eclipse

Mr. Ratcliffe — If *additional* movements are produced in region $F2$ during an eclipse, it might be that an eclipse effect could be observed outside the belt of the optical eclipse. It would be interesting to know if anything is observed even on the edge of the eclipse belt.

Dr. Davies — In the 1954 eclipse in Canada, Baker Lake was near the edge, and there was very little effect in the $F2$ layer.

Dr. Aono — In the recent eclipse, Wakkanai was located some 200 km outside the eclipse belt, and there were no clear effects.

Dr. Rawer — The large day-to-day variability in the $F2$ layer may make it difficult to be quite certain about anomalous movements during an eclipse. Has Dr. Thomas made an adequate analysis for *normal* days? I think that there are very few really quiet days for the $F2$ layer.

Dr. Thomas — We certainly have analysed data for normal days. I agree that it is very difficult to obtain reasonable control data for an eclipse in the $F2$ layer because of the large day-to-day fluctuations.

Mr. Ratcliffe — The normal days used by Dr. Thomas were selected as quiet days on the basis of the international *magnetic* character figure, not on the basis of the ionosphere. Furthermore, the movements deduced during the eclipses are noticeably larger than those present on normal days.

Dr. Thomas — The eclipses which we considered all occurred on quiet days.

Solar Eclipses and the Ionosphere (Special Suppl. Vol. 6 to J. Atmosph. Terr. Phys.), Pergamon Press ,London, 1956

SECTION IV

SOLAR RADIATION AND LAYER THEORY

Solar Sources of the Ionospheric Regions

C. W. Allen
University of London Observatory

Abstract—The first question considered asks which parts of the sun emit those radiations that produce the ionosphere. Our information comes mainly from correlations between ionospheric and solar data. For the *E* and *F* radiations there is a quiet component from the whole sun and a component associated with centres of activity. The level of emission is uncertain, but is probably the lower corona.

The second question concerns the intensity and spectrum of the radiation. We try to solve this from ionospheric studies, rocket observations, and solar theory, but conclusions reached are still somewhat tentative.

Introduction

It is now generally agreed that the various ionospheric regions are caused by ionizing radiation from the sun, and our attention is turned to the questions of the absolute evaluation of the radiation, its spectroscopic quality, its distribution in time and direction, and its physical effects. The ionosphere itself may be used as a photo-cell to measure the intensity and variations of the ionizing radiation, but such measurements refer to the radiation from the whole sun, and other devices have to be found to enable one to locate the individual sources, or to determine the intensity distribution. Some indications of this location may be obtained during a solar eclipse, when the sun is covered in a controlled manner by the moon; hence the study of solar sources of the ionospheric regions becomes a vital part of any discussion on solar eclipses and the ionosphere. However, solar eclipses are not the only means to this end, and in fact a large part of the present paper deals with information gained by solar correlations. These lead to certain conclusions which form a necessary background to any eclipse discussions.

The ionospheric regions caused by solar radiation are usually labelled *D*, *E*, *F1*, and *F2*, although some of these may be subdivided when observations permit. If a layer is produced in a region of constant scale-height H, recombination coefficient α, and absorption coefficient A, then the photon flux S producing the layer (in photons cm^{-2} sec^{-1} at earth) is given by

$$S = 4{\cdot}2 \times 10^8 \, H \alpha f_0{}^4 \tag{1}$$

where f_0 is the ordinary-ray critical frequency in Mc/s. This relation may be elaborated for more complicated ionospheric conditions, but is the main basis for determining the photon flux. The separation of the *F1* from the *F2* region is now generally regarded as being due to a change of α with height. This implies that *F1* and *F2* are both produced from the same radiation, and that for radiation studies we can use either. Naturally, we chose the more easily interpreted *F1*.

Regular daily and monthly values of f_0E and f_0F1 have been published, and by averaging a number of observing stations it is possible to determine the *E* and *F* variations of f_0 with considerable accuracy. Equation (1) shows S to be proportional to f_0^4. However, one cannot be certain that the effective value of the exponent of f_0 is precisely 4. For the $F2$ region the exponent is nearer 2 (ALLEN, 1948), and one might expect it to be somewhat less than 4 for $F1$. On the other hand, comparison with the sun's altitude leads to an $F1$ value 4·9 (ALLEN, 1948). For the D-region there are, unfortunately, no long series of observations and, therefore, this region cannot be studied by long-term correlations.

It will be found that the ionizing radiation comes both from centres of solar activity and from the general solar surface, and therefore to explain the radiation we must investigate both the special features and the normal quiet atmosphere. One of our purposes will be to find which of the phenomena associated with centres of activity are significant in the emission of radiation. Some phenomena to be considered are bright $H\alpha$ and Ca flocculi, dark flocculi (filaments) and prominences, faculae, sporadic flares, bright areas in the corona of λ5303 and other lines, coronal condensations. All are in some way associated with sunspot groups, but they have different mean lives—mostly longer than sunspots—and therefore their variations are inclined to lag behind those of sunspots. The filaments in particular do not usually develop until long after the sunspot has disappeared. By comparing the variation lags of ionospheric and solar phenomena, we can draw some conclusions about the solar sources.

2. Emission from Centres of Activity

It is clear that at least some of the ionizing radiation comes from centres of solar activity. The analysis of eclipse observations indicates emission from localized spots which can be identified with these centres (HIGGS, 1942). Furthermore, if monthly values of critical frequencies (preferably averaged over a worldwide set of stations) are plotted alongside monthly values of sunspot numbers, R, the detailed agreement makes it evident that the individual spot groups are responsible for some of the radiation.

The next question is: Which of the various phenomena associated with centres of activity can be held responsible for the ionizing radiation? The only way to discriminate between them observationally is by timing. Time-lags of the phenomena may be determined by studying correlation coefficients of monthly data, and although the process is not very accurate, it is at least objective and quantitative. An analysis (ALLEN, 1948) gave the following lags behind sunspot number variations:

Solar		Ionosphere	
Bright $H\alpha$ flocculi	0·15 month		
Ca flocculi	0·31 month	E region	0·43 month
Faculae	0·56 month	$F1$ region	0·66 month
λ5303 coronal line	0·87 month	$F2$ region	0·68 month

This table leaves the impression that (a) delay times increase for solar features in the order photosphere, chromosphere, corona; (b) $F1$ and $F2$ could come from the same source; and (c) the E and F regions are produced by different sources, with the E source lower than the F. Data that have accumulated since 1947 deserve analysis by this method to give independent checks on the above results. There is now abundant coronal material awaiting analysis.

The radiations from centres of activity are necessarily additive in their effects on the ionosphere, and the radiation will therefore be proportional to the number of centres. Individual centres, of course, vary enormously in their activity, but comparisons in size and complexity of sunspot groups at different phases of the sunspot cycle show that there is very little change in mean character during the cycle. It is only the numbers that change. Consequently the radiation measurements, when sufficiently smoothed, should be proportional to any measure of sunspot activity such as sunspot number. Results show this to be the case with considerable accuracy when the exponent of f_0 in (1) is known (cf. Fig. 6 of ALLEN, 1948), and in the $F2$ region this proportionality provides a means for determining the exponent. The proportionality also provides a means of extrapolating to zero sunspot activity.

3. EMISSION FROM QUIET SUN

The continuance of the E and F regions at considerable strength throughout the sunspot minimum makes it clear that there is a component of ionizing radiation that is *not* associated with centres of activity. The linear relation between radiation and sunspot number just mentioned enables us to extrapolate to zero sunspots to determine the radiation for a perfectly quiet sun. It would be difficult to prove from eclipse observations that this quiet radiation exists.

Quiet sun radiation is doubtless emitted throughout the sunspot cycle, but it is difficult to measure this quiet component in the presence of activity. No attempt has been made to do this up to the present. On the other hand, radio astronomers are attempting to solve the analogous problem of determining the quiet-sun emission of decimetre wave radiation throughout the sunspot cycle. The procedure is to correct the measured radiation by subtracting the active components. The amount of active radiation is to be determined by relating daily values of measured radiation with sunspot numbers and then extrapolating to zero sunspots. However, this does not allow for the radiation still emitted from a centre of activity after the sunspots have disappeared. Some estimate of this further correction may be made by using the lags mentioned in Section 2. The ionospheric data required for such an analysis are daily critical frequencies for the regions concerned, say E and $F1$. Random variations of f_0 at any one station will be considerable, and hence several stations should be averaged. Magnetic storms would have to be avoided (or studied as a by-product). It should be possible to carry out such an analysis for ionospheric u.v. radiation as well as for decimetre wave emission.

Eclipse observations can provide some indication of the height of emission. Thus PIDDINGTON (1951) has shown that neither the E nor $F1$ ionizing radiation is emitted *throughout* the corona, but the same information rather suggests that the emission might be concentrated in the lower corona.

4. COMPARISON OF ACTIVE AND QUIET SUN

We have seen that there are as yet no estimates of the variation of the quiet ionizing radiation during the sunspot cycle. We therefore assume it to be constant, and ascribe all the sunspot cycle variation to activity. Using the exponent 4 in (1), the ratio of the components becomes (ALLEN, 1948):

	Active (R = 100)	Quiet
E region	0·97	1·00
F region	1·24	1·00

The precise value of the ratio depends on the exponent used, and if this were smaller for $F1$ than E (say 3·4 for $F1$ and 4·0 for E), the ratios could become equal. In any case, it is evident that the ratios are not very different for the two regions. The interpretation we can put on this is that the spectrum of the ionizing radiation from the centres of activity is probably rather similar to that from the quiet sun.

The surface brightness of the active emission must be much greater than for the quiet emission. The mean sunspot area for $R = 100$ is 2×10^{-3} of the sun's disk, but the emission area is likely to be considerably greater. We will take it to be 0·01 of the disk, and the active emission brightness about one hundred times the quiet.

The characteristics of the ionospheric layers do not give any clear indication that there is any difference of quality of the radiation at sp-max and sp-min. It is possible that some of the sunrise stratification is associated with spectroscopic structure of the radiation, and it would be instructive to compare stratification at sp-max and sp-min with this in mind. I am not aware that this has been done. It is unlikely that change in radiation quality could explain the $F2$ region anomalies, since these are most severe at sunspot minimum when the quiet sun predominates (ALLEN, 1953).

5. IONOSPHERIC INDICATIONS OF SOLAR EMISSION

Estimates of the photon flux producing the ionospheric regions may be obtained from (1) when H, α, and f_0 are known. The following approximate values of the flux per cm² per sec at the earth for zenith sun and sunspot minimum are suggested as a working basis:

Region	Height	H	f_0	N_e	α	S
	km	km	Mc/s	cm^{-3}	cm^3 sec^{-1}	cm^{-2} sec^{-1} at earth
D	70	6	(0·15)	300	3×10^{-6}	4×10^5
E	120	10	3·3	$1{\cdot}3 \times 10^5$	$1{\cdot}1 \times 10^{-8}$	5×10^8
F	170	20	4·4	$2{\cdot}4 \times 10^5$	5×10^{-9}	$1{\cdot}6 \times 10^9$

The factor about which the greatest doubt exists is α, which could be ten times greater (PIDDINGTON, 1951).

If the absorption coefficient of the earth's atmosphere is known as a function of wavelength, one has a direct method of determining the possible wavelengths of the radiation from the height of the layer. In the simplest case we have

$$NH = 1/A$$

where A is the molecular absorption coefficient, H the scale height, and N the molecular density at the height of the layer. If N is known in relation to height h, a knowledge of the absorption coefficient enables one to determine a curve such as that given by GOLDBERG (1954b, Fig. 7) which connects the height of the layer formed with the wavelength of the ionizing radiation. From this diagram we read the following choice of wavelengths for the radiations:

Region	Height	Wave-length
D	70 km	1216 Å (Lα), X-rays (3 Å), or 1900 Å
E	120 km	1100 Å, or X-rays (20 to 50 Å)
F	170 km	Whole range 200–900 Å

This argument produces rather severe limits to the possible wavelengths of radiations producing the main layers.

6. ROCKET OBSERVATIONS

Recent rocket observations (FRIEDMAN and CHUBB, 1955) give 0·10 erg cm^{-2} sec^{-1} at earth as the normal value of Lα emission. This radiation is sufficient to produce the D layer, and the Lα wavelength is one of those indicated by absorption measurements for the D layer (Section 5). The measured intensity of Lα is small enough for it to be increased many-fold by the addition of a flare, and hence we would expect a flare to enhance the D region in the manner that is observed. These arguments, point rather convincingly to Lα as the chief cause of the D region. There may, however, be some short X-radiation during strong flares (FRIEDMAN and CHUBB 1955).

Opinions on the radiation producing the E region have been influenced by the rocket observation of soft X-rays suitable in wavelength and intensity to produce the ionization (BYRAM, CHUBB, and FRIEDMAN, 1954). Support for this identification of the E radiation is found in the calculations of ELWERT (1954), and the pre-ionization mechanism which enabled NICOLET (1945) to account for the E region by radiation near 1000 Å becomes unnecessary.

No rocket observations are available of the radiations between 200 Å and 900 Å which might be expected to produce the F region.

7. SOLAR PHYSICAL INDICATIONS OF EMISSION

First attempts at estimating the ionizing radiation from the sun (WOOLLEY and ALLEN, 1948, 1950; ELWERT, 1953, 1954; SHKLOVSKY, 1949) refer mainly to the quiet sun, since the physical condition is better understood for the normal solar atmosphere than for centres of activity. We have seen that the surface brightness of the emission from active areas is about one hundred times as great as for the quiet sun, but that the spectra are probably rather similar.

There are several layers in the solar atmosphere which might contribute ionizing radiation; we should consider particularly the photosphere, lower chromosphere, upper chromosphere or transition layer, and corona. These may emit black-body radiation, recombination continua, free-free continua, and line emission. We need consider only wavelengths less than the Lyman limit, since only these can ionize atmospheric gases. Lα can be omitted from the discussion, since the part it plays in the D region has already been mentioned, and its emission with suitable intensity from the chromosphere presents no theoretical difficulty.

As a first requirement for the calculations, we need a model of the sun's atmosphere; that is to say, the values of the atom density N, electron density N_e and temperature T as a function of height h. Models for the photosphere, lowest 1000 km of chromosphere (Böhm-Vitense, 1955; Athay, Menzel, Pecker, and Thomas, 1955), and corona (van de Hulst, 1950) are accurate enough for our purpose. The gap in our knowledge lies in the transition layer between the chromosphere and corona—sometimes called the outer chromosphere. Models available for this region are rather discordant (Woolley and Allen, 1950; Hagen, 1951; Piddington, 1954; Woltjer, 1954; Athay *et al.*, 1955). Furthermore, it is evident that the chromosphere is very inhomogeneous, and it may be erroneous to base calculations on mean conditions at each height.

As a second requirement we need methods for evaluating all the atomic factors concerned—optical absorption coefficients, excitation and ionization collision cross-sections, and recombination coefficients. Improved methods for obtaining some of these have been given recently by Elwert (1952, 1954).

The emission from the photosphere approximates to black-body radiation at the sun's surface temperature, about 4200°K, and is negligible.

The lower chromosphere is optically thick beyond the Lyman limit, and this too will emit mainly black-body radiation. However, somewhere in the chromosphere the temperature starts to increase outward, and the required black-body temperature will occur near unit optical depth. The level must be near the level of half-ionization, and it is found that the black-body temperature depends on atom density alone. With atom density 5×10^{10} cm^{-3}, Saha's equation leads to a radiation temperature of about 6200°K (Woolley and Allen, 1950; Goldberg, 1954a), and a photon flux about 5×10^9 cm^{-2} sec^{-1} at earth. This is more than enough to produce the F layer, and it is not easy to see how the radiation could be very much reduced. On the other hand, it would be difficult also to account for the hundred-fold increase required to explain the active component.

Similar arguments suggest that the upper chromosphere is optically thin. The free-bound emission of hydrogen may be evaluated then simply by determining the number of recombinations which depends on the recombination coefficients α and the ion and electron densities N_i, N_e. With $\alpha = 10^{-13}$, $N_i = N_e = 3 \times 10^9$, we find that a layer-thickness of 1000 km can produce about 10^9 photons cm^{-2} sec^{-1} at earth —almost enough for the F layer. A ten-fold increase in N_i and N_e could explain the active component.

An attempt to compute the line emission from the upper chromosphere (WOOLLEY and ALLEN, 1950) led to the value 7×10^9 photon cm^{-2} sec^{-1}. This is already too high for the observed *F* layer. ELWERT (1954) has shown that the formula used for collision cross-section gives values that are too high. Nevertheless, the error factor is not likely to be more than 10, and there is still almost enough line emission to produce the *F* layer. By increasing the transition layer thickness, the estimate could be a great deal higher.

Beyond the transition layer, the coronal temperatures and densities are fairly well known. The coronal calculations of WOOLLEY and ALLEN (1948) and ELWERT (1954) on emission by free-bound and free-free continua are in reasonable agreement, leading to a total emission of soft X-ray radiation of about 10^8 photons cm^{-2} sec^{-1}. This could produce the *E* region. ELWERT goes on to compute the line emission in the same spectral range, and again finds about 10^8 photons cm^{-2} sec^{-1}. There is, therefore, good reason to conclude that these radiations form a significant contribution to the *E* region. WOOLLEY and ALLEN's estimates of the intensity of lines of longer wavelength lead to the value 4×10^9 photons cm^{-2} sec^{-1}, but this is probably an over-estimate. In any case, this radiation would contribute to *F*, not *E*, ionization.

The calculations suggest that the *E* region is produced in the corona and the *F* in the chromosphere, which is the reverse of the impression mentioned in Section 2. The calculations throw very little light on the physical process of emission, because within the wide range of error all sources and processes (black-body lower chromosphere, free-free and free-bound upper chromosphere, lines in upper chromosphere and in corona for *F*; and free-free, bound-free, and lines for *E*) give about the same values. At least the values obtained are close to the ionospheric figures, and if they were added together the agreement would still be reasonably good. However, we have already seen in Section 3 that neither layer can be emitted entirely from the corona, and we might conclude that most of the *E* and *F* radiation comes from the upper-chromosphere to lower-corona region.

REFERENCES

ALLEN, C. W.	1948	*Terr. Mag. Atmos. Elect.*, **53**, 433.
	1953	*J. Atmosph. Terr. Phys.*, **4**, 53.
ATHAY, R. G., MENZEL, D. H., PECKER, J. C., and THOMAS, R. N.	1955	*Astrophys. J. Supp.*, **1**, 479.
BÖHM-VITENSE, E.	1955	*Z. Astrophys.*, **36**, 145.
BYRAM, E. T., CHUBB, T., and FRIEDMAN, H.	1954	*Rocket Exploration of the Upper Atmosphere*, London, Pergamon Press, p. 274.
ELWERT, G.	1952	*Z. Naturforsch.*, **7a**, 432.
	1953	*J. Atmosph. Terr. Phys.*, **4**, 68.
	1954	*Z. Naturforsch.*, **9a**, 637.
FRIEDMAN, H., and CHUBB, T. A.	1955	*Physics of the Ionosphere*, London, Phys. Soc, p. 58.
GOLDBERG, L.	1954a	*Astrophys. J.*, **120**, 185.
	1954b	*The Earth as a Planet*, Ed. KUIPER, Chicago, p. 434.
HAGEN, J. P.	1951	*Astrophys. J.*, **113**, 547.
HIGGS, A. J.	1942	*Monthly Not. Roy. Astron. Soc.*, **102**, 24.

NICOLET, M. 1945 *Mém. Inst. Roy. Met., Belgium*, **19**, 1.
PIDDINGTON, J. H. 1951 *J. Geophys. Res.*, **56**, 409.
1954 *Astrophys. J.*, **119**, 531.
SHKLOVSKY, I. S. 1949 *Crimea Mitt.*, **4**, 80.
VAN DE HULST, H. C. 1950 *B.A.N.*, **11**, 135.
WOLTJER, L. 1954 *B.A.N.*, **12**, 165.
WOOLLEY, R. v. d. R., and ALLEN, C. W. 1948 *Monthly Not. Roy. Astron. Soc.*, **108**, 292.
1950 *ibid.*, **110**, 358.

Solar Eclipses and the Ionosphere (Special Suppl. Vol. 6 to J. Atmosph. Terr. Phys.), Pergamon Press, London, 1956

The Origin of the *E*- and *F1*-Layer Ionizing Radiation

C. M. MINNIS

Radio Research Station, Slough

Abstract—The evidence obtained during earlier eclipses has tended to show that the eclipse-changes in the *E* and *F1* layers cannot, in general, be accounted for in terms of a uniformly bright solar disk. The data, however, did not permit any reliable conclusions to be reached regarding the identity of the emission areas.

During the February 1952 eclipse, the results obtained by the British expeditions to the Sudan and Nigeria allowed the locations and relative intensities of the radiation sources to be determined fairly accurately. The two sets of data are in good agreement and show that only two-thirds of the radiation on that occasion was emitted uniformly from the disk. The additional sources appeared to be associated with areas in which green coronal emission was known to be strong at the time.

The July 1954 eclipse occurred when solar activity was exceptionally low, but the measurements made both at Inverness and Slough show that nearly 10 per cent of the radiation came from sources other than the uniform disk. The accurate location of this residual radiation was not possible, but the best fit with the experimental data was obtained by postulating limb-brightening near the equator and limb-darkening near the poles.

A preliminary analysis of the *E*-layer data obtained during the June 1955 eclipse at Singapore shows that the distribution of the sources of ionizing radiation was not uniform, but the maximum obscuration was too small to allow much information to be gained about their locations.

1. INTRODUCTION

BETWEEN 1932 and 1937 the effects of solar eclipses on the ionosphere had been observed, with varying degrees of success, on six occasions. The results of these measurements left little doubt that the ionizing agent in the *E* and *F1* layers was electromagnetic rather than corpuscular radiation; the behaviour of the *F2* layer, on the other hand, appeared to be anomalous and no definite conclusions could be reached for this layer. During this period it was usual to assume that the sun's disk was uniformly bright, and to compute the effective recombination coefficients from the observed changes in electron density which occurred during an eclipse. There were, however, two difficulties: on some occasions the lag in the time of occurrence of minimum electron density behind that of the maximum phase of the eclipse was found to be zero or even negative, and it was also not unusual to find abrupt changes in the rate of rise or fall of electron density; these changes sometimes appeared to be related to the covering or uncovering of sunspots, and implied a non-uniform distribution of the sources of ionizing radiation.

The first attempt to derive the actual distribution appears to have been made by HIGGS (1942), who found that the data obtained during the 1940 eclipse in South Africa could not be explained in terms of a uniformly bright disk. Irregularities in the changes in the *E* and *F1* layers were assumed to be associated with the occultation of four isolated areas of calcium flocculi which were visible at the time; and these, in conjunction with a uniformly bright disk, were assumed to be the sources of ionizing radiation. The relative intensities of the different sources were then calculated, and it was found that the best fit with the *E*-layer data was obtained by assuming that the whole of the radiation was emitted from the four floccular areas. The *F1*-layer data led to the conclusion that a large fraction of the radiation could be attributed to the disk. HIGGS admitted that these were not unique solutions, and in the light of later knowledge it seems unlikely that they were correct.

WALDMEIER (1947) re-examined the eclipse data which had been obtained by RYDBECK (1946) in Sweden in 1945, and pointed out that the *E*-layer data suggested the existence of a link between the sources of ionizing radiation and the areas from which the coronal line at 5303 Å was emitted strongly. A similar suggestion was made by OSAWA (1950) after the 1948 eclipse in Japan.

2. FEBRUARY 1952 ECLIPSE

A preliminary examination of the 1952 data from Khartoum (MINNIS, 1952) gave partial confirmation of the relation between 5303 Å emission and *E*-layer ionizing radiation, but it was clear that the coronal intensity contours could not provide a basis for a quantitative explanation of the ionospheric data. However, the nature of the changes in the *E* layer was such that it seemed likely that a reliable estimate could be made, using the ionospheric data alone, of the distribution of the sources of ionizing radiation. This was done by computing, at frequent intervals, the function

$$J = \frac{A' q_0}{\alpha'} = \left(N^2 + \frac{1}{\alpha'}\frac{\mathrm{d}N}{\mathrm{d}t}\right) \sec^n \chi \, ,$$

in which the symbols have the usual significance and A' represents the fraction of the total radiation flux not obscured. It was assumed that the time-changes in the magnitude of J were due only to changes in A', and a model distribution of sources of ionizing radiation was constructed to fit these changes (MINNIS, 1955). The conclusion reached was that only 64 per cent of the radiation could be attributed to a uniformly radiating disk, while 25 per cent came from a thin bright annular area close to the west limb (Fig. 1). Some confirmation of this rather unexpected result was desirable, and a similar analysis was made of measurements which were made at the same time by PIGGOTT (1952) at Ibadan, where the eclipse was partial (Fig. 2). It is satisfactory to find that the agreement between the distributions derived from these independent sets of data is good, and some confidence in the reliability of the conclusions appears to be justified. It is also worth noting that the observed changes in the *F1* layer at Khartoum are consistent with the distribution derived using the *E*-layer data.

An examination of the usual visible features of the sun during the February 1952 eclipse shows that active areas existed near both the west and east limbs, but that over the remainder of the disk there were no important centres of local activity. These conclusions are confirmed by the intensity contours of the 5303 Å line, which reflect broadly the derived distribution of the ionizing radiation. There is, in particular, a marked correspondence between the position angles of the 5303 Å radiation at the limbs and of the two localized sources of ionizing radiation.

The results just discussed, although not conclusive, provide further evidence in favour of the hypothesis that areas from which 5303 Å radiation is emitted strongly are also strong sources of *E*- and *F1*-layer ionizing radiation.

3. JUNE 1954 ECLIPSE

Since the activity of the sun was still fairly high in 1952, it seemed likely that the isolated sources of ionizing radiation were temporary, and that a perfectly

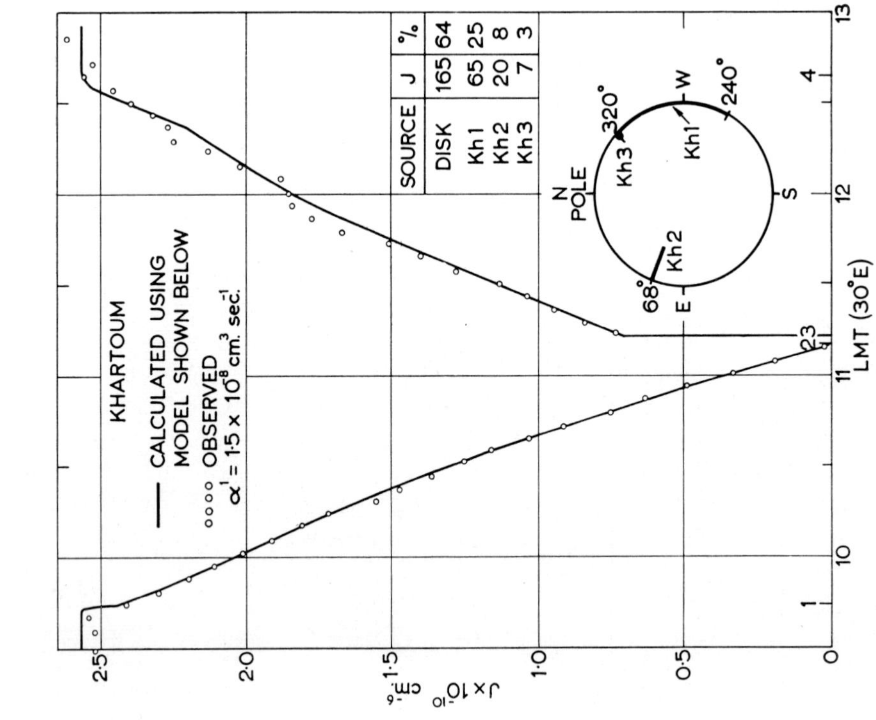
KHARTOUM
CALCULATED USING MODEL SHOWN BELOW
OBSERVED
α¹ = 1·5 x 10⁻⁸ cm.³ sec.⁻¹
SOURCE | J | %
DISK | 165 | 64
Kh1 | 65 | 25
Kh2 | 20 | 8
Kh3 | 7 | 3
N POLE
E
W
S
68°
320°
240°
Kh1
Kh2
Kh3
J x 10⁻¹⁰ cm.⁻⁶
LMT (30°E)

Fig. 1.

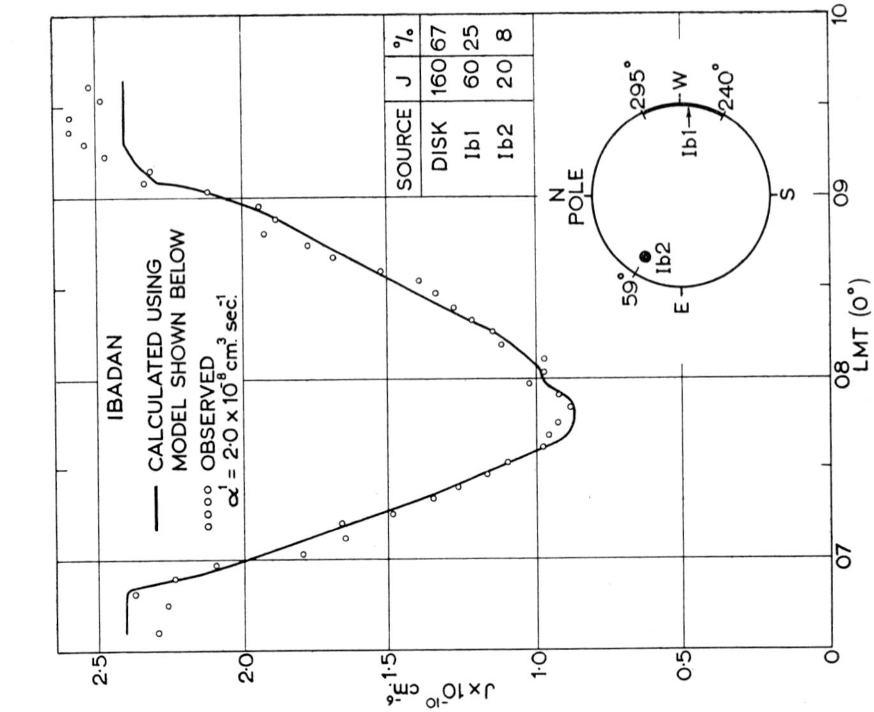
IBADAN
CALCULATED USING MODEL SHOWN BELOW
OBSERVED
α¹ = 2·0 x 10⁻⁸ cm.³ sec.⁻¹
SOURCE | J | %
DISK | 160 | 67
Ib1 | 60 | 25
Ib2 | 20 | 8
N POLE
E
W
S
59°
295°
240°
Ib1
Ib2
J x 10⁻¹⁰ cm.⁻⁶
LMT (0°)

Fig. 2.

quiet sun would exhibit a uniformly bright disk. For this reason the eclipse of June 1954 was of particular interest, because solar activity at that time was exceptionally low and the disk in visible light was almost featureless. Measurements of the changes in the E and $F1$ layers, which were made both at Inverness and Slough, indicate that about 92 per cent of the ionizing radiation can be attributed to a uniformly bright disk, while a very small area on the limb at $P = 255°$ contributes 4 per cent. It is not possible to determine unambiguously

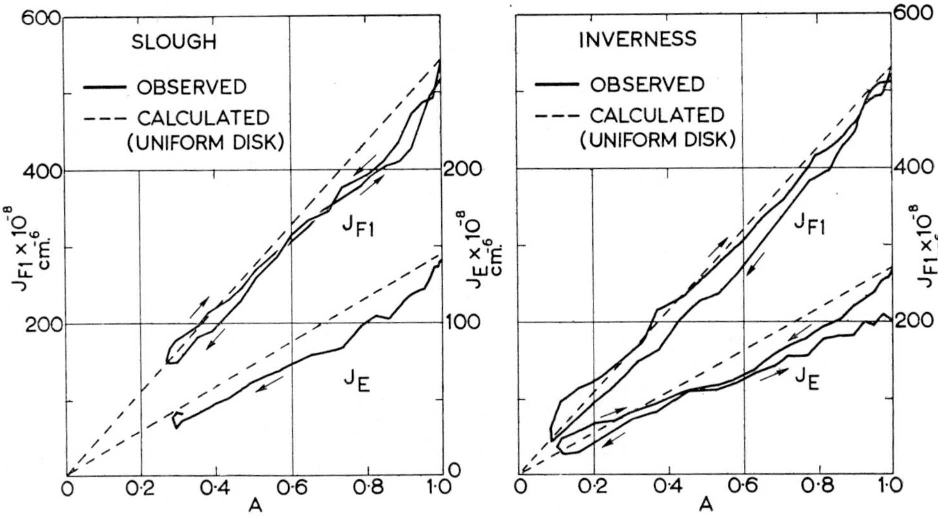

Fig. 3.

the location of the remaining 4 per cent but, when J is plotted as a function of the fraction A of the disk area not obscured, the increases in dJ/dA at the beginning and end of the eclipse point to the existence of additional bright sources near the west and east limbs (Fig. 3). The best fit is obtained by making the arbitrary assumption that there is brightening at the west and east limbs and darkening at the poles over arcs of 60° in all four positions (Table 1). It is interesting to

Table 1. Sources of *E*- and *F1*-layer ionizing radiation
30 June 1954

Source	Intensity per cent
Uniform disk	92
West limb	6
East limb	4
North pole	−2
South pole	−4
Spot ($P = 255°$ near west limb)	4
Total	100

note that the small spot is close to one of only two small areas in which very faint 5303 Å radiation was reported on the day of the eclipse.

4. June 1955 Eclipse

The maximum obscuration at Singapore during the eclipse of June 1955 did not exceed about 45 per cent; although this is too small to allow an unambiguous interpretation of the intensity distribution, an examination of the provisional values of f_0E, which have recently been received, suggests that at least 13 per cent

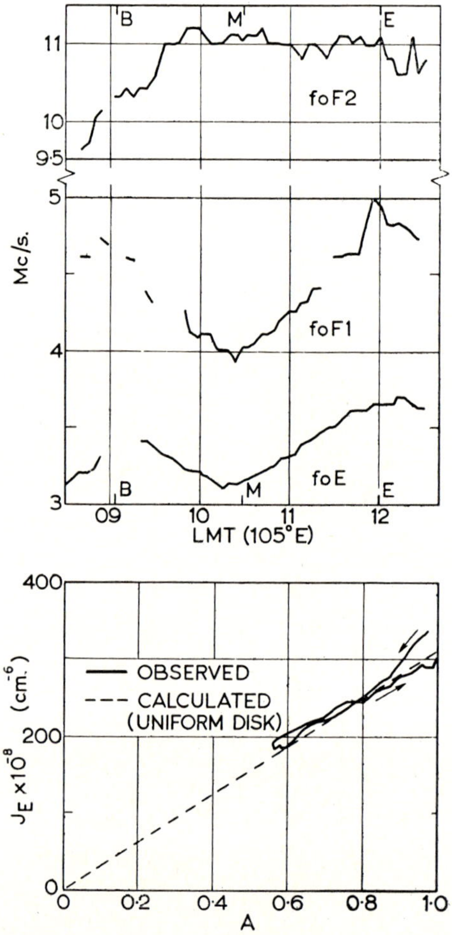

Fig. 4.

of the ionizing radiation must be attributed to sources other than a uniformly bright disk (Fig. 4). It is not possible to give any accurate information about their locations, since they were probably not obscured at Singapore. Owing to some irregularities in the provisional data, no conclusions can yet be reached concerning the possible existence of limb-brightening.

Conclusions

There is no doubt that the ionizing radiation responsible for the E and $F1$ layers is not always radiated with equal intensity from all parts of the sun's disk. From the evidence obtained during the eclipses of 1952, 1954, and 1955, it seems probable that important localized radiation sources are more likely to occur when solar activity is fairly high. However, even when activity is very low, a small contribution to the total radiation may come from sources other than the uniform disk radiation, and in 1954 this component appears to be consistent with limb brightening at the west and east limbs and darkening at the north and south poles. Evidence has also been obtained on the probable connection between isolated sources of ionizing radiation and areas from which 5303-Å coronal radiation is emitted.

Acknowledgement

This work was carried out as part of the programme of the Radio Research Board, and this account is published by permission of the Director of Radio Research of the Department of Scientific and Industrial Research.

References

Higgs, A. J. 1942 *Monthly Not. Roy. Astron. Soc.*, **102**, 24.

Minnis, C. M. 1952 *Nature*, **170**, 453.

1955 *J. Atmosph. Terr. Phys.*, **6**, 91.

Osawa, K. 1950 *Rep. Ionosph. Res. Japan*, **4**, 53.

Piggott, W. R. 1952 *Proc. Mixed Comm. Ionosph.*, Aug. 1952, p. 42.

Rydbeck, O. E. H. 1946 *Trans. Chalmers Univ. Tech.*, No. 53.

Waldmeier, M. 1947 *Terr. Mag. Atmos. Elec.*, **52**, 333.

Solar Eclipses and the Ionosphere (Special Suppl. Vol. 6 to J. Atmosph. Terr. Phys.), Pergamon Press, London, 1956

The Aspect of the Corona in Front of the Sun's Disk

M. WALDMEIER
Swiss Federal Observatory, Zurich

Abstract—Although a qualitatively correct picture of the corona over a given point of the solar disk, not on the limb, can be obtained by interpolation from the observed intensity over the point when it was on the limb, this method neglects the varying depth of coronal material penetrated along different lines of sight. The monochromatic corona thus determined will show pronounced limb brightening.

ONLY those regions of the corona which are outside the sun's limb may be observed in the case of a total solar eclipse. Hitherto no possibility of observing the corona in front of the sun's disk has been found at all. In investigating the behaviour of the ionosphere and the solar radio-emissions during a total or partial solar eclipse, the luminosity distribution of the corona in front of the sun's disk is at least as important as that outside the sun's limb.

For the first time a picture of the corona in front of the sun's disk was traced out by the author on the occasion of the eclipse of 9 July 1945 (WALDMEIER, 1947). Coronal observations from seven days before until seven days after the eclipse were used, under the assumption that the structure of the corona does not change much in the course of a few days. This assumption may be examined by comparing the observations made on one solar limb with those made fourteen days before or afterwards on the opposite limb. It proved to be well founded, especially for years of low solar activity. In the above mentioned picture of the corona we used for the intensity of a coronal emission-line over a given point of the sun's disk the intensity which was observed over the respective place when it was on the sun's limb. Certainly this process gives but a qualitative aspect of the distribution of the coronal intensity over the solar disk. The fact that the line of sight directed towards the centre of the sun's disk penetrates much less coronal material than one running tangentially to the sun is neglected.

The coronal emission-lines are, at least in the innermost part of the corona, excited by electron-collisions. Their intensity is therefore proportional to $\int N_e^2\, dx$, where N_e is the density of electrons, and the integral is to be extended over the whole line of sight. This integral was computed for a model corona which is in conformity with the minimum conditions. From this it follows, for example, that the corona over the equator, about $1/2'$ outside the sun's limb appears nearly seven times brighter than in the centre of the sun's disk. The monochromatic corona consequently shows quite distinctly the phenomenon of the bright ring. A slightly disturbed region on the sun's limb may appear much brighter than a heavily disturbed one in the centre of the sun's disk. Considering visual coronal lines, these conditions may be moderated through the contribution of excitation by radiation; for the short-wave emission-lines, the only ones of interest in ionospheric respects, the excitation by radiation is of no importance, on account of the small intensity of the sun's continuum in this region. Figs. 1 and 2 show the appearance of the corona in the region $\lambda < 1000$ Å during the total solar eclipses of 9 July 1945 and of 25 February 1952. In both cases the phenomenon of limb brightening is very pronounced. On 9 July 1945 the NW quadrant is

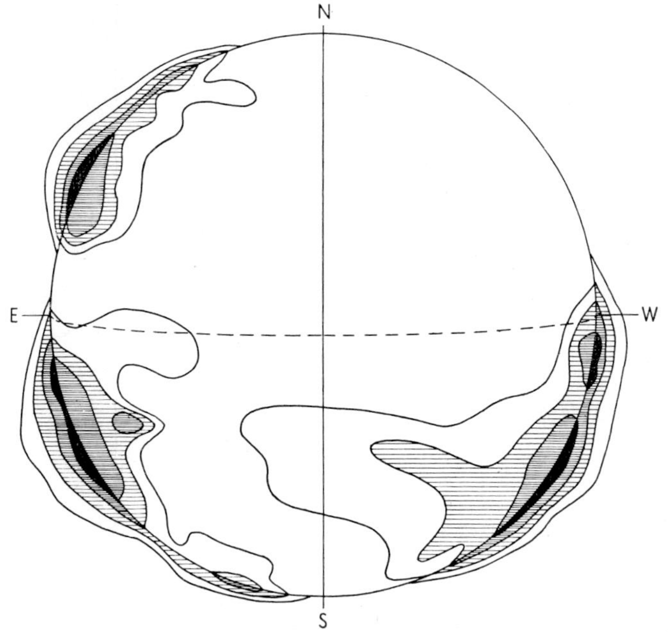

Fig. 1. The aspect of the corona during the total solar eclipse of 9 July 1945. The isophotes correspond to the intensities 5, 10, 20, and 40 . 10^{-6} Å in the centre of the sun's disk.

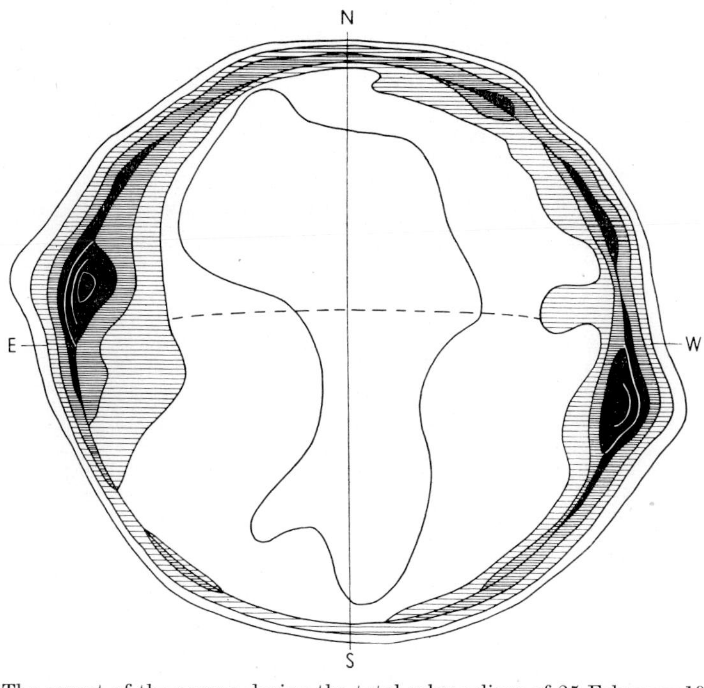

Fig. 2. The aspect of the corona during the total solar eclipse of 25 February 1952. The isophotes correspond to the intensities 5, 10, 20, 50, 100, 150, and 200 . 10^{-6} Å in the centre of the sun's disk.

practically free of emission. Especially intensive regions may result when disturbed regions are on the solar limb. This happened on 25 February 1952, when in both the NE and the SW quadrants a heavily disturbed region lay in the spot-zone just on the limb.

A more detailed publication appears in *Zeitschrift für Astrophysik* Vol. 38, p. 143 (1955).

REFERENCE

WALDMEIER, M. 1947 *Terr. Mag. Atmos. Elec.*, **52**, 333.

Solar Eclipses and the Ionosphere (Special Suppl. Vol. 6 to J. Atmosph. Terr. Phys.), Pergamon Press, London, 1956

The X-ray Radiation of the Solar Corona and Hot Coronal Condensations

G. ELWERT
Astronomisches Institut, University of Tübingen

Abstract—In earlier papers of the author, the soft X-ray radiation of the undisturbed sun was calculated and its effect on the atmosphere of the earth discussed. It was shown that this radiation could explain the formation of the normal *E* layer. The calculated height and thickness as well as the electron density in the case of equilibrium between ionization and recombination are in accord with the observed data.

This theory has been extended by calculations of the thermal radiation of coronal condensations, which are generally believed to be the origin of the slowly varying component of the radio emission of the sun. The radiation of the condensations was calculated assuming electron densities as derived by WALDMEIER and MÜLLER and temperatures of several million degrees as postulated by WALDMEIER, PIDDINGTON, and MINNETT. We find a remarkable increase of the X-ray radiation below about 20 Å and a short-wavelength limit in the neighbourhood of 7 Å. These results are in agreement with observations of the X-ray radiation by rocket exploration.

THE theory of the X-ray radiation of the undisturbed corona of the sun is based on data from the following sources: the density of the electrons, as given by the Allen-Baumbach formula; the intensity of the coronal lines; and the cosmic abundance of the elements. It is well known that the red and green coronal lines are produced by Fe-X and Fe-XIV. These ions contain only a few electrons in the *M* shell. In the state of equilibrium of ionization there is a balance between ionization by electronic collisions and by radiative recombinations. Considering the equilibrium of these processes, BIERMANN (1947), WOOLLEY and ALLEN (1948), and MIYAMOTO (1950) have derived ionization formulae. A more precise discussion of the cross-sections resulted in an ionization equation similar to that of WOOLLEY and ALLEN. These authors have also estimated the X-ray radiation of the corona. The maximum intensity of the green line is reached at about 10^6 degrees. For the same intensity of the green and the red line we get a temperature of $7 \cdot 10^5$ degrees, assuming both lines come from the same region of the corona.

With the help of the ionization formula it is possible to compute the abundance of the stages of ionization of the different elements. A general impression of the degree of ionization is obtained from Fig. 1, which shows ionization potentials in electron volts as a function of the nuclear charge *Z* for systems of a certain number of electrons. From right to left we give first the curves for systems with one and two electrons (the electrons of the *K* shell), then the curves for systems which have, in addition, one to eight electrons in the *L* shell, and finally the curves for those systems which still possess electrons in the *M* shell. In the corona the ions have only a few electrons left in the *M* shell; the first vertical line shows that the corresponding ionization potentials amount to roughly 350 eV. Electrons having a lesser potential are detached. This is the case to the left of the vertical line. Carbon, nitrogen, oxygen, and neon still possess their electrons in the *K* shell; magnesium, silicon and sulphur have only a few electrons in the *L* shell. The distribution of the ions in the stages of ionization can be calculated in detail from the ionization formula. Through the recombinations of the electrons with the ions, a continuous spectrum originates. Its short-wavelength limit lies at about 30 to 40 Å, as seen from the lower scale in Fig. 1. This scale gives the wavelengths corresponding to the ionization potentials in the upper scale. Besides this series-limit spectrum, a line spectrum is formed. Excited levels can be created by recombinations in higher shells and by electronic collisions. The latter processes are the more efficient. For the computation of the numbers of these processes one needs the corresponding cross-sections. These were calculated by quantum mechanics, with the help of BORN's approximation, for hydrogen-like atoms.

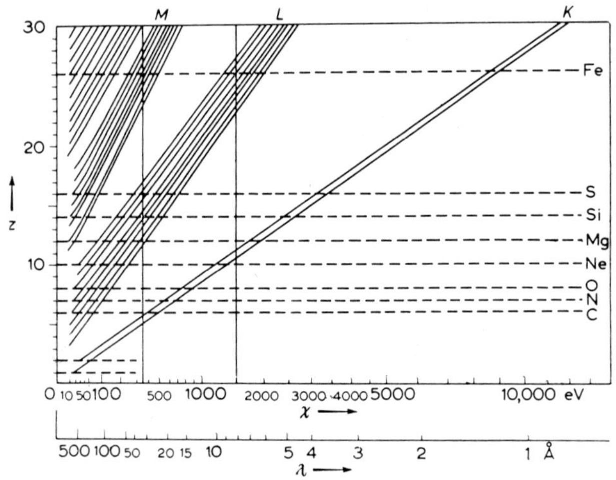

Fig. 1. Ionisation potentials and corresponding wavelengths of the series-limit continua as a function of the nuclear charge Z and the degree of ionization.

Still more refined computations would be desirable. The same may be said for the cross-sections for ionization and radiative recombination.

Fig. 2 shows the dependence on wavelength of the contributions to the line emission and the series-limit continuum for temperatures of $7 \cdot 10^5$ and 10^6 degrees. The lines are shown by crosses, the series-limit continua by circles which are given for the short-wavelength limits. The maximum intensity lies at about 80 Å for $7 \cdot 10^5$ degrees and at about 60 Å for 10^6 degrees. It is remarkable that, according to WIEN's law of black-body radiation, these maxima should occur almost at the same wavelength. The agreement is very good, if we take the limits of the continua. However, the intensities of black-body radiation should be very much larger. The intensities are given in erg cm^{-2} sec^{-1} for the distance of the earth. They must be multiplied by a "factor of uncertainty" Q' of value between 1 and 3, which is caused by fluctuations of density in the corona and by the uncertainty of the cross-sections. Using a value of 2 for this factor, the whole radiation has an intensity of about 1/10 erg cm^{-2} sec^{-1} at the earth.

Let us now consider the ionospheric consequences. BATES and HOYLE (1948) have already proposed a theory of the formation of the E layer by X-rays of about 40 Å. Meanwhile, HAVENS, KOLL, and LA GOW (1952) have determined the density of the air by means of rocket exploration up to a height of about 160 km, while values of the coefficient of absorption for different wavelengths are already known. By separating the spectrum into narrow wavelength ranges, the absorbed energy can be computed for different heights. From this the density of electrons existing in the equilibrium between ionization and recombination can be found. This electron

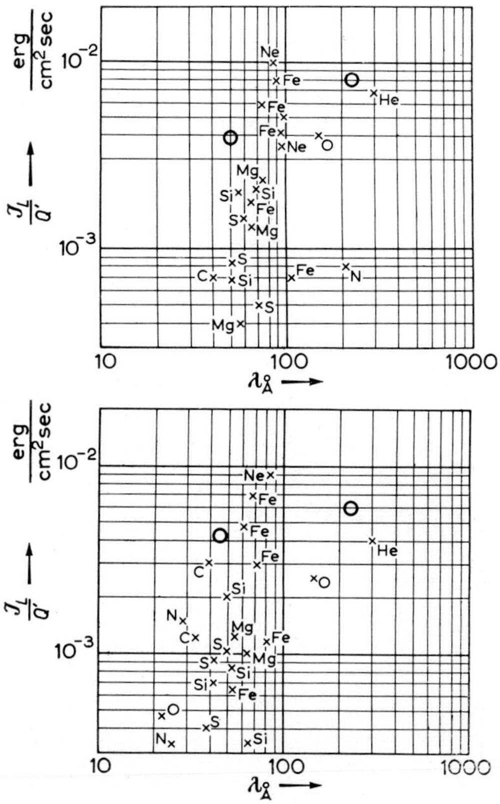

Fig. 2. Spectrum of the solar corona for 7 . 10^5 and 10^6 °K.

density as a function of height is shown in Fig. 3. The line on the right corresponds to a constant coefficient of recombination $\alpha = 10^{-8}$ cm³ sec⁻¹. The dotted line to the left is obtained if we assume (after Rawer, 1953) that the recombination coefficient is proportional to the square of the density of the molecules. By means of these curves the more complicated processes of recombination can be taken into account in addition to direct recombination. The maximum electron density is reached at a height of 115 to 120 km. This is the height of the normal *E* layer; the electron density and the semi-thickness are of the observed order. As the region between the corona and the chromosphere will contribute radiation at longer wavelengths, the electron density above 120 km will not be quickly diminished, but will probably remain nearly constant.

All the foregoing work relates to studies made some years ago. We shall now consider recent studies about coronal condensations. From the radio radiation of the quiet sun at dm wavelengths one can isolate a slowly varying component which is correlated with the area of sunspots. Observations of the eclipse in 1948 taken from three sites at a great distance from each other showed, according to CHRISTIANSEN, YABSLEY, and MILLS (1949), highly emitting areas. The average bright area occupied about 4/1000 of the solar disk; the mean brightness temperature was about $5 \cdot 10^6$ degrees. The most direct evidence concerning the localized bright areas was obtained by CHRISTIANSEN and WARBURTON (1953) with their interferometer consisting of thirty-two parabolic mirrors. The emission of these regions is interpreted as thermal

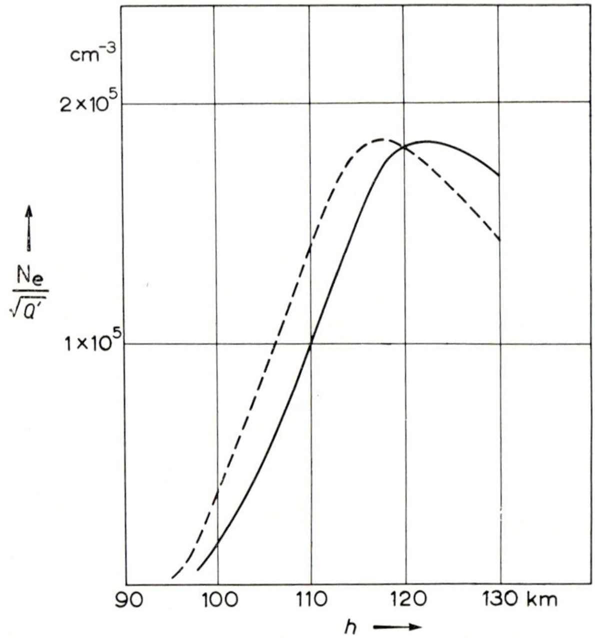

Fig. 3. Density of electrons as function of height.

radiation from regions having an exceptionally high electron density. These areas tend to occur in the lower corona over sunspots. Such regions were detected optically by their excessive brightness in the continuous spectrum and were given the name of coronal condensations. In a model of a typical coronal condensation after WALDMEIER and MÜLLER (1950), the maximum density in the condensation is about twenty times greater than in the innermost corona. As the condensations become optically thick for wavelengths greater than 20 cm, one obtains the observed intensity of the radiation if these condensations have a temperature of about $5 \cdot 10^6$ degrees. Computations of the spectrum by WALDMEIER and MÜLLER are in agreement with the observations. Similar calculations were given by PIDDINGTON and MINETT (1951), taking into account a magnetic field and a temperature of 10^7 degrees. We can conclude that this hypothesis of the origin of the slowly varying component is in close accord with the observations.

The calculation of the X-ray radiation of the undisturbed corona showed that the maximum intensity occurred at almost the same wavelength as it would according to WIEN's law of black-body radiation. The relation between temperature and wavelength of maximum intensity is shown in Fig. 4. For a temperature of 6 . 10^6 degrees the maximum would be at 5 Å corresponding to a potential of about 3000 eV. However, a correct calculation has to begin with the computation of the degree of ionization. We therefore have drawn in Fig. 4 an additional curve which gives, for different temperatures, the ionization potentials at which the quotient of the number-densities of the ions in the lower and upper stages of ionization is 1. To simplify the comparison, some approximations have been made. Both curves are very similar, and intersect at a temperature of about 10^6

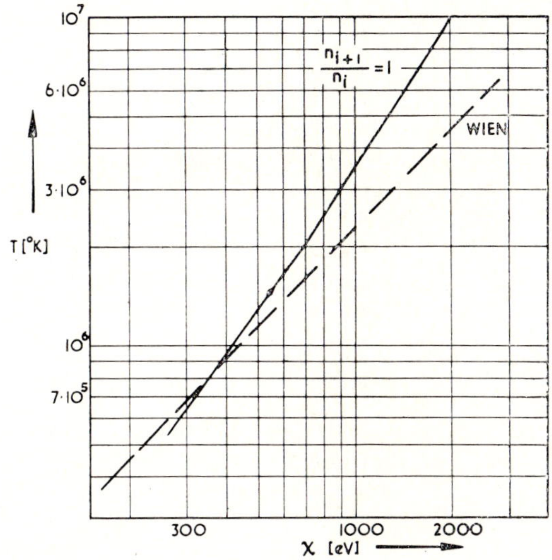

Fig. 4. Relation between temperature and wavelength of maximum intensity.

degrees. Therefore the wavelength of WIEN's law is nearly equal to the wavelength of the series-limits. At higher temperatures there are differences between the two curves. We get equal particle densities in the lower and upper stages of ionization at 6 . 10^6 degrees for ionization potentials of about 1500 eV.

We can obtain a general impression of the degrees of ionization from Fig. 1. In this figure a vertical line for the ionization potential of 1500 eV is drawn. We see that at 6 . 10^6 degrees all elements up to oxygen are completely ionized; neon will contain partly one electron in the K shell; magnesium, silicon, and sulphur still possess 2 K electrons, and iron has only a few electrons in the L shell. As we can see from the lower scale, the corresponding wavelength of the series-limits of iron is about 7 Å. Since the wavelengths of the lines are somewhat larger, a maximum of intensity at about 12 Å will result. In addition, we have the contributions of magnesium and silicon. The wavelengths of the series-limits are about 7 to 5 Å; the corresponding lines have wavelengths of 9 to 7 Å. Since the cosmic abundance of the heavier elements quickly decreases with increasing nuclear charge, no further essential contribution results from sulphur onwards. Therefore at about 7 Å there is a short-wavelength limit of the spectrum. This would also exist if the temperature of the condensation exceeded 6 . 10^6 degrees. Only if the speed of the particles were so great that the remaining K electrons of iron could be excited, could one obtain a radiation of the short wavelength of about 2 Å corresponding to the series-limit at 1·4 Å. It may be that this happens with flares. It is probable that this radiation makes a contribution to the formation of the ionospheric D layer. This would be in accordance with a hypothesis of Siedentopf (1950), who postulated a radiation of about 1 Å.

Fig. 5 shows the spectrum of the corona for a temperature of 10^6 degrees, together with the spectrum of a hot coronal condensation which has an electron density according to the model of Waldmeier and Müller (1950) and a temperature

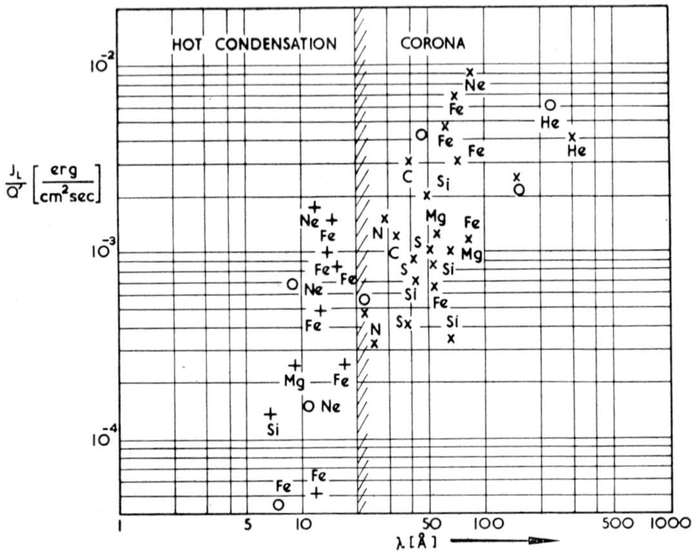

Fig. 5. Spectrum of the solar corona and a hot coronal condensation.

of 6 . 10^6 degrees. Below a wavelength of about 20 Å there is no essential contribution of the corona. On the other hand, for the condensations between 20 and 7 Å we have radiation of the intensity shown. For the model described, the total energy in this region is some 10^{-3} erg cm^{-2} sec^{-1}. Since the emission is proportional to the square of the electron density (and this is not constant in time) the intensity will be greatly variable. Below 10 Å our model gives nearly 10^{-3} erg cm^{-2} sec^{-1}. This value is in close accord with the data of Friedman, Lichtman, and Byram (1951) who have found 10^{-4} to 10^{-3} erg cm^{-2} sec^{-1} between 7 and 10 Å. These authors have already detected a short-wavelength limit near 7 Å.

Thus coronal condensations, owing to their exceptionally high electron density and temperature, are not only important for radiation in the dm-wavelength range, but also for solar X-rays. It will be a field of further research to look for a connection between both radiations and ionospheric data. On the lower side of the E layer an additional ionization should take place which ought to be correlated with the intensity of the radiations in the dm-wavelength range. Perhaps this ionization may be detected during solar eclipses, when the moon covers hot coronal condensations.

Details will be published elsewhere.

REFERENCES

BATES, D. R., and HOYLE, F. 1948 *Terr. Mag. Atmos. Electr.* **53,** 57.

BIERMANN, L. 1947 *Naturwiss.* **34,** 87.

BYRAM, E. T., CHUBB, T., and FRIEDMAN, H. 1954 *Rocket Exploration of the Upper Atmosphere,* Pergamon Press, London, p. 274.

CHRISTIANSEN, W. N., and WARBURTON, J. A. 1953 *Austr. J. Phys.,* **6,** 190, 262.

CHRISTIANSEN, W. N., YABSLEY, D. E., and MILLS, B. Y. 1949 *Austr. J. Sci. Res.,* **A2,** 506.

ELWERT, G. 1951 *Phys. Verhandlungen,* **2,** 112.

1952 *Z. Naturforschung,* **7a,** 202, 432.

1953 *J. Atmosph. Terr. Phys.,* **4,** 68.

1954 *Z. Naturforschung,* **9a,** 637.

FRIEDMAN, H., LICHTMAN, S. W., and BYRAM, E. T. 1951 *Phys. Rev.,* **83,** 1025.

HAVENS, R. J., KOLL, R. T., and LA GOW, H. F. 1952 *J. Geophys. Res.,* **57,** 59.

MIYAMOTO, S. 1950 *Publ. Astr. Soc. Pacif.,* I.

PIDDINGTON, J. H., and MINETT, H. C. 1951 *Austr. J. Sci. Res.,* **A4,** 137.

RAWER, K. 1953 *Die Ionosphäre,* P. Noordhoff N.V., Groningen.

SIEDENTOPF, H. 1950 *Archiv. el. Übertragung,* **4,** 92.

WALDMEIER, M., and MÜLLER, H. 1950 *Z. Astrophysik,* **27,** 58.

WOOLLEY, R. v. d. R., and ALLEN, C. W. 1948 *Month. Not. Roy. Astron. Soc.,* **108,** 292.

Solar Eclipses and the Ionosphere (Special Suppl. Vol. 6 to J. Atmosph. Terr. Phys.), Pergamon Press, London, 1956

The Distribution of Ly-α Radiation over the Solar Disk

C. DE JAGER
Sterrewacht Sonnenborgh, Utrecht

Abstract—The minimum intensity of the Ly-α line, measured with the rocket technique, yields a radiation temperature of 6800° for the centre of the line. The corresponding electron temperature of the chromosphere is 66,000° ± 10,000°; this temperature prevails in the layers about 9500 km above the photosphere. The variation of intensity of the Ly-α line over the disk is computed and compared with the observations. A qualitative discussion of the enhanced Ly-α radiation in plages and flares is given.

1. Lyman-alpha and the Temperature of the Upper Chromosphere

The first rocket measurements of the strength of the Ly-α line, the strongest emission line in the ultra-violet spectral region, yielded 0·10 ± 0·02 erg cm^{-2} sec^{-1} for the total emission of the line outside the earth's atmosphere (Byram, Chubb, Friedman, and Gailar, 1953). The corresponding solar emission is (4·6 ± 0·9) 10^3 erg cm^{-2} sec^{-1}. Later observations by Tousey *et al.* (1955) yielded values of 0·3 and 1·0 erg cm^{-2} sec^{-1}. This may mean that the strength of the line is variable. Since there is as yet no evidence of rapid fluctuations of the chromospheric structure, as far as the undisturbed chromosphere is concerned, these observations might be explained by assuming that the plages on the solar disk emit an enhanced radiation. The plages are known to have a higher radiation temperature (for $H\alpha$) than the normal chromospheric regions, so that it seems reasonable that also their Lyman-alpha radiation is stronger than the radiation emitted by the normal chromosphere. This suggestion has still to be followed up. Since the lowest measured value is likely to refer to the undisturbed chromosphere, this value (0·1 erg cm^{-2} sec^{-1}) will be used for our further computations. According to the most recent observations (Tousey, 1955), the width of the line is 0·3 Å or less.

The problem we are discussing in this section is whether these data can be brought into agreement with our current ideas about the structure of the chromosphere. Some authors (Goldberg, 1954; Miyamoto, 1954) have concluded, on the basis of the above observations, that the temperature of the upper chromosphere must be about 6500°, much lower than is, for example, adopted in Woltjer's (1954) model or in that of Athay, Menzel, Pecker, and Thomas (1955); but we want to show that this conclusion is incorrect.

We approximate the line by a Doppler profile with an area of (4·6 ± 0·9)10^3 erg cm^{-2} sec^{-1}. For the half-width of the line we take the observed maximum value, 0·30 Å. It will be shown below that the theoretically predicted half-width agrees with this observed value. For any Voigt profile the relation between the area S, the half-width h, and the central intensity c is given by

$$S = phc$$

with $p = 1{\cdot}06$ for a Doppler profile (van de Hulst and Reesinck, 1947), hence

$$c = 1{\cdot}45 \times 10^{12} \text{ erg cm}^{-2} \text{ sec}^{-1}$$

The corresponding black-body intensity, with Planck's law, is 6800° ± 100°.

On the basis of similar computations, it has been concluded that the temperature in the upper chromosphere is of the order of 6500°, but this conclusion would only be correct when we are dealing with a pure absorption line formed in an atmosphere in thermodynamic equilibrium. But the Ly-α line is produced by diffusion only, and the upper chromosphere, from where the Ly-α radiation emerges, is not in thermodynamical equilibrium, so that it is not permissible to speak of *the* temperature of the chromosphere. This should always be specified as electron temperature, ionization temperature, excitation temperature, etc. The relation between these various temperatures and the radiation temperature found above is rather complicated, but tables showing their relations as far as the hydrogen lines are concerned have been computed by GIOVANELLI (1948). These tables have been computed for a photospheric radiation temperature of 5000°, which may be considered as a happy choice, since, according to the rocket observations, the solar radiation temperature does not deviate much from 5000° in the whole spectral region that is important for the photo-excitation of the hydrogen levels (see Fig. 1, cf. DE JAGER, 1955). It seems, however, very important to repeat GIOVANELLI'S computations as soon as the far ultra-violet continuous spectrum near 1000 Å has been measured with some accuracy. The present knowledge about this radiation is not yet sufficient to justify such detailed computations. For the time being we have to use the existing tables.

To be able to apply the results of GIOVANELLI'S computations one should know first the values of N_e and N_1 (electron density and density of hydrogen atoms in ground-state) at the level where the hydrogen radiation is emitted. These values depend on the particular choice of the chromospheric model. At a height h^* the optical depth in the line centre is

$$\tau = \int_{h^*}^{\infty} N_1 \alpha_0 \, dh \, .$$

N_1 is the concentration of ground-level H atoms per cm^3,

$$\alpha_0 = \frac{\sqrt{\pi} e^2}{mc^2} \frac{\lambda^2 f}{\Delta \lambda_D} \, .$$

For the Doppler width $\Delta\lambda_D$ the observed maximum value is:

$$\Delta \lambda_D = \frac{h}{1{\cdot}655} = 0{\cdot}18 \text{ Å} \, .$$

We notice that this value is equal to the theoretical value; if we adopt that $T_{\text{el}} = 6{\cdot}6 \times 10^4$ °K, a value that will be justified below, and if we take for the macroturbulent velocity $\xi_t = 30$ km/sec (RUSH and ROBERTS, 1954), then

$$\Delta \lambda_D = \frac{\lambda}{c} \sqrt{2RT + \xi_t^2} = 0{\cdot}18 \text{ Å},$$

a value identical with the upper limit found above. Inserting this value into formula (1), we obtain $\alpha_0 = 1{\cdot}9 \times 10^{-14}$ cm^{-2}. With WOLTJER's model of the upper chromosphere (see, e.g., DE JAGER, 1955, Table 5), we find by numerical integration that an optical depth unity is reached at $h^* = 9400$ km above the surface of the sun. The integration has been performed outside the spicules, the contribution of the spicules above 8000 km being assumed negligible in WOLTJER's

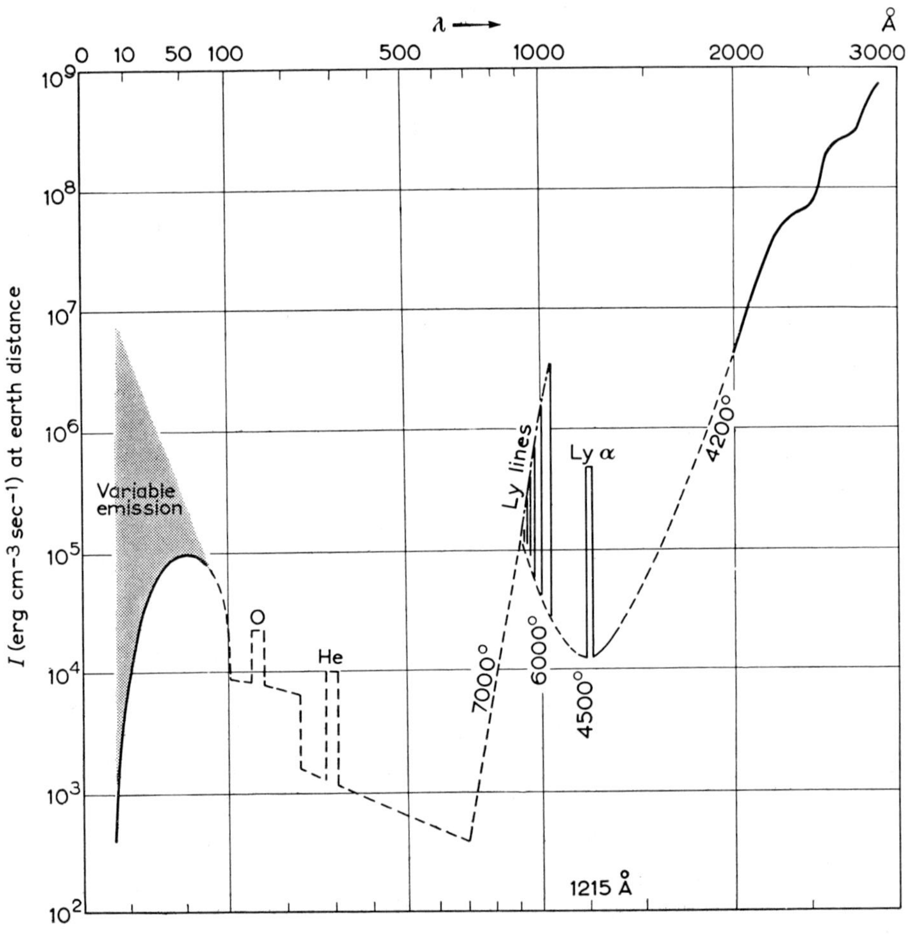

Fig. 1.

model. At 9400 km, according to WOLTJER: log $N_1 = 6{\cdot}1$ and log $N_e = 9{\cdot}6$; T_{el} is assumed to be 29,800°.

The next point to examine is whether these data agree with the computations made by GIOVANELLI. By interpolation from his tables we find for log $N_e = 9{\cdot}6$, and for a radiation temperature of 6800° $\pm$ 100° a corresponding electron temperature of $6{\cdot}6 \times 10^4$ °K, with an estimated error of 10^4 °K. This resulting T_{el} is higher than WOLTJER's value, but considering the great uncertainty of the existing models, and the steep increase of T in the region around 10,000 km (which is shown by all modern chromospheric models, cf. DE JAGER (1955), Fig. 2), the agreement cannot be regarded as unsatisfactory. Conversely, adopting WOLTJER'S $T_{el} = 29{,}800°$, we find with GIOVANELLI'S tables $T_{rad} = 6150°$.

The conclusion of this section is that the observed intensity of the Ly-α radiation points to a kinetic temperature of about 66,000° at the 9400-km level, and not to a temperature around 6500°.

2. INTENSITY VARIATION OF LY-α OVER THE DISK

The intensity variation of Ly-α over the disk is dependent on the variation of h^* with θ. Here h^* is the level where an optical depth unity is reached, and θ is the angle between the normal to the surface and the emergent ray of radiation. The

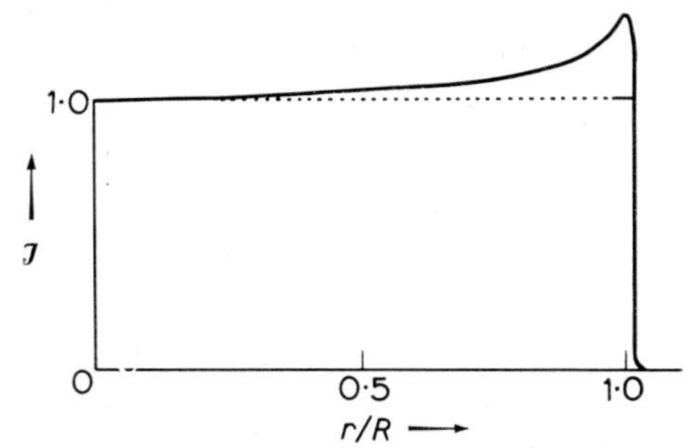

Fig. 2.

N_1 value increases so quickly with decreasing height that an increase of sec θ up to a value as high as 3·0 causes only a shift of the effective height of the layer of emission from 9400 km to 9700 km. The radiation temperature for the Ly-α radiation emitted at that layer can be read from GIOVANELLI'S tables if the values of T_{el} and N_e are known for $h = 9700$ km. These two values can be read from WOLTJER'S model. However, we then encounter the difficulty, already indicated above, that WOLTJER'S T_{el} value does not predict correctly the observed T_{rad} value, but yields a somewhat lower value. The observed T_{rad} value demands a higher T_{el} value (e.g., T_{el} at sec $\theta = 1$ is 29,800°, according to WOLTJER'S model, and it should be 66,000°, according to GIOVANELLI'S tables).

In order to make our computations internally consistent and to eliminate the above difficulty, we multiply all WOLTJER'S T_{el} values in the concerned h region by the ratio 66,000°/29,800°; this means that we submit WOLTJER'S T_{el} curve to a parallel upward shift on a logarithmic graph. This procedure, though being somewhat arbitrary, seems permissible, and seems the most appropriate one, because all log T_{el} curves deduced by various authors are more or less parallel in this region of the chromosphere, the absolute values of T_{el} being very uncertain.

With the T_{el} curve thus defined, the computations proceed easily: by numerical integration the value of h^* has been found as a function of sec θ; we read T_{el} and N_e, and deduce from this T_{rad} and compute with PLANCK'S law $B(T_{rad})$. The function $B(\sin\,\theta)$ is assumed to represent the distribution of the Ly-α radiation over the disk. The result is shown in Fig. 2.

We note that the variation of log N_e with h, though small, has a great influence on the resulting T_{rad} values. The assumption that N_e remains constant with height would double the limb brightening of the Ly-α radiation.

The figure shows that the regions outside the disk contributes only to the Ly-α radiation up to the heights where T_{el} shows its steep increase to values around 10^6 °K. In this particular model this occurs near 10,000 km; in other models this takes place at h values between 6000 and 10,000 km. But above these heights there are still a few spicules which in some cases may rise up to heights of 20,000 or 30,000 km. Their distribution with height has been studied by RUSH and ROBERTS (1954)–cf. their Fig. 1. These spicules have been observed in $H\alpha$, but their emission in Ly-α will probably be stronger. Our complete lack of knowledge of the structure of these high spicules makes it difficult, for the time being, to obtain an impression on their contribution to the chromospheric intensity in Ly-α. We presume that the high spicules will produce a short tail to the curve of Fig. 2, which tail goes to $r/R = 1{\cdot}03$, and has its greatest intensity near the poles (ROBERTS, 1954). From Fig. 2 it is clear that the Ly-α radiation shows only a faint limb-brightening, which agrees with the observational results of TOUSEY *et al.* (1955) which are smoothed by movements of 2′ to 5′ of the sun's image over the spectrograph slit, and which do not show any limb-brightening. The precise values of Fig. 2 might still be subject to changes (a) because of the uncertainties both in the chromospheric model and in the relations between T_{el}, N_e, and T_{rad} used in our computations; (b) because of the strong dependence of B on T in the far ultra-violet.

3. SOME REMARKS ON ENHANCED LY-α RADIATION

The above part of this paper is only concerned with the Ly-α radiation from the normal parts of the sun's disk. It seems probable that disturbed parts of the solar disk will emit an intenser Ly-α radiation, moderately enhanced in the plages, strongly intensified in the solar flares. It is premature to try to estimate the amount of Ly-α radiation emitted by plages and flares, because of our complete ignorance of the temperatures and pressures in the upper parts of these structures, but some guesses may perhaps be made.

In plages the $H\alpha$ radiation in the very centre of the $H\alpha$ line is increased generally by a factor of about 2 or 3, as compared to the normal spectrum (ELLISON, 1952), and corresponds to an increase of T_r from 3800° to 4600°. Since a mean plage has an area of about 1/25 that of the visible solar disk, the surface intensity of Ly-α in a plage should increase 50 to 250 times in order to explain the observed three- to ten-fold increase in the Ly-α radiation. This corresponds to a local increase in the radiation temperature from 6800° to 9000° respectively. It has still to be traced whether the electron temperatures and densities above plages are such that these radiation temperatures can be explained.

According to FRIEDMAN and CHUBB (1955), the largest flares are accompanied by sudden ionospheric phase anomalies, indicating a lowering of the reflecting ceiling for long waves by about 15 km. The necessary ionization could be produced either by an increase by a factor 10^4 of the Ly-α radiation or by the emission of strong X-rays with a wavelength near 1–2 Å. Since a strong flare occupies about 10^{-3} part of the visible solar hemisphere, the first assumption necessitates an increase in the Ly-α radiation per cm² by a factor 10^7. So we are confronted with the problem whether this enormous increase can be explained on the basis of our present knowledge of solar flares. A typical flare, though being transparent for visual continuous radiation, is opaque for continuous radiation in the 1000 Å region; hence it will emit a strong continuous radiation in the Ly-α region. (The absorption coefficient of the solar matter is much greater near 1000 Å than it is in the visual spectral region.) This continuous radiation probably has a radiation temperature of 12,000° (the generally adopted T_{el} value for typical flares). Then the continuous emission per cm^2 and per wavelength-unit is 10^4 times greater than that emitted by a black body of 6800°. The ionizing radiation may be stronger than the normal Ly-α radiation by another factor 10, since this is about the number of "windows" in the O_2 absorption spectrum near 1200 Å (WATANABE, ZELIKOFF, and INN, 1953), but even this is obviously still insufficient by a factor 10^2. In order to explain the factor 10^7, the radiation temperature of Ly-α should increase to 80,000°K. This value is obviously too high to make possible any neutral hydrogen emission. Nevertheless, although there are some indications that a flare might be colder in its upper regions than its lower regions (compare, for example, the self-reversal of Hα in some flares), there are also arguments supporting an upwards increase of T in the still higher regions above a flare: the top of a typical flare penetrates generally into the low corona and, moreover, the coronal layers above a flare are hotter than those of the normal corona. The coronal condensations above a flare-region are generally much larger than the flare itself and are comparable in size to the plages, which implies that the very high temperature computed above should possibly be decreased somewhat.

A detailed study of the D layer and its behaviour during eclipses may be useful for obtaining insight into these problems.

REFERENCES:

ATHAY, R. G., MENZEL, D. H., PECKER, J. C., and THOMAS, R. N.	1955	*Astrophys. J. suppl.*, series **1**, 505.
BYRAM, E. T., CHUBB, T., FRIEDMAN, H., and GAILAR, N.	1953	*Phys. Rev.*, **91**, 1278.
ELLISON, M. A.	1952	*Publ. Roy. Obs. Edinburgh*, **1**, 5.
FRIEDMAN, H., and CHUBB, T. A.	1955	*Physics of the Ionosphere*, London, The Physical Society, p. 58.
GIOVANELLI, R. G.	1948	*Austral. J. Sci.*, **1**, 189.
GOLDBERG, L.	1954	*Astrophys. J.*, **120**, 185.
HULST, H. C. V.D., and REESINCK, J.	1947	*Astrophys. J.*, **106**, 121.
JAGER, C. DE	1955	*Ann. Géophys.*, **11**, 330.
MIYAMOTO, S.	1954	*Publ. Astron. Soc. Japan*, **6**, 229.
RUSH, J. H., and ROBERTS, W. O.	1954	*Austral. J. Phys.*, **7**, 230.
TOUSEY, R.	1955	Communicated at the Dublin meeting of the I.A.U.
WATANABE, K., ZELIKOFF, M., and INN, E. C. Y.	1953	Geophys. Res. Paper No. 21, Air Force Cambridge Research Center.
WOLTJER, L.	1954	*B.A.N.*, **12**, 165.

Solar Eclipses and the Ionosphere (Special Supp. Vol. 6 to J. Atmosph. Terr. Phys.), Pergamon Press, London, 1956

Sudden Ionospheric Disturbances in Relation to Solar Flare Radiations

M. A. Ellison
Royal Observatory, Edinburgh

Abstract—A solar flare emits a burst of ultra-violet radiation which has immediate repercussions in the *D* and lower *E* regions. The operative wavelengths are still uncertain; but $L\alpha$ (1216 Å) radiation at the higher levels and X-rays ($\sim$5 Å) at lower levels seem probable.

Comparisons have been made between the development curves of the S.I.D.s and of the Hα radiation of the flares which caused them. On the assumption that the ultra-violet follows the same variation with time as the Hα light, some deductions can be made about the relaxation times and the maximum ionization at various levels.

1. Nature of Flares

A flare is a catastrophic event occurring on the sun which has immediate repercussions on the earth.

Flares are observed with filters of the Lyot type, transmitting the centre of the Hα line, or by means of the spectrohelioscope—Hale's instrument which has contributed more than any other to the study of these remarkable phenomena occurring in the solar chromosphere. Intense flares consist of complex patterns of "white-hot" filaments which reach their maximum brilliance within five or ten minutes of their first appearance, and then slowly decay within the next hour or so. Lesser flares are composed of bright patches, usually without any filamentary structure. But whether large or small, flares are confined to the bright hydrogen plages associated with sunspots, being most frequent in the central regions of the groups, and less so at increasing distances; beyond 100,000 km from a spot they are rarely observed. There is a tendency for flares to recur, with the same shape and in the same position, within an active region. The area of the largest (Class 3+) flare is less than one-hundredth part of the sun's visible hemisphere.

Although the occurrence of an individual flare is quite unpredictable, the probability of flare occurrence is much greater in the magnetically complex sunspots of β- and γ-type than in the α-type.

2. Spectral Characteristics

The light of a flare is confined chiefly to individual spectral lines. In the visible region the hydrogen lines of the Balmer series are outstanding in emission, Hα being the strongest. The *H* and *K* lines of Ca-II are comparable in intensity with Hα, and other less intense lines arise from He-I, Fe-II, and a few other metallic atoms. The spectrum is essentially one of low temperature, that is to say it is representative of conditions in the chromosphere ($T \sim 10{,}000°$K) rather than in the corona ($T \sim 10^6 °$K). From the nature of the visible spectrum we can infer the presence in the ultra-violet of the strong resonance lines $L\alpha$ (1216 Å) and He (584 Å); and it seems probable that X-rays ($\sim$5 Å), generated in the corona above the flare, may also be present. Any, or all, of these might give rise to the sudden increase of free electrons in the lower ionosphere during a flare, but it is still uncertain which wavelengths are operative and to what extent.

3. Development of a Flare

The most outstanding feature of the flare spectrum is the sudden flash of radiation which occurs, usually within five to ten minutes after the first visual onset. This burst of radiation is best exhibited by plotting either the line-width, or the central intensity, of the Hα line versus time. Such plots are known as *development curves*. In an intense flare (Class 3 or 3+) the line-width may increase from 2 Å to 20 Å in a period which may be as short as one minute. At the same time the central intensity of the line increases to a value about twice the brightness of the continuous spectrum of the surrounding photosphere. Such increases in brightness have never been observed to take place gradually. The flash of radiation is immediately followed at first by a rapid and then by a slower rate of decay, the full duration of the decline being one hour or more.*

4. Sudden Ionospheric Disturbances

The extra ionization generated by a flare is believed to be confined to the level 60–100 km, that is in the D region and lower E region. One or more new layers of ionization may be formed.

The resulting S.I.D.s are recorded in the following ways:

(1) Crochets in terrestrial magnetism (conductivity effect).
(2) Short-wave radio fade-outs (S.W.F.) 5–20 Mc/s, and cosmic-noise fade-outs (S.C.A.) 18 Mc/s (absorption effects).
(3) Sudden-phase anomalies (S.P.A.) of very long radio waves 16 kc/s (lowering of the D-layer reflecting ceiling).
(4) Sudden enhancements of atmospherics (S.E.A.) 27 kc/s (improved reflectivity of D at oblique incidence).
(5) Sudden anomalies in field strength (S.F.A.) of a distant 16-kc/s transmitter (interference between ground-wave and sky-wave from D).

Delay Times

At Edinburgh we have been comparing the development curves of the S.I.D.s with those of the flares which generate them (Ellison, 1950, 1953). The times of commencement of the two curves are in close agreement, and the sudden increase in ionization of the D region is found to coincide with the flash of Hα radiation from the flare. Thereafter the two development curves follow separate paths. The ionization reaches its maximum value on the average about seven minutes after the peak of the flare, and in many cases it does not return to its pre-flare value for half an hour, or more, after all visible radiation in Hα has ceased.

The following results have been obtained for the various time delays:

	S.I.D. max.–flare max.	Number of cases
S.P.A.	5 min.	8
S.E.A.	7	21
S.F.A.	5	8
Crochets	2·5	13

* For a general survey of the properties of flares, see *The Sun and its Influence* by M. A. Ellison (London, Routledge and Kegan Paul, 1955).

It is natural to suppose that the intensities of the visible and ultra-violet wavelengths in the flare radiation follow the same course of rise and decay with time,* even though only the former can be directly observed. If this is so, we must interpret the time-delays as "relaxation times," due to the sluggishness in the response of the D layer to the increase in ionizing radiation. In other words, if q (number of ionizing quanta) passes through a maximum, the maximum in N (number of free electrons) lags behind. Following Appleton (1953), the relaxation time τ is given by $\tau = 1/(2\alpha N)$, where α is the recombination coefficient. Thus, if τ_f is the observed relaxation time following the maximum of q in a flare, and τ_n that which follows the maximum of q at midday ($\sim$60 min at 75 km), we have

$$N_f/N_n = \tau_n/\tau_f = 60/7 = 8{\cdot}5.$$

It therefore appears that the number of free electrons at this height increases during a flare by five to ten times the normal value—a result which is in good agreement with that derived from absorption measurements made during a short-wave fade-out (Appleton and Piggott, 1954).

6. Magnitude of S.I.D.s Compared with Flare Importance

In those S.I.D.s which can be measured quantitatively (e.g., S.P.A.s and S.C.A.s) there is abundant evidence that the magnitude of the S.I.D. is in general proportional to the estimated importance (Classes 1, 2, 3, and 3+) of the flare in Hα light (Bracewell and Straker, 1949; Shain and Mitra, 1954). Even the smallest flares of Class 1 are found to generate S.P.A.s and S.C.A.s. The magnitudes of the S.E.A.s, although influenced by other factors than purely D-layer ionization, increase likewise with flare importance (Ellison, 1953).

The details of this relationship need further study, because hitherto quantitative flare information has been inadequate. During the International Geophysical Year we hope to have a Lyot Hα heliograph in operation at the Cape Observatory, along with several others well distributed in longitude. The photographic records of the chromosphere, taken automatically by such instruments at 30-sec intervals during the day, will enable us to plot photometric light curves for an increasing number of flares. We hope by this means to place the classification of flares by importance on a much surer footing than is at present possible by means of visual observations.

7. Crochets

The crochets, or magnetic solar flare effects, differ in many respects from those S.I.D.s which are recorded by radio methods. The following points are of special interest:—

(1) If we regard the crochets as arising from an augmentation of the normal daily variation currents (S_q) flowing in the ionosphere, we find that the magnitude of the crochet represents an increase in ionospheric conductivity by a factor of 2. This is very much less than the increase by a factor of 10 in the number of free

* This is to be expected if $L\alpha$ is the ionizing agent, but it does not necessarily follow if coronal X-rays are involved.

electrons which has been deduced from the fade-out data (APPLETON and PIGGOTT, 1954).

(2) Crochets are recorded only in association with intense flares (Class 3+, 3, and occasionally Class 2). They are thus a much less sensitive indicator of flare occurrence than the S.P.A. and S.C.A. effects.

(3) When a crochet does occur, its duration is always less than that of the other types of S.I.D. recorded during the same flare. Its duration is sometimes less than that of the flare itself as seen in Hα light.

(4) The maximum of the crochet lags behind the maximum of the flare by $2\frac{1}{2}$ min on the average. This figure may be compared with the time-lag of 7 min for the S.E.A.s which are representative of conditions at a height of 70–75 km.

These differences strongly suggest that the crochet currents are confined to a level, or levels, different from those in which the fade-out and radio-reflexion effects are operative. The precise level has not yet been established. A possible location is in the E layer at a height of 100–120 km—the same as that in which the normal daily variation (S_q) currents are believed to flow. The currents responsible for the crochets are, in fact, found to be in the same general sense as the ionospheric currents the existence of which is inferred in order to account for the daily variation-departures existing at the time of the flare. This does not necessarily mean, however, that both current systems are located at the same height. Indeed, the contrary seems to be indicated by the absence of crochets in association with Class 1 flares. Class 1 flares are known to generate S.P.A.s and S.E.A.s at heights of about 70 km. To cause these effects, the radiation must also have traversed the E region, yet no crochets occur at these times.

In consequence, I have suggested elsewhere (ELLISON, 1950) that the crochets may represent the establishment of an independent current system at a much lower level of about 60–70 km—a region which is appreciably ionized only during an intense flare.

REFERENCES

APPLETON, E. V.	1953	*J. Atmosph. Terr. Phys.*, **3**, 283.
APPLETON, E. V., and PIGGOTT, W. R.	1954	*J. Atmosph. Terr. Phys.*, **5**, 141.
BRACEWELL, R. N., and STRAKER, T. W.	1949	*Monthly Not. Roy. Astron. Soc.*, **109**, 28.
ELLISON, M. A.	1950	*Pub. Roy. Obs. Edinburgh*, **1**, 53.
	1953	*J. Atmosph. Terr. Phys.*, **4**, 226.
SHAIN, C. A., and MITRA, A. P.	1954	*J. Atmosph. Terr. Phys.*, **5**, 316.

Solar Eclipses and the Ionosphere (Special Suppl. Vol. 6 to J. Atmosph. Terr. Phys.), Pergamon Press, London, 1956

Formation of the Ionized Layers

D. R. Bates
Department of Applied Mathematics, Queen's University, Belfast

Abstract—A short summary is given of current views regarding the processes responsible for the formation of the ionized layers. It seems probable that *D* is due to the photo-ionization of nitric oxide by the Ly-α line; that *E* is due mainly to the action of X-rays in the wavelength region centred about 40 Å; and that both *F1* and *F2* may be due partly to the action of longer X-rays and partly to the photo-ionization of atomic oxygen by ultra-violet radiation.

1. Introduction

Many hypotheses concerning the origins of the various ionized layers have been advanced. Indeed, a by no means exhaustive survey of the literature reveals that the number for each layer averages at least three. In the past, much of the work has necessarily been somewhat speculative in character. It is only recently, with the exploitation of the rocket as a research tool, that steady progress has been possible.

Early theorists were presented with a very difficult problem, for though general formulae describing the formation of an ionized layer had been developed by Chapman (1931), the detailed inferences that could be drawn from them were rather limited, owing to the lack of reliable information on the structure of the upper atmosphere, on the absorption cross-sections of its constituents, and, most especially, on the intensity distribution through the solar spectrum.

Potentially the most useful of the general formulae is that relating to the location of a layer produced by incident monochromatic radiation. For simplicity we take the atmosphere to be isothermal. If the active radiation is being absorbed in the photo-ionization of a constituent X, and in addition is being absorbed in some other way by a constituent Y, then the peak photo-ionization rate occurs where

$$n(X) = \cos \chi / H(X) A(X) \left\{ 1 + \frac{A(Y) n(Y)}{A(X) n(X)} \right\} \tag{1}$$

in which χ is the zenith angle of the sun and the n's, H's, and A's are respectively the number densities, scale heights, and absorption cross-sections of the constituents indicated. In principle $n(X)$, and hence the corresponding altitude, can thus be calculated. The photo-ionization rate might, however, be quite insignificant. Without knowing the incident photon flux, it is not possible to say whether the process under consideration is responsible for one of the ionospheric layers. Unfortunately, the estimates of the photon flux were usually grievously in error.

If the chemical constitution of the atmosphere were not as it is, spectroscopic studies during the twilight periods might be of invaluable assistance by providing information on the species of ion occurring at different altitudes. However, with the constitution as it is, there is only a single species which might be expected to be important and which can readily be detected by spectroscopic methods. This species, N_2^+, is found to be remarkably rare, but as it is rapidly removed by dissociative recombination its virtual absence during the twilight periods does *not* mean that photo-ionization of molecular nitrogen proceeds at a negligible rate—though it is very strongly suggests that such photo-ionization cannot give rise to one of the layers (BATES, 1954).

Having indicated some of the factors which made it possible to put forward erroneous but plausible hypotheses about the processes yielding the layers, we shall now briefly summarize the current views.

2. *D* REGION

As has long been realized, ultra-violet radiation of photon energy in excess of 12 eV cannot reach the *D* region. This severely restricts the number of processes that need be considered, for 12 eV is less than any of the ionization potentials of the main gases in the atmosphere.

NICOLET (1945) has suggested that the *D* region arises from the photo-ionization of nitric oxide

$$NO + h\nu(> 9{\cdot}5 \text{ eV}) \rightarrow NO^+ + e. \tag{2}$$

His suggestion has been widely accepted. The following are amongst the arguments which may be adduced in its favour.

(i) All other hypotheses which have been proposed have been examined and shown to be untenable (BATES and SEATON, 1950). It is unlikely that the correct hypothesis can have been overlooked.

(ii) Photochemical studies have established that nitric oxide must be a constituent of the atmosphere at *D*-region altitudes (cf. NICOLET, 1954; BATES, 1954).

(iii) Laboratory measurements by PRESTON (1940), WATANABE, MARMO, and INN (1953), and DITCHBURN, BRADLEY, CANNON, and MUNDAY (1954) show that Ly-α (λ1216) must penetrate just about as deeply into the atmosphere as is required, the absorption cross-section presented by an *air* molecule being $1{\cdot}7 \times 10^{-21}$ cm^2. This has been confirmed directly by BYRAM, CHUBB, FRIEDMAN, and LICHTMAN (1952), using rocket-borne photon-counters.

(iv) The photon flux in Ly-α that is received from the quiet sun is found to be approximately 6×10^9 cm^{-2} sec^{-1} (BYRAM, CHUBB, FRIEDMAN, and GAILAR, 1953). At λ1216 the cross-section for (2) is $2{\cdot}0 \times 10^{-18}$ cm^2 (WATANABE, 1954). Using the absorption cross-section cited in (iii), it may hence be shown that, when the sun is overhead, the photo-ionization rate is about $1 \times 10^6 f(NO)$ cm^{-3} sec^{-1} at 90 km, about $3 \times 10^6 f(NO)$ cm^{-3} sec^{-1} at 80 km, and about $2 \times 10^6 f(NO)$ cm^{-3} sec^{-1} at 70 km, where $f(NO)$ is the volume abundance of nitric oxide relative to the other constituents. These rates may well be considerable. Unfortunately, neither $f(NO)$ nor the rates actually required are known. All that can be said with certainty about $f(NO)$ is that it is less than some 2×10^{-4}, for if it were not, certain bands of nitric oxide would have been detected in absorption (BYRAM, CHUBB, FRIEDMAN, and GAILAR, 1953). If α is the effective recombination coefficient, if $n(e)$ is the number-density of electrons, and if λ is the negative ion-to-electron ratio, then the photo-ionization rate required is at least $\alpha n(e)^2$ in the case of electronic recombination, and $(1 + \lambda)\alpha n(e)^2$ in the case of ionic recombination. At 80 km α may perhaps be 2×10^{-7} cm^3 sec^{-1}, and $n(e)$ may perhaps be 8×10^2

cm^{-3} (cf. MITRA and JONES, 1954). The corresponding value of $\alpha n(e)^2$ is 0·1 cm^{-3} sec^{-1}; that of $(1 + \lambda)\alpha n(e)^2$ may be up to about an order of magnitude greater.

According to NICOLET (1954), the enhancement of D during fade-outs is due to flare activity increasing the Ly-α intensity and thus the rate of photo-ionization of nitric oxide. An alternative hypothesis has been proposed by FRIEDMAN and CHUBB (1955), who point out that X-rays at 1–2 Å might be responsible for the enhancement. Such X-rays have not been observed under quiet solar conditions, but may be emitted with sufficient intensity at times of flare activity. Doubtless the issue will be decided before long by rocket measurements during a flare.

3. E LAYER

In 1938 HULBURT and VEGARD independently suggested that the E layer might be formed by the action of X-rays emitted by the sun. The hypothesis must now be regarded as firmly established, for rocket scientists have verified that the sun is a sufficiently intense source of X-rays in the necessary wavelength range centred about 40 Å (BYRAM, CHUBB, and FRIEDMAN, 1953, 1954; HAVENS, FRIEDMAN, and HULBURT, 1955). The intensities reported show a rather wide scatter, but are in general well in excess of the intensity that might naively have been inferred from the values of $n(e)$ and α obtained by radio measurements. However, it must be borne in mind that α is the *effective* and not the *average* recombination coefficient— that is, α determines the variation of $n(e)$, and is *not* such that $\alpha n(e)^2$ must necessarily be the rate at which electrons are liberated (BATES, 1951, 1955).

In addition to the X-rays producing ionization in the main part of the E layer, there are shorter X-rays producing ionization at slightly lower levels (FRIEDMAN and CHUBB, 1955). The wavelength of the shortest X-rays so far detected is 6 Å. For this wavelength the absorption peak is at about 90 km. It is found that the intensity of the shorter X-rays is strongly dependent on coronal activity.

WULF and DEMING (1938) ascribed the formation of the E layer to the photo-ionization of molecular oxygen. The transition involved appears to be rather weak, but as NICOLET (1945) has pointed out, pre-ionization

$$\left.\begin{array}{r} O_2 + h\nu \begin{pmatrix} >12\cdot 0 \text{ eV} \\ <13\cdot 6 \text{ eV} \end{pmatrix} \quad \rightarrow O'_2 \\ O'_2 \rightarrow O_2^+ + e \end{array}\right\} \tag{3}$$

is also possible. It cannot be said whether the contribution is appreciable until the intensity of the incident radiation is determined. The results of some attempted measurements on the intensity were published in 1951, but these are now believed to be erroneous.

4. *F1* and *F2* LAYERS

The hypothesis that the radiation producing *F1* also produces *F2*, the double layer arising because of the decrease with altitude of the effective recombination coefficient, was put forward in 1938 by BRADBURY. It is probable that the hypothesis is essentially correct (BATES, 1954; RATCLIFFE, SCHMERLING, SETTY, and THOMAS, 1956). This does not of course mean that different photo-ionization processes are of precisely the same relative importance in the two layers.

Adopting the intensity measurements at shorter wavelengths as a guide to the conditions prevailing in the solar corona, HAVENS, FRIEDMAN, and HULBURT (1955) estimate that the continuum between 100 Å and the He II threshold at 228 Å carries about 0·05 erg cm^{-2} sec^{-1}, and that so do the He II and He I resonance lines at 304 Å and 584 Å (taken together). They claim that the *F* layers are due to the general ionization produced by this radiation (and by radiation of rather shorter wavelengths). However, the ionization rates which they compute (see table) are

Table

Altitude (km)	150	200	250	300	350
Ionization rate on hypothesis of HAVENS *et al.* (cm^{-3} sec^{-1})	200	80	20	5	1·6

low. Furthermore, they do not take cognizance of the fact that much of the available energy must be used in producing rapidly recombining N_2^+ ions. Except at great altitudes (where the effect is unimportant owing to diffusive separation), the computed ionization rates should therefore exceed the ionization rates deduced by the customary radio methods. In view of this, the older hypothesis that the *F* layers are at least partly due to the photo-ionization of atomic oxygen

$$O + h\nu\,(>13{\cdot}6\text{ eV}) \rightarrow O^+ + e \tag{4}$$

should not be abandoned lightly. The distribution of the ionization resulting from (4), which is of course influenced by absorption by molecular nitrogen, appears to be in fair accord with the observed distribution. Once again rocket measurements are required to determine if the radiation concerned is intense enough.

REFERENCES

BATES, D. R. 1951 *Proc. Phys. Soc.*, **B64**, 805.

1954 *The Earth as a Planet* (Ed. G. P. KUIPER), Chicago, University of Chicago Press, p. 576.

1955 This volume, p. 191.

BATES, D. R., and SEATON, M. J. 1950 *Proc. Phys. Soc.*, **B63**, 129.

BRADBURY, N. E. 1938 *Terr. Mag. Atmos. Elect.*, **43**, 55.

BYRAM, E. T., CHUBB, T., and FRIEDMAN, H. 1953 *Phys. Rev.*, **92**, 1066.

1954 *Rocket Exploration of the Upper Atmosphere*, (Eds. R. L. F. BOYD and M. J. SEATON), London, Pergamon Press, p. 274.

BYRAM, E. T., CHUBB, T., FRIEDMAN, H., and GAILAR, N. 1953 *Phys. Rev.*, **91**, 1278.

BYRAM, E. T., CHUBB, T., FRIEDMAN, H., and LICHTMAN, S. W. 1952 *J. Opt. Soc. Amer.*, **42**, 876.

CHAPMAN, S. 1931 *Proc. Phys. Soc.*, **43**, 26, 283.

DITCHBURN, R. W., BRADLEY, J. E. S., CANNON, C. G., and MUNDAY, G.	1954	*Rocket Exploration of the Upper Atmosphere,* (Eds. R. L. F. BOYD and M. J. SEATON), London, Pergamon Press, p. 327.
FRIEDMAN, H., and CHUBB, T. A.	1955	*Physics of the Ionosphere,* London, Physical Soc., p. 58.
HAVENS, R. J., FRIEDMAN, H., and HULBURT, E. O.	1955	*Physics of the Ionosphere,* London, Physical Soc., p. 237
HULBURT, E. O.	1938	*Phys. Rev.*, **53,** 344.
MITRA, A. P., and JONES, R. E.	1954	*J. Geophys. Res.*, **59,** 391.
NICOLET, M.	1945	*Mém. Instit. Roy. Met., Belgium,* Vol. 19.
	1954	*The Earth as a Planet* (Ed. G. P. KUIPER), Chicago, University of Chicago Press, p. 644.
PRESTON, W. M.	1940	*Phys. Rev.*, **57,** 887.
RATCLIFFE, J. A., SCHMERLING, E. R., SETTY, C. S. G. K., and THOMAS, J. O.	1956	*Phil. Trans. Roy. Soc.*, **A248,** 609.
VEGARD, L.	1938	*Geofys. Publ. Oslo,* **12,** No. 5.
WATANABE, K.	1954	*J. Chem. Phys.*, **22,** 1564.
WATANABE, K., MARMO, F., and INN, E. C. Y.	1953	*Phys. Rev.*, **90,** 155.
WULF, O. R. and DEMING, L. S.	1938	*Terr. Mag. Atmos. Elect.*, **43,** 283.

DISCUSSION OF PAPERS IN SECTION IV

Chairman: Prof. S. CHAPMAN

Dr. RAWER — I should like to make some remarks on the paper of Professor ALLEN. In general, for ionospheric work, critical frequencies are considered to be the most reliable parameters, and that is correct—except for f_0F1, which is not exactly a critical frequency. On the other hand, in addition to critical frequencies, a lot of information exists on heights. This may be useful despite the fact that the question of true heights has yet to be resolved. Now, ionospheric observations also give fairly reliable estimates of layer thickness, and this thickness enters into the calculations. For example, consideration of layer thickness introduces difficulties for the Nicolet theory of the formation of the E layer, which are not encountered on the X-ray theory. Information on the solar radiation cannot be obtained directly from critical-frequency measurements alone—one also needs to measure layer thickness and so determine the total electron content. This applies to the E layer in addition to higher layers.

Dr. BARTELS — In Professor ALLEN's paper it was supposed that the additional radiation coming from the active sun had the same spectrum as that of the quiet sun. I should like to say that there is some geomagnetic evidence which seems to show that for the E region (where the ionospheric currents responsible for the S_q variations flow) the additional radiation coming from the active sun has a slightly different spectrum from that of the quiet sun.

Mr. RATCLIFFE — Professor ALLEN, I think, suggests that a large portion of the extra effect at sunspot-maximum is due to radiations from the especially active regions, and this, it seems to me, should make corresponding variations in the day-to-day overall density of the E region. Now, on the curves which he showed from his own papers there is, of course, a big variation, and I am wondering whether a lot of this is purely experimental and due to inaccuracies in reading properly the cusps on the original records. I believe, but am not sure, that BIBL has carefully studied actual records, and concluded that the day-to-day variation is very small. Was the E region critical-frequency variation sufficient for us to conclude that irregular ionizing regions on the sun are *really* as important as the curves of Professor ALLEN suggest? Additionally, if they are, are they bigger at sunspot-maximum than sunspot-minimum?

I should also like to question Dr. ELLISON's use of a relaxation time. It is not difficult to show that the response of the ionosphere, considered as a decaying system of electrons with a relaxation time, depends markedly on the nature of the disturbing force, and I doubt very much whether one can really deduce anything about time-constants of the ionosphere for the sort of perturbing force to be expected in solar flares.

Dr. GOLDBERG — With regard to Dr. ELLISON's discussion of the time-lag of S.I.D.s and Hα, I should comment that there is nothing in the solar evidence that requires one to look for an explanation of this time-lag in the ionosphere. Recently my colleague, Miss HELEN DODSON, has been investigating the time-lag for a large number of flares, and finds indeed that the lag is dependent on the position of the flare on the solar disk. This leads one to suppose that the time-lag between maxima in Lyα and in Hα arises in the solar atmosphere. Now, there is much evidence that flare-layers are not isothermal, but that lower temperatures occur near the top. The major evidence for this is that when one observes flares near the limb, the rate of change of the Hα intensity is much steeper than when observed at the centre of the disk. This suggests that there are two separate layers with quite different radiating characteristics. Now, I should think that the Lyα radiation, which has a very high optical depth, would always come from the very top of the flare-layer, whether the flare concerned is at the centre of the disk or at the limb. On the other hand, Hα radiation may come from *any* point within the flare.

Dr. ELLISON Can you say what sort of differences in time-lag are involved?

Dr. GOLDBERG The difference is of the order of several minutes.

Dr. ELLISON I should have thought that the optical depth in Hα is so great that there will be very little difference between optical depths in Hα and Lyα.

Dr. GOLDBERG Not if the flare has on top of it a relatively cool layer of low optical depth. Near the limb the observations may refer to a cool layer on top, but this will not necessarily be true at the centre of the disk.

Solar Eclipses and the Ionosphere (*Special Suppl. Vol. 6 to J. Atmosph. Terr. Phys.*), *Pergamon Press, London*, 1956

SECTION V

RECOMBINATION

Recombination in the Ionosphere

D. R. Bates

Department of Applied Mathematics, Queen's University, Belfast

Abstract—A survey is given of the hypotheses concerning recombination in the ionosphere: amongst the collision processes considered are radiative recombination, charge transfer, ion-atom interchange, dissociation recombination, attachment, ionic recombination, and complex ion formation.

1. Introduction

Recombination in the ionosphere is due, not to a single process, but to a variety of processes. Though the more important of these have probably been recognized, a complete theory cannot yet be presented. In the present survey we shall discuss in particular how one group of processes gives place to another as we proceed downwards through the atmosphere.

We shall denote the effective recombination coefficient by the symbol α followed by the letter used for the layer or region concerned; the recombination coefficient for a particular process by the same symbol with the appropriate equation number as an identifying subscript; and other rate coefficients by k similarly identified.

2. Upper Ionosphere

At sufficiently great altitudes the constituents of the atmosphere are atomic. Because of this there are few processes by which free electrons can disappear. Indeed the only possibilities that need be considered are *radiative recombination*

$$\left.\begin{array}{l} O^+ + e \rightarrow O' + h\nu \\ N^+ + e \rightarrow N' + h\nu \end{array}\right\} \quad (1)$$

and perhaps *dielectronic recombination*

$$\left.\begin{array}{l} O^+ + e \rightarrow O'', \; O'' \rightarrow O' + h\nu \\ N^+ + e \rightarrow N'', \; N'' \rightarrow N' + h\nu \end{array}\right\} \quad (2)$$

the single primes indicating either the ground or a singly excited state, and the double primes a doubly excited state. Quantal calculations show that α_1 is about 10^{-12} cm^3 sec^{-1} (Bates, Buckingham, Massey, and Unwin, 1939), and that α_2 is at most comparable (Massey and Bates, 1943).

The observed effective recombination coefficients in the ionized layers are very much greater than the value just cited: thus $\alpha(E)$ is some 10^{-8} cm^3 sec^{-1} (cf. Waynick, 1955), $\alpha(F1)$ is probably of the same order (cf. Bates, 1951), and

$\alpha(F2)$ may be expressed by the approximate formula

$$\alpha(F2) = \left[\frac{9 \times 10^{-5}}{n(e)}\right] \exp\left\{\frac{300 - z}{35}\right\} \text{cm}^3 \text{sec}^{-1} \tag{3}$$

where $n(e)$ is the number of electrons per cm^3 and z is the altitude in km, so that, for example, if $n(e)$ is 10^6 cm^{-3} and z is 300 km, $\alpha(F2)$ is about 10^{-10} cm^3 sec^{-1} (Ratcliffe, Schmerling, Setty, and Thomas, 1956).*

After having examined various alternatives and shown them to be unacceptable, Bates and Massey (1946, 1947) proposed that the process responsible for the rapid disappearance of electrons might be *dissociative recombination*, which in the general case may be written

$$\text{XY}^+ + e \rightarrow \text{X}' + \text{Y}'. \tag{4}$$

When it was put forward, the proposal was highly speculative, for laboratory studies on dissociative recombination had not been carried out and the mechanism of the process was not properly understood. Since then considerable progress has been made experimentally (Biondi and Brown, 1949; Biondi, 1951a; Richardson and Holt, 1951; Dandurand and Holt, 1951; Radfield and Holt, 1951; Faire, Fundingsland, and Aden, 1954), and some has also been made theoretically (Bates, 1950). There is now no doubt but that α_4 can indeed be extremely large. Values even exceeding 10^{-7} cm^3 sec^{-1} have been reported. However, much remains to be done, particularly regarding the identification of the *active* ions in the various gases, and regarding the temperature dependence.

It is not possible to explain the results of the radio scientists by invoking (4) alone, since atomic ions (which of course cannot take part in such a process) are certainly formed in the F layers and, as may easily be seen, it is in general the most slowly disappearing species of ion which determines the effective recombination coefficient.

Bates and Massey (1947) advanced the view that the atomic ions are first converted into molecular ions and that dissociative recombination follows. They suggested that the conversion takes place through *charge transfer*

$$Z^+ + XY \rightarrow Z + XY^+ \tag{5}$$

where Z is either O or N, and XY is some molecule such as O_2 having a *lower ionization potential* (so that the process is *exothermic*).

The possible identification of XY as O_2 did not at the time seem at all plausible, for in 1947 atmospheric oxygen was believed to be almost completely dissociated above about the 100-km level. Bates and Massey indeed mentioned that the actual O_2 distribution might differ from what it would be if there were local photochemical equilibrium: but they made no attempt at a quantitative treatment. It was not until 1954 that the problem was resolved by the important researches of Nicolet and Mange, who showed that, because of mixing and

* The author wishes to thank Mr. Ratcliffe and his associates for letting him see a first draft of their paper.

diffusion, O_2 must in fact be quite abundant even in the *F* layers. Their conclusion is in harmony with certain measurements that have been made from rockets (cf. BYRAM, CHUBB, and FRIEDMAN, 1955).

BATES (1955) has questioned whether the charge-transfer process suggested can be sufficiently effective in view of the fact that an electronic transition involving a considerable change in the internal energy is rarely brought about in a thermal encounter (cf. MASSEY and BURHOP, 1952). He suggested that the original hypothesis be slightly modified by replacing (5) by the *ion-atom interchange* process

$$Z^+ + XY \rightarrow ZX^+ + Y \qquad (6)^*$$

which, as he pointed out, does *not* require an electronic transition and might have a rate coefficient as high as 10^{-10} cm^3 sec^{-1}. Recent measurements by STEVENSON and SCHISSLER (1955) confirm that in at least some instances ion-atom interchange occurs very readily.

Supposing, for the sake of definiteness, that O^+ ions are the primary ions formed, and accepting the hypothesis described, then the recombination sequence operative is

$$O^+ + O_2 \rightarrow O_2{}^+ + O \qquad (7)$$

$$O_2{}^+ + e \rightarrow O' + O'. \qquad (8)$$

At great enough altitudes (7) is clearly the rate-limiting process. The critical altitude must lie below the altitude of *F2*, so that

$$\alpha(F2) = k_7 n(O_2)/n(e) \qquad (9)$$

which is similar to formula (3) of RATCLIFFE, SCHMERLING, SETTY, and THOMAS (1956). Equating the two yields,

$$n(O_2) = \frac{9 \times 10^{-5}}{k_7} \exp\left\{\frac{300 - z}{35}\right\} \text{cm}^{-3} . \qquad (10)$$

If we provisionally take k_7 to be 10^{-10} cm^3 sec^{-1}, we see that we require $n(O_2)$ to be about 10^6 cm^{-3} at the 300-km level. The work of NICOLET and MANGE (1954) suggests that the requirement is not unreasonable.

It is apparent that in equilibrium

$$n(O_2{}^+)/n(O^+) = k_7 n(O_2)/\alpha_8 n(e) . \qquad (11)$$

Hence at low enough altitudes $O_2{}^+$ ions must predominate, and (8) must become the rate-limiting process, so that the usual recombination law is obeyed. This must be the situation in the *F1* and *E* layers.

---

* It might be thought that the apparently somewhat similar process

$$WZ^+ + XY \rightarrow WY + XZ^+$$

might provide an efficient means of transforming one species of molecular ion into another. This is probably not the case, for the process would be expected to have considerable activation energy.

The level above which (7) is rate-limiting and below which (8) is rate-limiting is approximately the level where

$$n(O^+) = n(O_2^+) , \qquad (12)$$

that is, where

$$n(O_2) = \alpha_8 n(e)/k_7 . \qquad (13)$$

If, again provisionally, we take α_8 to be 10^{-8} cm^3 sec^{-1}, $n(e)$ to be 2×10^5 cm^{-3}, and k_7 to be as before, we thus obtain that at this level $n(O_2)$ is about 2×10^7 cm^{-3}. According to NICOLET and MANGE (1954), such a value occurs just above the *F1* layer. The hypothesis thus seems to be in accord with the radio evidence. However, as need scarcely be emphasized, the values chosen for k_7 and α_8 may be seriously in error. It is most desirable that these coefficients should be properly determined.

Rocket scientists could provide a useful test of the hypothesis by measuring the variation with altitude of the $n(O_2^+)/n(O^+)$ ratio. By (11), this ratio should be large in the *E* layer and small in the *F2* layer.

It should be noted that ionization of the molecular nitrogen must result from the action of solar ultra-violet radiation and X-rays. The ions formed are removed through

$$N_2^+ + e \rightarrow N' + N' . \qquad (14)$$

If this is very rapid, as it probably is, radio scientists would normally be unaware of the additional ionization (BATES, 1951). Their effective recombination coefficient can therefore only be used to derive a *lower limit* to the incident photon flux. The additional ionization might be detected from measurements made during an eclipse.

A further complication is possible. Owing to the disposition of the relevant potential energy curves, dissociative recombination may be an utterly negligible process for some species of molecular ion, perhaps proceeding even more slowly than radiative recombination. Such a species might conceivably be responsible for the persistence of the *E* layer through the night. NICOLET (1955) has tentatively put forward an alternative explanation of this persistence. According to him, it may be due to the presence of Na^+ and similar atomic ions which, owing to the conservation of energy, cannot be converted into rapidly recombining molecular ions by charge transfer or ion-atom interchange.

There is no need for us to repeat the potent arguments that have been advanced (BATES and MASSEY, 1946, 1947) against the possibility that negative ions of atomic oxygen, which are continually being produced by *radiative attachment*

$$O + e \rightarrow O^- + h\nu, \qquad (15)$$

and destroyed by *photo detachment*

$$O^- + h\nu \rightarrow O + e \qquad (16)$$

and *associative detachment*

$$O^- + O \rightarrow O_2 + e, \qquad (17)$$

might contribute appreciably through *ionic recombination*

$$O^- + O^+ \rightarrow O' + O'. \tag{18}$$

It may, however, be mentioned that these arguments have been somewhat strengthened by the discovery (BRANSCOMB and SMITH, 1955) that the electron affinity of atomic oxygen, instead of being 2·2 eV, is only 1·45 eV. Thus the adoption of too high an affinity naturally led to an overestimate for the radiative attachment rate and to an underestimate for the photodetachment rate. Furthermore, a certain resonance effect, which was helpful to the negative ion hypothesis, can now be dismissed as non-existent.

3. LOWER IONOSPHERE

The effective recombination coefficient does not remain constant as we continue on downwards into the D region. According to MITRA and JONES (1954), it is about 10^{-7} cm^3 sec^{-1} at 90 km, about 10^{-6} cm^3 sec^{-1} at 75 km, and about 10^{-5} cm^3 sec^{-1} at 60 km. It would seem that some other mechanism enters.

Negative ions must eventually become important, but it is difficult to calculate the altitude at which this happens. BATES and MASSEY (1952) have considered the collision processes involved. Radiative attachment

$$O_2 + e \rightarrow O_2^- + h\nu \tag{19}$$

yields negative ions of molecular oxygen, the rate coefficient probably being $1{\cdot}4 \times 10^{-15}$ cm^3 sec^{-1} (BIONDI, 1951b). These negative ions may be destroyed by a number of processes including photodetachment

$$O_2^- + h\nu \rightarrow O_2 + e, \tag{20}$$

associative detachment

$$O_2^- + O \rightarrow O_3 + e, \tag{21}$$

and ion-atom interchange

$$O_2^- + O \rightarrow O^- + O_2 \tag{22}$$

(after which (16) might occur). Ionic recombination

$$O_2^- + X^+ \rightarrow O_2' + X' \tag{23}$$

in which X may, for example, be NO, also takes place. Its contribution to the effective recombination coefficient in the D region is $\alpha_{23}\lambda$, where λ is the $n(O_2^-)/n(e)$ ratio. A quantitative treatment is not feasible at present; quite apart from uncertainty as to the value of α_{23} it is not possible to estimate λ with any assurance, for with the exception of k_{19}, none of the rate coefficients required have been determined. Nevertheless it is instructive to make certain numerical substitutions. If τ is the life of an O_2^- ion at a certain level, then at that level

$$\lambda = k_{19}\tau n(O_2) = 1{\cdot}4 \times 10^{-15}\tau n(O_2). \tag{24}$$

At the 75-km level $n(O_2)$ is 2×10^{14} cm^{-3}. Hence λ is there about (0·3) τ, so that if $\alpha_{23}\lambda$ is to be 10^{-6} cm^3 sec^{-1}, τ must be as long as 30 sec even if α_{23} is as much as 10^{-7} cm³ sec⁻¹. For comparison it may be noted that the life of an O^- ion towards photo-detachment alone is less than 3 sec (Bates and Massey, 1946). A life of 30 sec for an O_2^- ion cannot be excluded, but it would scarcely have been predicted. The hypothesis that ionic recombination is of dominant importance in D may be correct; but it should be viewed with caution.

In considering D it must be remembered that the positive ions in it are probably not the same as those in the E layer. The main species of ion may for example be NO^+, and it is conceivable that the dissociative recombination coefficient for this species is unusually high. Again *complex ions* must be being formed in three-body collisions

$$X^+ + 2M \rightarrow X^+ . M + M \tag{25}$$

X^+ being perhaps an NO^+ ion and M being perhaps an N_2 or O_2 molecule. Should k_{25} be as high as 10^{-30} cm³ sec⁻¹, the life of a simple ion would only be about 1 sec at the 75-km level. The complex ions may be destroyed in ordinary two-body collisions

$$X^+ . M + M \rightarrow X^+ + 2M \tag{26}$$

if the available kinetic energy of relative motion is sufficient. If ε is the energy required, we have approximately

$$k_{26} = 3 \times 10^{-10} \exp(-\varepsilon/RT) \text{ cm}^3 \text{ sec}^{-1}. \tag{27}$$

The table gives the value of $\exp(-\varepsilon/RT)$ for some selected cases. It may well be that k_{26} is so small that many ions in D are complex rather than simple. Unfortunately, the vital parameter ε is unknown.

Table

ε (kcal/mole)	2	4	6	8	10
T(°K)	$\exp(-\varepsilon/RT)$				
200	$6{\cdot}5 \times 10^{-3}$	$4{\cdot}3 \times 10^{-5}$	$2{\cdot}8 \times 10^{-7}$	$1{\cdot}8 \times 10^{-9}$	$1{\cdot}2 \times 10^{-11}$
250	$1{\cdot}8 \times 10^{-2}$	$3{\cdot}2 \times 10^{-4}$	$5{\cdot}7 \times 10^{-6}$	$1{\cdot}0 \times 10^{-7}$	$1{\cdot}8 \times 10^{-9}$
300	$3{\cdot}5 \times 10^{-2}$	$1{\cdot}2 \times 10^{-3}$	$4{\cdot}3 \times 10^{-5}$	$1{\cdot}5 \times 10^{-6}$	$5{\cdot}2 \times 10^{-8}$

References

Bates, D. R. 1950 *Phys. Rev.*, **77**, 718; **78**, 492.
1951 *Proc. Phys. Soc.*, **B64**, 805.
1955 *Proc. Phys. Soc.*, **A68**, 344.

Bates, D. R., Buckingham, R. A., Massey, H. S. W., and Unwin J. J. 1939 *Proc. Roy. Soc.*, **A170**, 322.

BATES, D. R., and MASSEY, H. S. W.	1946	*Proc. Roy. Soc.*, **A187**, 261.
	1947	*Proc. Roy. Soc.*, **A192**, 1.
	1952	*J. Atmosph. Terr. Phys.*, **2**, 1, 253.
BIONDI, M. A.	1951a	*Phys. Rev.*, **83**, 1078.
	1951b	*Phys. Rev.*, **84**, 1072.
BIONDI, M. A., and BROWN, S. C.	1949	*Phys. Rev.*, **76**, 1697.
BRANSCOMB, L. M., and SMITH, S. J.	1955	*Phys. Rev.*, **98**, 1127.
BYRAM, E. T., CHUBB, T. A., and FRIEDMAN, H.	1955	*Phys. Rev.*, **98**, 1594.
DANDURAND, P., and HOLT, R. B.	1951	*Phys. Rev.*, **82**, 278, 868.
FAIRE, A. C., FUNDINGSLAND, O. T., and ADEN, A. L.	1954	*Phys. Rev.*, **93**, 650.
MASSEY, H. S. W., and BURHOP, E. H. S.	1952	*Electronic and Ionic Impact Phenomenon*, Clarendon Press, Oxford, p. 441.
MASSEY, H. S. W., and BATES, D. R.	1943	*Rep. Progr. Phys.*, **9**, 62.
MITRA, A. P., and JONES, R. E.	1954	*J. Geophys. Res.*, **59**, 391.
NICOLET, M.	1955	*Meteors* (Ed. T. R. KAISER), Pergamon Press, London, p. 99.
NICOLET, M., and MANGE, P.	1954	*J. Geophys. Res.*, **59**, 15.
RATCLIFFE, J. A., SCHMERLING, E. R., SETTY, C. S. G. K., and THOMAS, J. O.	1956	*Phil. Trans. Roy. Soc.*, **A248**, 609.
REDFIELD, A., and HOLT, R. B.	1951	*Phys. Rev.*, **82**, 874.
RICHARDSON, J. M., and HOLT, R. B.	1951	*Phys. Rev.*, **81**, 153.
STEVENSON, D. P., and SCHISSLER, D. O.	1955	*J. Chem. Phys.*, **23**, 1353.
WAYNICK, A. H.	1955	*Physics of the Ionosphere*, Physical Society, London, p. 1.

Solar Eclipses and the Ionosphere (Special Suppl. Vol. 6 to J. Atmosph. Terr. Phys.), Pergamon Press, London, 1956

A Critical Survey of Eclipse Information on Recombination Data

K. RAWER
Service de Prévision Ionosphérique Militaire, Freiburg

Abstract—The more important assumptions underlying the analysis of ionospheric eclipse measurements are critically examined and proposals made for future work in this field.

THE PROBLEM

ECLIPSES of the Sun afford occasions on which the recombination coefficient in the different ionospheric layers can be determined. One uses the equation

$$\frac{dN}{dt} = I - R \tag{1}$$

where

$$I = g(t) \cdot I_0(t) \qquad \text{and} \qquad R = \alpha_m \cdot N(t)^m \tag{2}$$

N = electron density, I = electron production, I_0 = undisturbed production, g = screening ratio of ionizing radiation during the eclipse, α_m = the recombination coefficient. Normally it is believed that $m = 2$. N and dN/dt are determined from the observations, generally from the critical frequency. By comparison with the undisturbed observations, $I_0(t)$ may be eliminated. g, in general, is calculated from the optical eclipse data. Thus from a set of observations during the eclipse, α may be obtained for any instant.

This determination assumes a constant radiation intensity during the control period. This assumption will not be discussed here, but other assumptions will be discussed in more detail. These are:

(a) The ionizing radiation has a constant intensity over the whole solar disk; the effective solar diameter is equal to the optical one (otherwise a geometrical calculation of $g(t)$ is not justified).

(b) There is no change, during the eclipse, of the height of the maximum of the layer (otherwise the values of N deduced from the critical-frequency measurements are not consistent).

(c) The effective electron density is not influenced by either slow or rapid transport or contraction phenomena (otherwise additional terms must be introduced into (1)).

(d) The *theory* used in (1) and (2) is true, especially the simple recombination formula (otherwise the effective coefficient α must depend on further parameters, e.g., N).

DISCUSSION

(a) A number of observations which can hardly be explained on the assumption of a constant intensity have been presented. The most striking proof is found with the eclipse of 25 February 1952, in which rapid changes were observed during the eclipse, and the maximum of the ionospheric effect was observed long *before* the optical maximum. Fig. 1 shows the variation of f_0F1 at Djibouti (10-min time-delay); Fig. 2 that of f_0E at Lwiro* (4-min time-delay). Clearly any recombination equation cannot be applied here, since negative α values are obtained. These observations can be explained if we assume centres of intense radiation

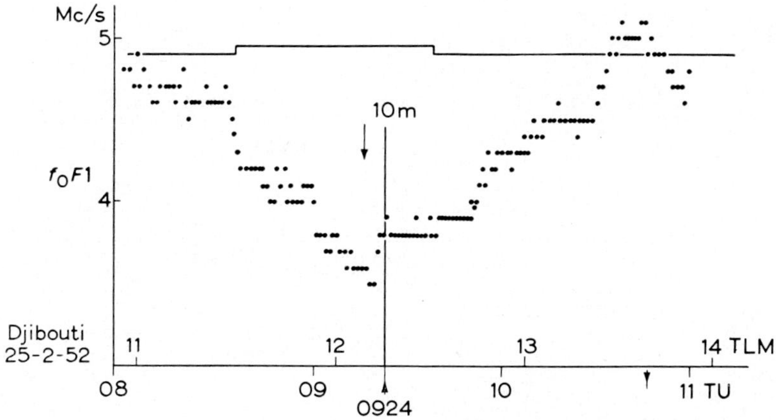

Fig. 1. Variation of f_0F1 at Djibouti, 25.2.52

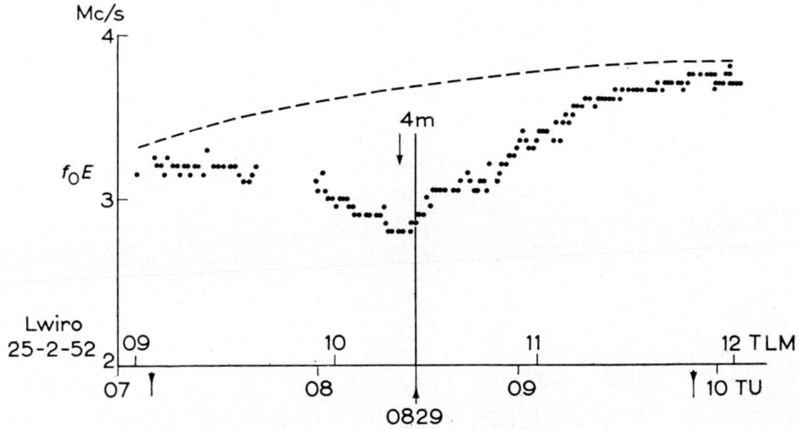

Fig. 2. Variation of f_0E at Lwiro, 25.2.52

on the solar disk (PIDDINGTON, 1951). Such rapid variations as are shown in Fig. 1 can hardly be explained otherwise (except in the case of transitory movements).

There seems to be no doubt that such irregularities exist in the distribution of the radiation. It is very difficult to deduce their real distribution on the solar disk by comparing different ionospheric and solar observations, and in many cases it may be impossible to obtain it with sufficient accuracy. In the presence

* By kind permission of Dr. NICOLET and the I.R.S.A.C.

of important irregularities, no satisfactory evaluation of α can be made. An estimation may be obtained in these cases from the apparent time-constant of the rapid variations; and for the results shown in Fig. 1 this has been done (MÉRIAU and RAWER, 1954).

In addition to such irregularities, a limb effect (darkening or brightening) is quite probable. Finally, a radiation from the corona must have a larger effective diameter than the optical diameter of the sun. In order to deal with these effects, we need further information concerning the ionizing radiation both from theory and experiments.

(b) The height corresponding to N_{max} is probably very variable for the *F1* layer. We meet here with the problem of the definition of a "characteristic

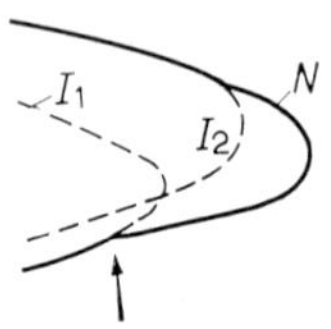

Fig. 3. Composition of two independent layers

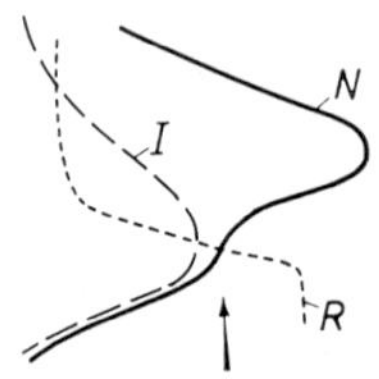

Fig. 4. Layer formation with variable recombination term

frequency" for a layer which has no maximum of electron density. As recently pointed out (RAWER, 1955), different physical interpretations of f_0F1 values must be applied, according to the supposed origin of the layer. Two different theories exist. According to the first, different radiations give rise to the *F1* and *F2* layers. In this case the "cusp" in real height is seen at the lower edge of the upper layer (Fig. 3). The second theory (MOHLER, 1937, 1940) suggests that both layers have the same ionizing source, but that the separate layers are a consequence of a rapid decrease of the recombination coefficient with height (Fig. 4). Such a variation may be found at the upper limit of the existence of catalyzer molecules, e.g., with a dissociative recombination process. Different observations seem favourable to this latter theory. In this case f_0F1 is effectively determined by the height at which the catalyzer disappears.

For a Chapman *E* layer the height of the maximum is found to be independent of the intensity of the radiation. This probably holds also if the layer is produced by X-rays (ELWERT, 1952, 1953; ARGENCE and RAWER, 1954). If, on the other hand, we had a specific ionization of molecular oxygen, as is believed by some workers in this field (CHOUDHURY, 1952), the height of the ionized layer should increase during the eclipse in consequence of a decreased dissociation rate. Apart from noon conditions, the height is in any case variable with cos χ.

It seems worth while for the *F1* layer, at least, to base the evaluation on real rather than on virtual height. This has been done for the 1952 eclipse by MINNIS (1955). Fig. 5 gives some results for the recent eclipse of 20 June 1955 at Nha-Trang (Vietnam).* The variations of N at a constant height show some slight

* A film showing the variation of real as well as virtual height has been produced.

irregularities, partly due to the method of reduction (Kelso) and partly to transitory phenomena. We note that the eclipse-effect decreases in amplitude, but increases in delay-time with increasing altitude. This is to be expected if the effective recombination decreases with increasing height.

(c) There can now be no doubt that transport phenomena play an important role in the *F2* layer. Thus a reliable determination of the recombination coefficient

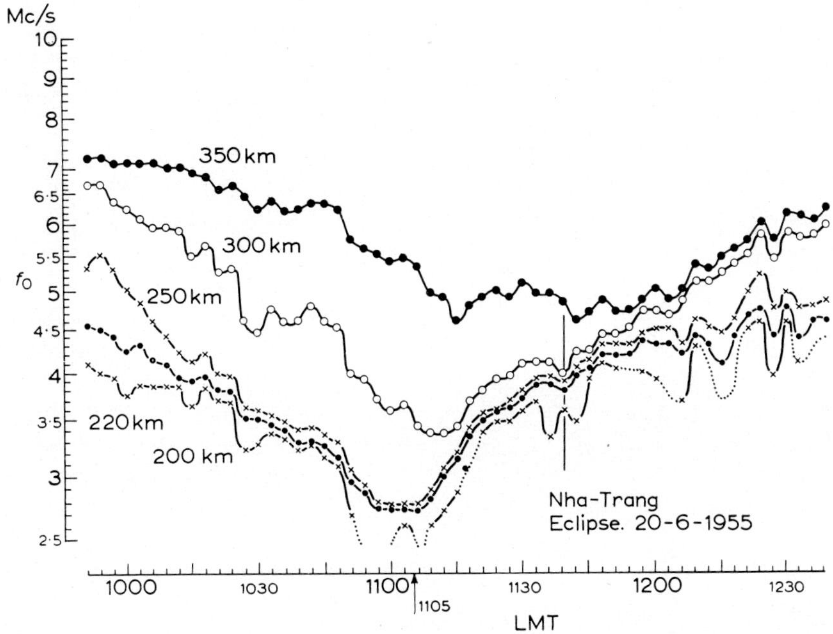

Fig. 5. Variation of the electron density N at different real heights (Nha-Trang, 20.6.55).

is not possible for this layer. On the other hand, transitory changes (generally rapid downward movement from the high *F* region) are quite normal phenomena during the daytime. Observed f_0F1 values are often affected by such rapid changes. These phenomena, also observed during eclipse conditions (Bibl and Delobeau, 1955), are an element of instability which can only be detected and eliminated with high-speed $h'f$ sweeps.

(d) We continue to apply twenty-year-old theories to the fundamental ionospheric processes. Although we know, from theory as well as from observations, that direct (radiative) recombination cannot explain the effective disappearance of electrons in the ionosphere, the formula (2) with $m = 2$ is nearly always used. No other simple recombination or attachment formula is consistent with the theory. The dissociative recombination proposed by Bates and Massey (1947) is a two-step process (charge transfer and recombination-dissociation). Only if the first process is rapid enough will the second give something like an attachment law (but with the catalyzer ions as second partner). Thus this sort of recombination depends even on an unknown factor (the catalyzer density) and one which is variable with height. This objection to existing eclipse analyses is most important.

One way out of these difficulties would appear to be the following: in equation (2) we should take an empirical exponent m instead of the "theoretical" one. In fact, neither for E nor for $F1$ does the daily variation give $m = 2$, but $m = 5/3$ and $5/2$ respectively. This method of obtaining an effective m is better; the value has also been checked with eclipse results (minimum electron density). With this method an α_m is obtained which may easily be translated into an effective "α_2", if this is thought necessary ($\alpha_2 = \alpha_m \cdot N^{m-2}$).

One serious objection to this method may be mentioned: the recombination itself may be influenced by the eclipse, so that normal and eclipse recombination coefficients are different. This possibility can hardly be excluded.

In this situation we may adopt an alternative approach. When we determine one value of α only from all the readings of an eclipse, we really reject an important part of the information acquired and it would be better to retain this information instead of losing it by applying a doubtful theory. A complete set of eclipse data may become very useful for future work. Thus we may one day obtain precise information concerning the actual catalyzer, in which case complete data will be required to check the theory. With the present method of putting together all information in one parameter, we shall not be able to do this. It is proposed that during the whole of an eclipse two parameters which can easily be obtained should be evaluated. They may be the time-constant (N/R) and the corresponding electron density N; or, if it is thought better to continue with effective α_2 values, we may take R/N^2 and N. Thus for every eclipse we should tabulate g, R/N^2 and N.

Conclusions

The following proposals are made for further work in this field (the reasons are indicated in parentheses, see the paragraphs above).

(1) A reliable determination of the $F2$-recombination coefficient is not at present possible, (c).

(2.1) Rapid sweeps are desirable, (a, c).

(2.2) Only the most regular (continuous) eclipses may be expected to give reasonable values, (a).

(2.3) The corona, as well as the limb-effect, must be studied theoretically, (a).

(3.1) Evaluations should be based on real heights (taking account of the magnetic field for medium and high latitudes), (b).

(3.2) The E layer should be observed with a sufficiently large height-scale (in order to apply a real height-analysis to it), (b).

(4.1) The results must not be evaluated in terms of a doubtful theory, (d).

(4.2) Results should be given as a two-parameter function (e.g., R/N^2 and N) versus the screening factor g, (d).

References

Argence, E., and Rawer, K. 1954 *Phys. Rev.*, **94**, 253.

Bates, D. R., and Massey, H. S. W. 1947 *Proc. Roy. Soc.*, **A192**, 1.

Bibl, K., and Delobeau, F. 1955 *Z. Geophysik.* (in press).

Choudhury, D. C.	1952	*Phys. Rev.*, **88,** 405.
Elwert, G.	1952	*Z. Naturforschung*, **7a,** 202.
	1953	*J. Atmosph. Terr. Phys.*, **4,** 68.
Mériau, P., and Rawer, K.	1954	*Ann. Géophys.*, **10,** 9.
Minnis, C. M.	1955	*J. Atmosph. Terr. Phys.*, **6,** 91.
Mohler, F. L.	1937	*Bur. Stand. Journ. Res.*, **19,** 447, 559.
	1940	*Bur. Stand. Journ. Res.*, **25,** 507.
Piddington, J. H.	1951	*J. Geophys. Res.*, **56,** 409.
Rawer, K.	1955	*Compt. Rend. Acad. Sci.* (Paris), **240,** 331.

Solar Eclipses and the Ionosphere (Special Suppl. Vol. 6 to J. Atmosph. Terr. Phys.), Pergamon Press, London, 1956

The Effective Recombination Coefficients in the E and $F1$ Layers

C. M. MINNIS
Radio Research Station, Slough

Abstract—The realization that the sun cannot always be regarded as a uniformly radiating disk has meant that it is no longer possible to deduce α from the time-difference between the electron-density minimum and the eclipse maximum. Other methods may sometimes be used if conditions are appropriate.

During the total eclipse at Khartoum, dN/dt could be measured up to about 30 sec before second contact; the legitimacy of the extrapolation to second contact has been questioned, but a consideration of d^2N/dt^2 shows that it is probably valid. Assuming $q = 0$ at second contact, the evaluation of $(dN/dt)/N^2$ gives $\alpha_E = (1{\cdot}51 \pm 0{\cdot}12) \times 10^{-8}$ and $\alpha_{F1} = (8{\cdot}2 \pm 1) \times 10^{-9}$ cm^3 sec^{-1} at estimated true heights of 125 and 170 km respectively.

For a partial eclipse, the solution of CHAPMAN's equation for various values of α and the matching of one of the resulting $N(t)$ curves with the experimental one is unreliable unless $q(t)$ is similar to the geometrical form. However, under certain conditions an approximate value of the component of q_0 due to the uniform disk radiation can be obtained by using the observed values of dN/dt at times when N or $N^2 \sec \chi$ are equal; α can then be computed, using the noon value of q_0/α. Such values are not accurate, but they provide confirmation that the very high values of α which are sometimes proposed to explain zero time-lag in minimum N are unlikely to be correct.

The close correlation between eclipse changes in f_0E and f_0F1 shows that the locations and relative intensities of the sources of ionizing radiation for the two layers are the same. The relation between f_0E and f_0F1 can be used to compute α_E/α_{F1}, which is equal to q_0F1/q_0E.

INTRODUCTION

IF the changes in the incident radiation intensity during a solar eclipse followed the simple geometrical law, it would, in principle, be possible to deduce the effective recombination coefficients in the E and $F1$ layers from the variation of electron density with time. Unfortunately, the sun's disk can seldom be regarded as uniformly bright, and the method cannot be applied with any confidence. Although several alternative methods for determining α'_E, α'_{F1} are available, the circumstances peculiar to any particular eclipse will determine which, if any, are appropriate.

KHARTOUM 1952

When an eclipse is total, it is probably correct to assume that electron production falls to a very low level during the interval between second and third contacts. If electron dissipation takes place by an effective recombination process, it then follows that at the instant of second contact we can write

$$\alpha' = -\frac{dN}{dt} \Big/ N^2 . \tag{1}$$

Since both dN/dt and N can be found from the changes in critical frequency with time, this relation allows an estimate of α' to be made.

At Khartoum in 1952, N was measured at intervals of about 4 min. No measurements were made during the interval from 30 sec before second contact to 45 sec after third contact. It was found that dN/dt increased rapidly and in a regular manner up to the beginning of this interval, and it seemed reasonable to extrapolate dN/dt and N up to the instant of second contact. This allowed equation (1) to be used, giving:

$$\alpha'_E = (1{\cdot}51 \pm 0{\cdot}12) \times 10^{-8} \text{ cm}^3 \text{ sec}^{-1}.$$

The height of the E layer was estimated to be 125 km.

This value for α'_E is rather higher than that usually assumed, and it has been suggested that the extrapolation of dN/dt is not valid and that $(-dN/dt)$ may have begun to fall before second contact. Since dN/dt changed from -41 to $+47$ during the interval mentioned above, this argument is a plausible one, but on further consideration it does not seem likely to be correct.

If the electron-balance equation for a Chapman layer is differentiated, we have

$$-\left|\frac{d^2N}{dt^2}\right| = -\left|\frac{dq}{dt}\right| + 2\alpha' N\left|\frac{dN}{dt}\right| . \tag{2}$$

The signs are those appropriate to the period up to second contact at Khartoum. A fall in the magnitude of $(-dN/dt)$ during the 30-sec interval covered by the extrapolation mentioned above would imply that d^2N/dt^2 had become positive; this, in turn, would require a decrease in the magnitude of $(-dq/dt)$. For a uniformly bright disk, dq/dt is constant just before second contact, and hence a decrease would imply some darkening of the east limb relative to the uniform disk. Bright sources of ionizing radiation were found near both the east and west limbs during the 1952 eclipse, and that near the west limb was very close to the limb and appeared to be associated with 5303-Å coronal radiation which was visible during the eclipse. A similar coronal condensation was observed near the east limb, but the location and intensity of the ionizing radiation is such that its contribution to dq/dt is small. The conclusion is reached, therefore, that most of any postulated decrease in dq/dt would have to be attributed to a considerable deficit in the uniform disk radiation over a narrow annular arc on the east limb; this does not seem very likely. If the extreme but unlikely assumption is made that q fell to zero 30 sec before second contact, the decrease in the value of α', computed using equation (1), would be only $0{\cdot}09 \times 10^{-8}$ cm^3 sec^{-1}, and the effect of the extrapolation is evidently not very great in any case.

A similar procedure was used to determine α'_{F1} at Khartoum. Owing to slight irregularities in the changes in f_0F1, the measured values of dN/dt are less reliable than those for the E layer, and the possible error in computing α'_{F1} is greater. The final value obtained is

$$\alpha'_{F1} = (8 \pm 1) \times 10^{-9} \text{ cm}^3 \text{ sec}^{-1},$$

and the height of the layer is about 170 km.

The Khartoum E-layer data were also used to obtain an approximate value for the rate of electron production, using a method which does not require a knowledge of α'. If the electron production is divided into two parts; $q_0AD \cos \chi$, due to the fraction D of the total ionizing radiation attributable to uniform disk radiation, and q', due to the remaining sources, then

$$\frac{dN}{dt} = q_0\, AD \cos \chi + q' - \alpha' N^2 \tag{3}$$

where A is the fraction of the disk area not obscured. If times are selected before and after the maximum phase of the eclipse such that the electron densities are equal, then:

$$\left(\frac{dN}{dt}\right)_2 - \left(\frac{dN}{dt}\right)_1 = q_0 D\,(A_2 \cos \chi_2 - A_1 \cos \chi_1) + q_2' - q_1' \tag{4}$$

The slope of a plot of $(dN/dt)_2 - (dN/dt)_1$ against $(A_2 \cos \chi_2 - A_1 \cos \chi_1)$ allows $q_0 D$ to be obtained, providing $(q_2' - q_1')$ remains constant. A similar result can be obtained by choosing times when $N^2 \sec \chi$ has the same value.

At Khartoum, changes in $(q_2' - q_1')$ did not allow a unique value of the slope to be obtained, but it was found that $q_0 D \simeq 260$ cm^{-3} sec^{-1} as compared with $q_0 D = 248$ cm^{-3} sec^{-1} calculated from the independently derived values of q_0/α', α', and D. The good agreement is probably partly fortuitous, because the difference $(A_2 \cos \chi_2 - A_1 \cos \chi_1)$ is usually small compared with its component terms.

INVERNESS AND SLOUGH 1954

The difference $(A_2 \cos \chi_2 - A_1 \cos \chi_1)$ applicable to the Inverness and Slough results during the partial eclipse of 1954 is so small that it is not possible to obtain a value for $q_0 D$ on which any reliance can be placed. Since the eclipse was not total at either station, equation (1) cannot be used to determine α'.

Solar activity was very low in June 1954, and it was expected that the distribution of sources of ionizing radiation over the disk would be nearly uniform, and the experimental results show that to a first approximation this assumption is correct. This allows α' to be determined approximately, using equation (5) and assuming that the effective fraction A' of the ionizing radiation unobscured is given by the simple geometrical relation.

$$\alpha' = \frac{dN}{dt} \Big/ \left[\left(\frac{q_0}{\alpha'}\right) A' \cos \chi - N^2\right]. \tag{5}$$

Both at Inverness and Slough, the *F1*-layer results gave

$$\alpha'_{F1} \simeq 2 \times 10^{-8} \text{ cm}^3 \text{ sec}^{-1},$$

which is much too large and suggests that the effective radiation intensity differed from the calculated value. A lower limit can be obtained by computing $(dN/dt)/(-N^2)$ when dN/dt is negative, and this gave $\alpha'_{F1} > 0{\cdot}6 \times 10^{-8}$ cm^3 sec^{-1} and $\alpha'_E > 1{\cdot}2 \times 10^{-8}$ cm^3 sec^{-1} at Inverness. Since N^2 did not fall to such a low value as at Inverness, the lower limits obtained at Slough are too small to be informative.

The distribution of sources of ionizing radiation described in a separate paper were deduced by putting $\alpha'_{F1} = 2 \times 10^{-8}$ cm^3 sec^{-1}. If this distribution is used to compute A' for substitution in equation (5), a lower value for α'_{F1} would result, but, since the derived value is very sensitive to the assumed value of A', it is not possible to obtain a reliable value for α'_{F1}. The best information available on recombination coefficients at Inverness may be summarized:

$$2{\cdot}0 \times 10^{-8} > \alpha_{F1} > 0{\cdot}6 \times 10^{-8} \text{ cm}^3 \text{ sec}^{-1}$$

$$\alpha_E > 1{\cdot}2 \times 10^{-8} \text{ cm}^3 \text{ sec}^{-1}$$

THE RELATION BETWEEN f_0E AND f_0F1

(a) *Diurnal variations*

It is well known that the diurnal variations in the E- and $F1$-layer critical frequencies do not conform strictly to the $\cos^{1/4}\chi$ law. If we put:

$$f_0E = a\cos^m\chi \qquad \text{and} \qquad f_0F1 = b\cos^n\chi \tag{6}$$

then, providing χ is not too large, we can write

$$f_0E \simeq \frac{am}{bn} f_0F1 + a\left(1 - \frac{m}{n}\right) = Af_0F1 + B. \tag{7}$$

Typical values of the parameters in equation (7) are given in Table 1, and the large negative values of B indicate that, in general, m is considerably greater than n. Table 1 has been based on monthly mean values, but any fuller investigation ought preferably to be based on carefully measured data for selected days.

Table 1. Constants in equation (7) (diurnal variations)

Station	Date	A	B	a/b	m/n
Singapore	May–July 1950–52	1·16	−1·92	0·79	1·48
Singapore	May–July 1954	1·59	−3·54	0·76	2·07
Slough	May–July 1950	1·03	−1·70	0·70	1·47
Slough	May–July 1954	1·18	−1·87	0·74	1·60
Khartoum	February 1952	1·04	−1·33	0·76	1·37

(b) *Eclipse variations*

During an eclipse which occurs not too far from noon, the variations in f_0E and f_0F1 are due mainly to the partial obscuration of the respective sources of ionizing radiation which are responsible for the E and $F1$ layers. Since the identity of neither radiation is known, it is not possible to assume that the intensity-time relation is the same for both layers. However, during several recent eclipses, it has been found that there is a close correlation between the changes in f_0E and f_0F1, and it seems reasonable to infer from this that the radiation sources for the two layers are similarly distributed over the sun's disk, and that they have similar relative intensities. It has also been found that the constants A and B measured under eclipse conditions differ considerably from those typical of the normal diurnal variation. The constants determined during several eclipses are given in Table 2, and represent lines passing much closer to the origin than those defined in Table 1.

Table 2. Constants in equation (7) (eclipse variations)

Station	Date	A	B
Khartoum	February 1952	0·75	+0·17
Inverness	June 1954	0·66	+0·18
Slough	June 1954	0·73	−0·12
Singapore	June 1955	0·72	+0·25

Since B is very small during an eclipse, it is concluded that both layers respond in almost the same way to changes in the intensity of the ionizing radiation. This is obviously equivalent to making $m = n$ in equation (7) for the diurnal variation.

If it is assumed that during an eclipse

(i) the fractional obscuration of the ionizing radiation is the same for both the E and $F1$ layers,

(ii) $B = 0$ in the f_0E/f_0F1 relation,

then, as explained in the Appendix, it can be shown that the recombination coefficients, the rates of electron production, and the ratio a/b as defined in equation (6), are related simply as follows:—

$$R = \alpha'_E/\alpha'_{F1} = q_0F1/q_0E = \sqrt{(q_0/\alpha')_{F1}/(q_0/\alpha')_E} = b^2/a^2$$

The ratio b^2/a^2 may be derived from the data given in Tables 1 and 2, and the values of R calculated in this way are given in Table 3. During eclipses, it is sometimes possible to measure α'_E and α'_{F1} independently, and the values of R derived from results quoted by several eclipse observers since 1940 are also included in Table 3.

Table 3. Determinations of R

Method of measurement		R
Relation between f_0E and f_0F1	Diurnal (Table 1)	1·78
	Eclipse (Table 2)	1·88
Independent measurements of α'_E and α'_{F1} during eclipses	GILLILAND, Brazil, 1940	(2)
	PIERCE, S. Africa, 1940	2·00
	RYDBECK, Sweden, 1945	1·83
	DENISSE *et al.*, Brazil, 1947	(2·5)
	MINNIS, Khartoum, 1952	1·85

At Inverness during the 1954 eclipse, sporadic E ionization made it difficult to read the E-layer critical frequencies. A small increase in group retardation which was tabulated as f_0E appears, in fact, to have been a ledge below the peak of the E layer. This conclusion is based on the small value of A at Inverness (Table 2). The height of this ledge is only a few km below the peak of the E layer, but the recombination coefficient implied by $A = 0{\cdot}66$, as compared with $A = 0{\cdot}73$, appears to show that, at the level of the ledge, the recombination coefficient is about 1·2 times that in the normal E layer.

CONCLUSIONS

Values of the recombination coefficients in the E and $F1$ layers have been deduced from measurements made during recent solar eclipses. These are:—

Khartoum 1952. $\alpha'_E = 1{\cdot}5 \times 10^{-8}$, $\alpha'_{F1} = 8 \times 10^{-9}$ cm³ sec⁻¹

Inverness 1954. $\alpha'_E > 1{\cdot}2 \times 10^{-8}$ cm^3 sec^{-1},

$$2{\cdot}0 \times 10^{-8} > \alpha'_{F1} > 0{\cdot}6 \times 10^{-8} \text{ cm}^3 \text{ sec}^{-1}.$$

At Khartoum it was possible to obtain a direct measure of the rate of electron production which was in agreement with α' and q_0/α' measured separately.

A relation between the recombination coefficients and the critical frequencies in the E and $F1$ layers allows the ratio α'_E/α'_{F1} to be determined from the changes in critical frequencies which occur either diurnally or during an eclipse.

Appendix

Let the electron density in the E and $F1$ layers during an eclipse be given by an expression of the form

$$\frac{dN}{dt} = A^x q_0 \cos^y \chi - \alpha N^2 \tag{1}$$

where N = electron density,
q_0 = rate of production of electrons,
α = effective recombination coefficient,
A = fraction of ionizing radiation not obscured by the moon,
χ = solar zenith angle.

For a Chapman layer, $x = y = 1$, but, owing to causes which are not yet fully understood, it is found that $y \simeq 0{\cdot}9-1{\cdot}4$ in the E and $F1$ layers. The value of x specifies the relation which would connect A and N for a constant value of χ, and it is much nearer to unity than y.

Rearranging the terms in equation (1), we obtain:

$$A^x \left(\frac{q_0}{\alpha}\right) = \left(N^2 + \frac{1}{\alpha}\frac{dN}{dt}\right) \sec^y \chi \tag{2}$$

Further, at noon when $dN/dt = 0$, equation (1) gives:

$$\left(\frac{q_0}{\alpha}\right) = N_0^2 \sec^y \chi_0 \tag{3}$$

From equations (2) and (3):

$$A = \left(\frac{N}{N_0}\right)^{2/x} \left(\frac{\sec^y \chi}{\sec^y \chi_0}\right)^{1/x} \left(1 + \frac{1}{\alpha N^2}\frac{dN}{dt}\right)^{1/x} \tag{4}$$

This expression can be used to compute A as a function of time during an eclipse for comparison with the simple geometrical function. In principle it could also be used to compare the ratio of the fractional obscurations in E and $F1$ layers (A_E/A_{F1}), but in practice the accuracy of measurement of dN/dt is not usually sufficiently high to allow firm conclusions to be reached. An indirect argument can, however, be used to throw some light on this point. During several recent eclipses, in which simultaneous accurate measurements of f_0E and f_0F1 have been made, it has been found that the correlation between the critical frequencies is very high, and this can only be so if there is a correspondingly high correlation between A_E and A_{F1}. It seems reasonable, therefore, to conclude that the locations and relative intensities of the E- and $F1$-layer sources of ionizing radiation are the same, and consequently to assume that during an eclipse $A_E/A_{F1} = 1$.

For the eclipses of 1952, 1954, and 1955, it has been found that the relations between f_0E and f_0F1 can be represented by straight lines passing very close to the origin, and it can be shown that such relations have important implications concerning the ratio of the recombination coefficients in the layers.

Let

$$N_E = \mu N_{F1} + n \tag{5}$$

$$\alpha_E = \lambda \alpha_{F1} \tag{6}$$

$$x = r, \qquad y = s \; (E \text{ layer}) \tag{7a}$$

$$x = u, \qquad y = v \; (F1 \text{ layer}) \tag{7b}$$

where: the subscripts E and $F1$ denote respectively the E and $F1$ layers,
x and y are defined in equation (1),
μ, λ are constants,
and $n/\mu N_{F1} \ll 1$.

If second-order quantities are neglected, it is possible to represent the ratio A_E/A_{F1}, using equations (4), (5), (6), and (7) as follows:

$$\frac{A_E}{A_F} = C\,\frac{\left\{1 + Z\Big/\left[\lambda\mu\left(1 + \dfrac{2n}{\mu N}\right)\right]\right\}^{1/r}}{(1 + Z)^{1/u}} \tag{8}$$

where

$$Z = \frac{\mathrm{d}N}{\mathrm{d}t}\Big/\alpha N^2 \tag{9a}$$

$$C = \frac{(N^2 \sec^s \chi)^{1/r}\,(N_0{}^2 \sec^v \chi_0)^{1/u}}{(N^2 \sec^v \chi)^{1/u}\,(N_0{}^2 \sec^s \chi_0)^{1/r}}\left(\frac{1 + n/\mu N}{1 + n/\mu N_0}\right)^{2/r} \tag{9b}$$

and $N = N_{F1}$, $N_0 = N_{0F1}$.

Since $s \simeq v$, $u \simeq r$, and $n/\mu N \ll 1$, C can differ little from unity. Also, since $\sec \chi$ does not change much during the eclipses under consideration, C remains almost constant.

If we now put $C = 1$ and assume, for reasons explained earlier, that $A_E/A_F = 1$ throughout the eclipse, then from equation (8), assuming $r = u$

$$\lambda\mu\left(1 + \frac{2n}{\mu N}\right) = 1. \tag{10}$$

The experimental evidence on the magnitude of n is insufficient to show whether the true value of n differs from zero, but the observed magnitude of $2n/\mu N < 0{\cdot}1$. The conclusion is therefore that $\lambda\mu = 1 \pm 0{\cdot}1$, and it follows from equations (5) and (6) that:

$$\alpha_E/\alpha_{F1} = q_0F1/q_0E = \sqrt{(q_0/\alpha)_{F1}/(q_0/\alpha)_E} = b^2/a^2$$

where a and b are defined by

$$f_0E = a \cos^m \chi; \quad f_0F1 = b \cos^n \chi.$$

Addendum

The values of α'_E and α'_{F1} derived from the eclipse measurements made at Khartoum are based on the assumption that the intensity of the ionizing radiation falls to zero during totality. Following the discussion, at the symposium, on the structure of the sun's atmosphere, it now seems to be necessary to admit at least the possibility that complete obscuration may not occur during totality. If there is some residual radiation at second contact, equation (5) ought to be used to compute α' instead of equation (1), but, since A' is unknown, it is not possible to obtain a unique value for α'

Table 4. Effective recombination coefficients, Khartoum, 25 February 1952

Residual radiation at second contact (A')	$\times 10^8$ cm³ sec⁻¹ α'_E	α'_{F1}
0	1·5	0·8
0·05	2·6	1·6
0·10	9·2	(14)
0·11	—	∞
0·12	∞	—

In Table 4 several different values of the residual radiation at second contact have been assumed, and it can be seen that, provided this does not exceed 5 per cent of the total radiation, the order of magnitude of the α' values for $A' = 0$ remains correct. The α' values for total obscuration at second contact must now be regarded as lower limits; it is not possible to quote upper limits at present, but a further inter-comparison of the results obtained at several stations in 1952 might allow more information to be obtained on the minimum possible intensity of the bright sources and hence on the maximum value of α'. The correction of the results obtained during the partial eclipse at Inverness requires a more detailed investigation.

Solar Eclipses and the Ionosphere (Special Suppl. Vol. 6 to J. Atmosph. Terr. Phys.), Pergamon Press, London, 1956

Recent Laboratory Studies of Recombination Cross-sections

J. SAYERS
Electron Physics Department, University of Birmingham

Abstract—A special tube has been developed which, in addition to measuring the rate of disappearance of ions or electrons in an afterglow following ionization, incorporates an R.F. mass spectrograph which identifies the ions at the same time. Provisional results of rate coefficients are given for oxygen and nitrogen for electron temperatures of a few thousand degrees.

WE have very recently been making laboratory studies of recombination, using equipment which has gone through a long period of development.

It appeared necessary for any reliable information on recombination in laboratory work, to identify the ions as well as measure their rate of disappearance,

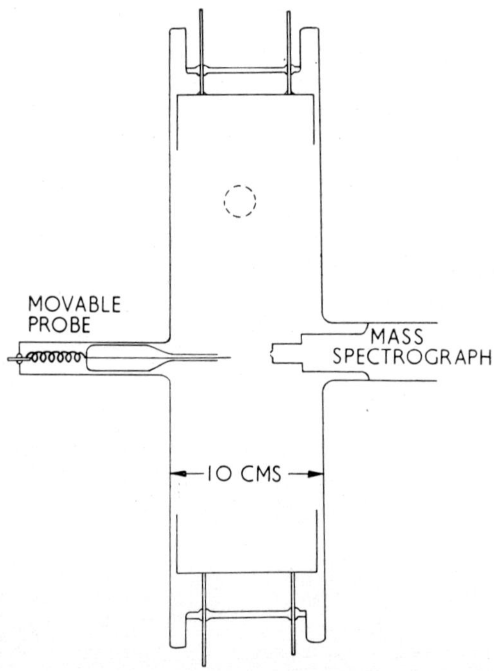

Fig. 1.

and this required the building of a tube which contained not only a probe for sampling ion density but also a miniature mass spectrograph for analysing the ion types. The tube is illustrated diagramatically in Fig. 1. The movable probe is made of platinum wire 0·002 inch in diameter, and is used for recording Langmuir current-voltage characteristics from which we can calculate both the electron temperature and the electron density. A miniature radio frequency mass spectrograph on the opposite side from the probe enables the ion-mass spectrum to be determined at the same time as the ion or electron density measurements.

In making experimental studies relative to the upper atmosphere, it is desirable to operate at the appropriate pressure, of the order of 10^{-3} mm. At this pressure, however, diffusion of ions and electrons will be very rapid, so to reduce diffusion,

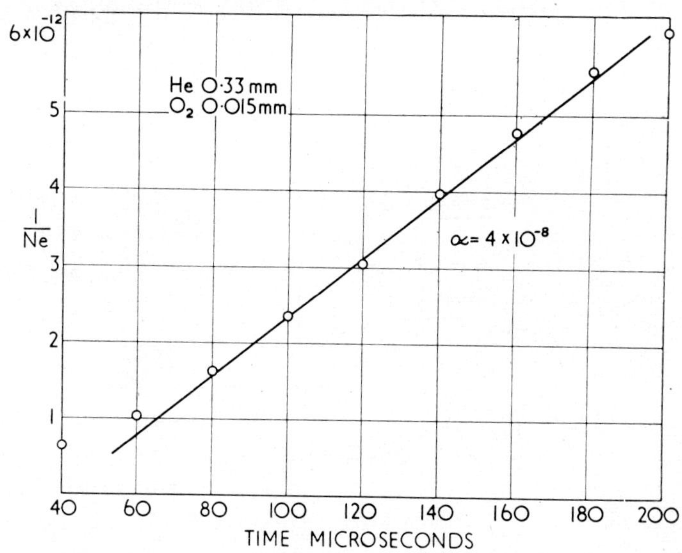

Fig. 2.

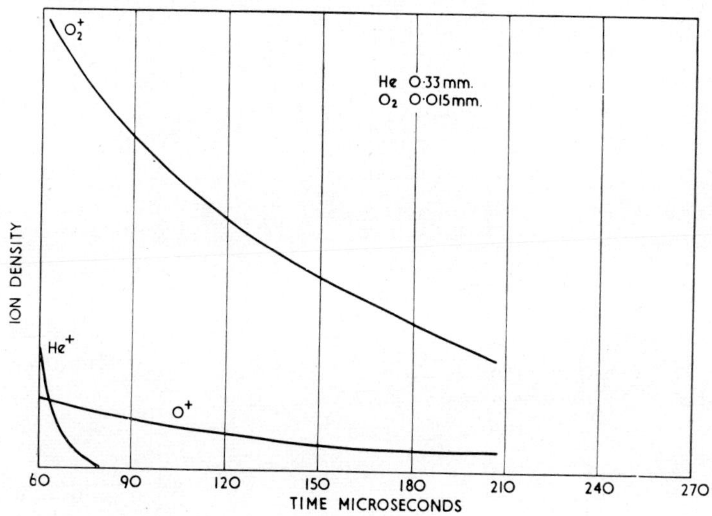

Fig. 3.

the atmospheric gases at the appropriate pressure (oxygen and nitrogen have been studied so far in the present series) are mixed with an inert gas, helium or argon, and we are able to discover by the use of the mass spectrograph what part the inert gas is playing in the ionic processes.

The gas mixture in the tube is subjected to a 20-μsec pulse of R.F. power at 50-cycles repetition frequency, which produces the ionization. The electron- or ion-density and ion-mass analysis is made in the afterglow following such pulses. Now, if the rate of decay of electron density follows a recombination law, this will give a linear relation between the reciprocal of the electron density and time in the afterglow. The relation is illustrated in Fig. 2, for oxygen at a partial pressure of 0·015 mm and an electron temperature of approximately 2500°K. Fig. 3 shows the results of the mass spectrograph analysis in the same afterglow. The ions are mainly O_2^+ with a small quantity of O^+ and a short-lived He^+ population. The electron and O_2^+ ion populations are therefore approximately equal (since the medium must be electrically neutral) and apart from the small effect of diffusion, the ion decay follows the recombination law as shown in Fig. 2.

If we now reduce the oxygen partial pressure, the proportion of O^+ ions increases, and at a partial pressure of 0·005 mm oxygen we get O^+ as the main ion type at the beginning of the afterglow. The O^+, however, disappears by ion exchange, according to the reaction

$$O^+ + O_2 \rightarrow O_2^+ + O$$

with the formation of O_2^+. The O_2^+ meanwhile is being removed by dissociative recombination, and on balance the O_2^+ population changes slowly in the afterglow. The electrons, recombining with a nearly constant O_2^+ population, will decay according to the equation

$$\frac{dNe}{dt} = -\alpha Ne\, O^+$$

which will simulate an attachment law. In this case, a graph of $1/Ne$ against time is no longer a straight line, but if we plot log Ne against time a linear relation is obtained. This is indeed the ion-exchange reaction modifying the electron decay to an attachment-like law which BATES has mentioned in relation to the *F2* layer in his paper. The accuracy of the results for the O^+ decay is not yet as good as we would wish, but the rate-coefficient is undoubtedly of the order of 10^{-10}. Nitrogen has also been investigated. Here the analysis of results is much more accurate. In the first place, we have no negative ions present, and there appears to be no ion exchange with the N^+—as indeed we would expect, since the ionization potential of N_2 is higher than that of N.

For N_2^+ we have measured a dissociative recombination coefficient of $1{\cdot}1 \times 10^{-7}$ cm^3 sec^{-1} for an electron temperature of 3200°K.

DISCUSSION OF PAPERS IN SECTION V

Chairman: Mr. J. A. RATCLIFFE

Mr. LYON — It is well known that α can be obtained from the diurnal curve for fE or the electron density N using the formula $\alpha = \frac{dN/dt}{N \cdot \Delta N}$ given by APPLETON many years ago. ΔN is the difference between the values of N at two times symmetrical about noon. We find that because of irregularities in the value of ΔN it is very difficult to get values of α by this method. However, we have found that the irregularities seem to be due mainly to tidal perturbations associated with the S_q system, which are different in sign north and south of the current focus round about 35° latitude. If we take curves near the current focus, then the variation of ΔN and the other parameters in this equation are fairly regular, and we can get consistent values of α near the current focus. The average value of α obtained is $1{\cdot}0 \times 10^{-8}$ cm^3 sec^{-1}. I shouldn't place very great reliance on the figure after the decimal point, but I would certainly say that it is not very far out.

Prof. BATES — I think the essential trouble is that the symbol α is used both for "effective recombination coefficient" and for "average recombination coefficient." Now, the α that is obtained by the method indicated by Mr. LYON is an *effective* recombination coefficient, that is, it describes the variation of electron density throughout the day. If we consider two ions—we call them A and B, and denote the concentration of A by the symbol A, etc.—then we have:

$$dA/dt = q(A) - \alpha(A) \,.\, A \,.\, n(e)$$

$$dB/dt = q(B) - \alpha(B) \,.\, B \,.\, n(e)$$

Suppose that these two ions are being formed at approximately the same rate, but that for some reason $\alpha(B)$ is very much larger than $\alpha(A)$. Then calculation shows that, provided the difference between $\alpha(B)$ and $\alpha(A)$ is large enough, the coefficient obtained by this method is, in effect, $\alpha(A)$ and the q is $q(A)$. This is as far as the diurnal variation is concerned. If one considers the eclipse effect, then if the measurements are sufficiently precise and if there is no complication due to active areas on the sun, there is a possibility of being able to get $\alpha(A)$ and $\alpha(B)$. At the moment I think it is misleading to speak of anything but an *effective* recombination coefficient. Many of the solar physicists have been misled into thinking that *effective* recombination coefficient and *average* recombination coefficient are the same. I think it is probable that the average recombination coefficient may well be two or three or even more times larger than the effective recombination coefficient.

Mr. RATCLIFFE — What is known about the recombination coefficients of O_2^+ and N_2^+? It has been suggested that both these ions may exist in the E region.

Prof. SAYERS — The afterglow measurements which I described indicated that the N_2^+ recombination coefficient is round about 1×10^{-7}. We are not quite so sure about the value for O_2^+—the experimental work has only recently been carried out, and the analysis is not quite complete.

Mr. RATCLIFFE — It appears that the values of α that Professor SAYERS has given are high compared with the usual ionospheric values.

Prof. BATES — I believe the temperatures are very different.

Prof. SAYERS In our laboratory experiments, the electron temperatures were around 3000°K. From our results so far we are not able to estimate the variations with temperature. We might be able to do this, but it would be technically very difficult.

Dr. RAWER It should be noted that, in the evaluation of α' from eclipse results, it is incorrect to apply the simple theory (equations 1 and 2 in my paper), since we know from the usual diurnal measurements that in practice the exponent m does not have the theoretical value 2.

Dr. MENZEL We have done a lot of work on the value of the exponent n (in $fE^n \propto \cos \chi$.) This is described in a paper submitted to this symposium. The value of 4 for a simple Chapman isothermal layer is altered if there is a temperature-gradient. If the temperature increases upwards, then n is less than 4; and if the temperature decreases upwards, n is greater than 4. Our analysis suggests a value of about 3 for the equator, and that this increases to about 4 at the poles.

Mr. RATCLIFFE But in that analysis you are assuming there are no movements in the region.

Dr. MENZEL That is so. Our treatment is empirical.

Dr. BEYNON At Swansea Mr. G. M. BROWN and I have made an extensive study of region-E data, and I might mention one point concerning this index n. In the complete continuity equation

$$\partial N / \partial t = q - \alpha N^2 - \text{div}\,(N\boldsymbol{v})$$

the effect of movement is represented by the term div $(N\boldsymbol{v})$, where $\boldsymbol{v}$ is the vector drift-velocity. The assumption $fE^n \propto \cos \chi$ is strictly only applicable to noon or near-noon conditions, and to cases in which the movement term div $(N\boldsymbol{v})$ is negligibly small. In our analysis we adopted the approach that some of the known variations in the exponent n might result from neglect of the movement term. This is an indirect way of looking at it. We have examined n, using monthly mean noon fE data for the three years near sunspot minimum 1952–54 for more than fifty stations, and obtained some quite interesting results. Thus, if we consider the variations in n along a line of longitude, we find that although there is a general increase from $n \sim 3$ at the equator to $n \sim 4$ at high latitudes (similar to that found by Dr. MENZEL for n determined from the diurnal variation), there are marked singularities around latitudes $\pm 30°$. We suggest these singularities may be associated with the S_q current systems which have foci near these latitudes. We think that the index may indeed be a good index of movement in region E, and that these results indicate very clearly that one must take account of movement-terms in considering anything to do with region E.

Dr. MANGE I should like to emphasize that we *do* know that in the E region there is an increasing scale-height with altitude, and we must consider this result in making further calculations on the subject.

Mr. RATCLIFFE It is clear that these points are important as far as eclipse results are concerned. In dealing with eclipses we are tending to use the power 4 always, and we need to consider how far this is going to lead us astray in our estimates of α'. Scale-heights may, or may not, change; we might possibly have to revise our theoretical ideas on that, but unless the experiments have gone wrong, the fact is that the value is not 4.

Mr. LYON I am extremely interested in Dr. BEYNON's results obtained for the index n, because in Edinburgh we obtained a very similar result which I think must be equivalent to this, by using ΔN, the difference between values of N at two points symmetrically placed with respect to noon. We find that the values of ΔN show a latitude variation with a reversal of sign near the current foci. I think that the results presented by Dr. BEYNON and our own must in effect be the same. The values of n, as calculated by Dr. BEYNON, are determined in such a way that they include the effect of movements. We tried to eliminate the effect of movements from N by taking not fE itself, but the average $(\overline{fE})$ of the values at two times symmetrically placed with respect to noon. Now, when we do that, we find that the values of n are very consistent at all latitudes. This is something that other people have not found, but I think the reason must lie in this fact that we have taken $\overline{fE}$. Whereas other people seem to have found that n varies with latitude, we find that the value is very consistently $\frac{1}{3}$. It may have slight variations between 0·30 and 0·36, but not consistent ones so far as latitude is concerned, and no very large variation over the sunspot cycle either. I think it is about 0·02 less at sunspot-maximum than at sunspot-minimum.

Dr. BEYNON By using the value $\overline{fE}$ instead of dealing with a.m. and p.m. separately you try to eliminate dN/dt?

Mr. LYON Yes.

Dr. BEYNON But I think that this method does not get rid of it completely, since dN/dt is not of exactly the same magnitude at times symmetrical about noon.

Mr. LEPECHINSKY At noon the layer is lower than at other hours, and we may need to consider the effect of collisional frequency.

Mr. PIGGOTT It has always seemed to me that n really depended on how the layer moved during the day. Hence, in so far as eclipses are concerned, the crucial question is: is the layer during an eclipse always at the same height or not? If the levels to which measurements refer remain unchanged, I don't see any reason why one should not accept the simple αN^2 type of law. I feel that probably the errors due to ignoring the changes of height would be at least as great as, if not greater than, the troubles involved in the breakdown of the diurnal cos χ variation.

Mr. RATCLIFFE There is, of course, another point which I suppose we must not forget; the movements in the S_q region are due to the interaction of the magnetic field of the earth and the current in the region. The eclipse alters the current, and so the eclipse possibly alters the movements, and the problem is very complex.

Dr. BEYNON We find that n determined from the diurnal variation of fE varies regularly with season, showing maxima near the solstices and minima near the equinoxes. This may again find an explanation in terms of the S_q current system.

Mr. RATCLIFFE In his paper, Dr. BURKARD suggested that there might be considerable temperature changes during the eclipse and that these might affect the exponent.

Dr. BURKARD I find that one can explain the diurnal variation and also the variation during an eclipse on the assumption that the recombination coefficient is a function of pressure and of temperature.

Dr. MENZEL I have also carried out a general analysis assuming that the recombination coefficient varies with pressure.

Mr. RATCLIFFE In Dr. BURKARD's paper, temperature changes of 20% are suggested to occur during an eclipse. What is known about the thermal time-constants of the atmosphere at these levels?

Prof. BATES I think the time-constant will be much too long to expect anything but a very small effect during an eclipse.

Prof. CHAPMAN It seems to me that the temperature change required is too large and unlikely to occur.

Dr. RAWER I think an eclipse (such as that of 1954) when solar activity is very low, provides one of the best occasions for the determination of α'.

Mr. MINNIS I haven't seen Dr. RAWER's results, but even for the 1954 eclipse, when solar activity was at a minimum, it was not possible to assume a perfectly uniformly radiating disk.

Mr. RATCLIFFE I think it would be important to discuss for a short time whether we can rule out an important contribution from the corona. Does anyone feel that they could demonstrate from eclipse data that there is *not* 10% radiation coming from the corona?

Mr. PIGGOTT In my opinion the radio evidence, in so far as it can be used at all to discuss the corona (and this is by no means too sound), gives two limiting conditions. You can assume a uniform sun with intense bright patches on it or a more uniform sun with a bright ring round it. In the 1952 eclipse the best solar model indicated about 13 to 14% of the total photon emission coming from a bright ring overlapping the visible edge of the sun. I do not feel that the corona proper is involved. I think that the ionospheric information is consistent with there being real limb-brightening.

Mr. MINNIS Mr. PIGGOTT's analysis is based on the $J(t)$ relation as determined using a small value of α', which was consistent with the assumption that at Khartoum no ionizing radiation came from outside the disk. Now, if we assume that there is such a source of radiation, we must use a much greater value of α', and this would lead to different conclusions regarding the amount of limb-brightening at Ibadan.

Mr. RATCLIFFE Your analysis, of course, assumes there is no corona to get your α'; if you had not assumed that, you would have got a different value of α'.

Mr. MINNIS Yes, the Khartoum value of α' ($1{\cdot}5 \times 10^{-8}$) assumes no radiation at totality.* Mr. PIGGOTT's analysis is sound in so far as the existence of a bright ring is concerned, but since the eclipse was partial, it cannot be used to decide whether this ring is on the disk or in the corona. What we are discussing is the existence of coronal radiation beyond the limb.

Mr. PIGGOTT The Khartoum data are crucial. The maximum intensity of a ring whose radius is equal to that of the visible sun is 15 per cent of the total radiation, whereas the maximum intensity for a ring of slightly greater radius is considerably more. The greater part of this ring must, however, be eclipsed or there are difficulties in matching the results at the two stations. I am, therefore, considering a bright ring which can only be slightly larger than the visible sun.

Mr. RATCLIFFE A survey of all eclipses, as in my paper, shows an unexpectedly large number in which the minimum of electron density comes during visible totality. The visible curve goes really to zero, and the electron-density minimum

* *Note added November* 1955.

To explain the Khartoum and Ibadan eclipse changes in the E layer it is necessary to postulate the existence of three sources of ionizing radiation, viz.: the uniform disk of the sun, and the two isolated bright sources near the E and W limbs which may or may not include coronal radiation. The fraction (I/I_0) of the total radiation attributable to each source can only be stated if α'_E is known, but it has been found possible to obtain expressions of the form $I/I_0 = a + b/\alpha'_E$ for each source both at Khartoum and at Ibadan. Equating the pair of expressions corresponding to each source gives three simultaneous equations, which can be solved for α'_E (Khartoum) and α'_E (Ibadan). The difference between these is found to be negligible, and the final value for both stations is $\alpha'_E = 1{\cdot}2 \times 10^{-8}$ cm^3 sec^{-1}.

Such a low value implies that there can have been no residual radiation during totality at Khartoum. This was the assumption made earlier which, using another method, gave $\alpha'_E = 1{\cdot}5 \times 10^{-8}$ cm^3 sec^{-1}, and this value is therefore justified. However, the new method contains a smaller experimental error and, further, it makes use of a larger proportion of the experimental data at both stations; for these reasons, the new value of α'_E is preferred.

occurs during that time and ought to drop to zero, but it fails to do so by about 10%. This happens so often that one suspects that, as PIDDINGTON has suggested, there is always 10 to 15% of the total ionizing radiation coming from beyond the visible disk. Now, an analysis of the type given by MINNIS and PIGGOTT involves values of α' which are calculated on the assumption that there is no residual ionizing radiation at totality.

Mr. LANDMARK I should like to say that in our calculations we also made the assumption of no radiation at totality, and deduced a value of $\alpha'_E \sim 3$ to 4×10^{-8}.

Dr. BECKER All the observations discussed are dependent on obtaining accurate values of fE, and one questions whether it is always possible to do this. The effects of sporadic E are often difficult to eliminate.

Dr. RAWER I think that BIBL has shown how one can get a fairly satisfactory value of fE. For the eclipse of 1954 some difficulty arose due to the presence of Es ionization. During that eclipse, measurements at Lindau and at our stations showed a decrease of fE before first contact. Now, we cannot be sure about this, because of the presence of sporadic E making it necessary to extrapolate the trace.

Dr. BARTELS From our experience with geomagnetic data it is always important to consider possible lunar influences. This applies to phenomena observed near the time of solar eclipses. It may sometimes be wrong to attribute all the observed effects to the sun.

Mr. RATCLIFFE As I understood it, the analysis of Professor BATES suggests that, when different sorts of positive ions are present, each one determining a recombination coefficient, those having the largest recombination coefficient will disappear first and then there will be a reduction in the net recombination coefficient. Therefore, one would expect the recombination coefficient to vary with time. This is an extremely important point. In the past a variation of recombination coefficient with time has generally been considered in terms of λ (the ratio of electrons to ions). Furthermore, since the work of BATES and others has shown that λ is probably very small, most ionospherists concluded that it was no good thinking about an α varying with time. Professor BATES now suggests another explanation whereby the recombination coefficient might vary with time.

Prof. BATES For a typical eclipse one would expect that in the initial stage the most rapidly recombining ions presumably are the N_2^+ ions. These would disappear first, leaving perhaps O_2^+ ions or O^+ ions, which disappear more slowly. An analysis of the complete decay-curve (assuming no complications due to possible variations of activity on the solar disk) would show that the recombination coefficient began at a large value and ended with rather a small value. That is the case in the E layer. In the $F1$ and $F2$ layers the position is perhaps rather more complicated, because in these layers there are two types of decay mechanism. The two decay laws are entirely different—the one is proportional to N_e and the other is proportional to N_e^2. The one case gives an exponential decay, and the other gives a $1/\alpha N_e$ type of decay. Has anyone any indication at all that α appears to be greater at the beginning of an eclipse than at later stages?

Mr. RATCLIFFE Several workers have pointed to the fact that the predominant peculiarity about eclipses is that the ionospheric eclipse starts too quickly. Many years ago PIERCE and HIGGS and others tried to analyse results assuming an α' varying with time. I wonder if anyone recently has considered a similar approach.

Mr. PIGGOTT We tried changing α' for the Ibadan eclipse, but it did not help in the least. The rates of change are so small that no certain conclusions concerning this point could be drawn.

Mr. MINNIS That applies also to eclipses which are nearly total, because it is only near the centre part of the eclipse that one can determine anything about α'.

Mr. RATCLIFFE I think that we might get all those workers who feel that they have a really good result on the E region to consider three extreme standpoints: (a) To what extent could their results be explained in terms of an α' which is changing during most of the eclipse? (b) How far could they get by assuming a corona contributing something of the order of 10%? (c) How far could they get by assuming α' constant and spots only on the visible disk? Consideration has already been given to (c), but the results need to be re-examined in terms of (a) and (b).

Mr. MINNIS If we assume that the E layer is partly caused by radiation from the corona, then according to Dr. WALDMEIER's diagrams we would expect a bright ring at coronal heights. This would imply a much larger value of α' at Khartoum than that calculated, assuming no coronal radiation. This shows that, if both the intensity of the coronal radiation and the value of α' are unknown, there is a wide range of possibilities in the interpretation of eclipse results.

Mr. RATCLIFFE There is a real distinction between limb-brightening, however intense, which *is* covered at totality and coronal radiation, however weak, which is *not*. I am sure this is a real difference which the solar physicists want us to disentangle.

Mr. MINNIS Even during a total eclipse, as at Khartoum, it is impossible to say how much comes from the corona unless you have some idea of the magnitude of α'. During a partial eclipse it appears to be impossible to distinguish between a bright ring on the disk and one in the corona.

Mr. RATCLIFFE When the disk is really covered and there is a minimum at totality, one can then find out something about the radiation at totality.

Mr. MINNIS In the Khartoum case, if you assume that there was, say, 10% during totality, then the brightness of the patch on the disk drops to something negligibly small, and that does not tally with Professor WALDMEIER's curves.

Solar Eclipses and the Ionosphere (Special Suppl. Vol. 6 to J. Atmosph. Terr. Phys.), Pergamon Press, London, 1956

SECTION VI

GEOMAGNETIC ECLIPSE PHENOMENA

Survey of Geomagnetic Eclipse Phenomena

S. Chapman

Geophysical Institute, College, Alaska,
also High Altitude Observatory, Boulder, Colorado*

Abstract—Consideration is first given to the so-called corpuscular eclipse effect. In view of recent evidence that the charged particles responsible for aurorae penetrate to the E level by means of some accelerating process near the earth, dependent on the magnetic field of the earth, a corpuscular eclipse due to neutral particles is not to be expected in the E layer. Any (weak) corpuscular eclipse effects are likely to be confined to the high layers of the atmosphere.

Secondly, the effects of an eclipse on the earth's magnetic field are considered. It is shown that, on an idealized picture of the S_q current sheet under eclipse conditions, a substantial reduction of electric conductivity in a limited part results in only a small change of field at the surface of the earth.

The Corpuscular Eclipse

In 1932, Professor (now Sir Edward) Appleton came to discuss with me some observations of the English eclipse of 29 June 1927. He found that the minimum of the electron-density peak occurred about four minutes after the optical centre of totality. At that time it was uncertain whether ultra-violet radiation or corpuscular radiation was the chief cause of ionization. He noticed the coincidence between the time 4 min, or 240 secs, and the time it would take for particles moving with the speed of 1,000 miles per sec to traverse the distance from the moon to the earth (240,000 miles). This suggested that corpuscles might be the principal cause of the ionization in the E layer. It led me to look into the question of what happens to a corpuscular stream occulted by the moon, and I briefly recount what was found (Chapman, 1932).

The earth-moon system travels round the sun in a nearly circular orbit, with a speed of 29·80 km/sec. The moon describes a less-nearly circular orbit relative to the earth, with a mean orbital speed 1·02 km/sec. Thus even when the moon is on the sunward side of the earth, and relative to the earth is moving backward (opposite to the earth-moon system's orbital motion), the moon's motion is still forward. (The proportions of the sun-earth-moon system are such that the moon's orbit is always concave to the sun.) Just as an umbrella carried by a man walking in rain provides a dry "shadow" sloping backwards, so the moon will cast a shadow devoid of corpuscles pouring outward from the sun, and this shadow will slope backwards, lagging behind the moon's onward motion. But relative to the earth, when the moon is moving backwards towards its position at new moon, when it may eclipse the sun, this lagging shadow free from corpuscles will reach the earth before the optical shadow does, because the latter lags scarcely at all, so great is the speed of light. Hence the moon will cause what I called a "corpuscular eclipse," before it causes the optical eclipse. With the corpuscular speed at that

* Now engaged in a joint research programme with the High Altitude Observatory and the National Bureau of Standards, at Boulder, Colorado.

time (1932) thought probable, namely 1000 miles or 1600 km/sec, the moon's mean forward speed, reduced to 28·8 km/sec near new moon, would cause the corpuscular shadow to lag by 28·8/1600 radian, or 1·03°. If the corpuscular speed is only 500 km/sec, as may be so for the supposedly slower particles that cause moderate magnetic disturbance, the lag would be 3·3°. The distance between the optical and corpuscular shadow-centres at the earth's distance (384,400 km) from the moon would be 6920 km for the faster particles; hence the shadow in the beam (for the faster corpuscles) would reach the earth one hour and fifty-three minutes before the optical shadow; for the slower particles the time would be about six hours.

As Dr. Jackson pointed out in the 1932 Geophysical Discussion on this subject at the Royal Astronomical Society, the difference in time would also be accompanied by a considerable change in the position on the earth where the eclipse occurred, because of the considerable amount by which the earth turns during the two hours or more by which the optical eclipse is delayed after the corpuscular eclipse. So the fact that the eclipse of the ionosphere is observed during the optical eclipse is clear evidence that a major part of the ionization is due to ultraviolet radiation, without any concern about small details such as a lag of the minimum ionospheric electron density by a few minutes or seconds after the centre of optical totality.

At the time, however, much attention was paid to the corpuscular eclipse (APPLETON and CHAPMAN, 1935), and then and since that time there have been some who thought that they had identified it (NETHERLANDS P.T.T., 1955). In 1932 it was thought that the corpuscular eclipse would affect the E layer, because the particles (neutral particles, which would come unimpeded onto the earth) were supposed to be travelling at the same speed as the charged particles from the sun, which produce aurorae and magnetic storms; and we know that these in the auroral zone come down to the level of the E layer. As early as 1919, E. A. MILNE and I had pointed out that the speed of 1600 km/sec seemed quite inadequate for particles to get down to that level, and in 1932 this was an unsolved problem (and really still is). The speed required for penetration of auroral particles down to the E level exceeds by a factor of 10 or more the speed reckoned from the travel-time for particles to pass from the sun to the earth (as inferred from the lag of magnetic-storm commencements behind solar events, when one can correlate outbursts on the sun and sudden commencements of magnetic storms). But although this problem was unsolved, it was nevertheless thought that the neutral particles would have the same penetrating power as the charged particles, and so one might expect a corpuscular eclipse in the E layer. Since that time views have changed, and now we know (through the work of MEINEL) that the particles come into the auroral zone with speeds greatly in excess of 1600 km/sec: hence it is now thought that there must be some accelerating process in the neighbourhood of the earth, which raises the speed of the charged particles that come into the earth's atmosphere to several times beyond that with which they travel from the sun.

Any such process will certainly depend on the magnetic field of the earth and the electric charges of the particles; but the magnetic field of the earth will not affect the neutral particles at all, and so, if there should be a substantial quantity of neutral particles in the solar stream along with the charged particles that produce magnetic storms, these particles will not be so accelerated, and will impinge on the atmosphere with the speed of 1600 to 500 km/sec; hence they will be stopped high up in the atmosphere, and so one would not expect a corpuscular eclipse at all in the E layer. Since that time KAHN has further discussed the question of the degree of ionization of the solar streams, and has found reason to suppose that the proportion of neutral particles in the stream will be extremely low; therefore, I think it is likely that, if there are corpuscular eclipses, they are very weak, and only affect a high layer in the atmosphere.

Effect of an eclipse on the earth's magnetic field

This is a matter which has been studied for several decades; one of those most active in this field was the late Dr. L. A. BAUER, the first director and inspirer of that rightly celebrated institution, the Department of Terrestrial Magnetism of the Carnegie Institution of Washington. BAUER sent many expeditions to places where there were solar eclipses, to try to observe the effect of eclipses on the earth's magnetic field. He thought he had found eclipse effects, but other people questioned whether they were indeed eclipse effects, and not ordinary disturbances of the field, which happened to be going on at this period. It was a difficult question to answer, and at the time there was no theoretical guidance as to what one might expect for the effect of a solar eclipse on the magnetic field. In 1933, my attention having been called to ionospheric eclipses by the discussions with APPLETON, I looked into the matter (CHAPMAN, 1933). It was clear that the reduction in the electron and ion content of the atmosphere during a solar eclipse must reduce the conductivity of the upper atmosphere, and therefore must affect the intensity of those electric currents which produce the transient magnetic variations. The main magnetic field and the secular magnetic variation, a long-continued one, proceed from within the earth; but the transient magnetic variations (those that occur within times of the order of a day or less) are known from harmonic and spherical harmonic analyses (and in other ways) to be produced by electric currents mainly above the earth—although the changing magnetic field of these electric currents above the earth also induces earth currents inside the earth, which make a minor contribution to the observed magnetic variations.

At ordinary times (times not of magnetic disturbance) the main electric currents of which we are aware in the upper atmosphere are those associated with the solar daily magnetic variation. By studying the daily magnetic variations at many observatories over the earth by means of spherical harmonic analysis, the pattern and intensity of the electric currents in the upper atmosphere can be obtained, with only slight uncertainty—the uncertainty being associated with the doubt as to the exact height at which the currents flow. The magnetic data do not indicate this height; it has to be inferred from radio data and the theory of electric conductivity of the ionosphere. The currents probably flow in the E layer. The magnetic data tell us the pattern and the total intensity of the electric current integrated through the ionosphere at any time. Fig. 1 shows this pattern as drawn by Dr. BARTELS from a spherical harmonic analysis which I made. It shows two views—one of the daytime hemisphere as viewed from the sun, and one of the night hemisphere—at the equinoxes, and two similar views at the solstices. The curves indicate the direction of current-flow at each point, and they are drawn so that between adjacent curves 10,000 amperes flow. The currents are strongest over the day hemisphere, and circulate round points, one north and one

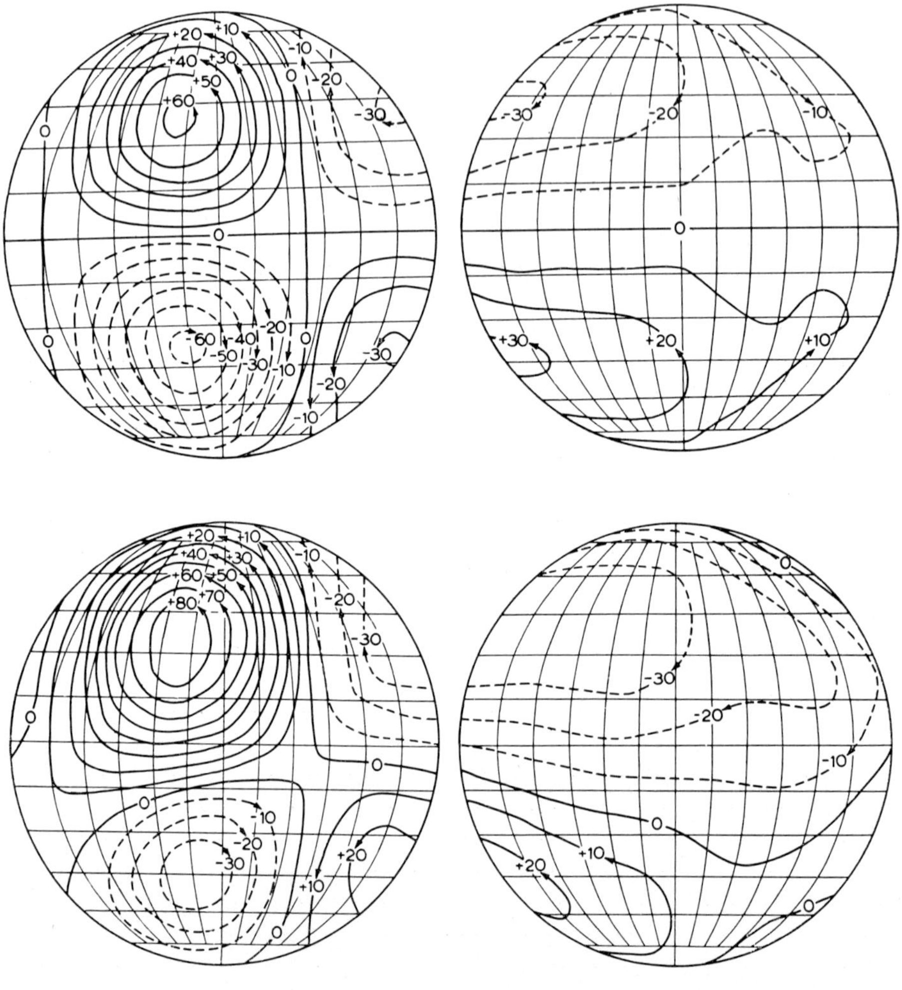

Fig. 1.

south of the equator, in middle latitudes and located on a meridian about an hour before noon; over the night hemisphere they die away considerably. A total of about 60,000 amperes flows round each of the two daytime circuits at the equinoxes. In the northern summer the system is intensified in the northern hemisphere and weakened in the southern hemisphere. There is a similar but less intense system of currents due to the moon; their pattern is governed by the meridian on which the moon lies, and also by the ionization over the sunlit hemisphere.

In order to consider what would be the effect on a current system of the kind described, whose origin we can explain by the dynamo theory, it seemed to me adequate for the first step greatly to idealize the problem in the following way:—

Let the atmosphere be treated as a plane, thin, infinite conducting sheet, and in the first instance let it be supposed to be uniformly conducting all over. This is to illustrate any part of the actual current system that may be affected by an eclipse. The eclipse affects only a small area compared with any one circuit of the current system; over this part of the current system I treated the currents as uniform and flowing through a uniformly conducting medium. Now if in this medium there is an eclipse, then, whereas the conductivity outside is uniformly K, the eclipse reduces the conductivity to some lower value K', again treated as uniform. The radius of the circle of reduced conductivity will correspond to something greater than the circle of totality, which has a maximum diameter of 163 miles; but of course we know that the reduction of conductivity is a continuous process (the case treated is an idealization of what actually occurs, in order to make a simple calculation of order of magnitude). We suppose that there is uniform current-flow until the eclipse occurs and reduces the conductivity over an area treated as circular. Hence, before the eclipse, one supposes that there is a uniform electric field present, driving this current. Supposing that the forces driving the current persist, this uniform electric field will, during the eclipse, have to drive current through a sheet which is uniformly conducting except over the circular eclipsed region. This is a very simple problem, worked out in standard textbooks on electricity and magnetism. If i_0 is the current intensity over the whole sheet before the eclipse, the current when the conductivity is reduced becomes

$$i' = i_0 \frac{2K'}{K + K'},$$

a very simple relation.

If K', instead of being a reduced conductivity, was suddenly made infinite (this does not apply to an eclipse, of course), then i' would be $2i_0$. Thus a very great *increase* in the conductivity would not increase the current by more than a very moderate factor; the area of increased conductivity would be a favoured channel for the current, but still it has to flow into the area and out again, which lengthens the path, and so prevents the increase of current being very great. This has a bearing on the fact that the electrojet over the magnetic equator at Huancayo is only about twice as intense as the currents in the neighbourhood. At Huancayo near the magnetic equator the conductivity is enhanced perhaps tenfold, but this does not quite double the electric current which would otherwise flow.

In the case of an eclipse, K is *reduced* to K'. Suppose for example we take $K' = \frac{1}{2}K$; then $i' = (2/3)i_0$. So if the conductivity is reduced by one-half, the current is reduced by two-thirds. If K'/K is $\frac{1}{4}$, then we get $i'/i_0 = 2/5$, again a moderate reduction. This reduction of the current will be accompanied by a corresponding reduction of the magnetic field of this part of the current at the surface of the ground below; hence the change of the magnetic field under this eclipsed region will be a moderate fraction of the field of these currents themselves. Now the field of the currents themselves is in most places of the order of 40 gammas, and so if this is reduced by say three-fifths, the effective change of the magnetic field will be of amount 24 gammas—modified a little by the fact of the internal currents. Thus we get an estimate of the magnetic change to be expected. It is quite a small change, and certainly not easy to detect in a field which is always liable to small changes otherwise. Mr. EGEDAL is presenting a paper at this symposium dealing with what is actually observed, and making a comparison with my simplified theory. I should stress again that the theory is a highly idealized one, the supposed change of K being taken as discontinuous and uniform over a circle, whereas really there is a gradual change, as we know from the electron-density measurements.

We expect, then, that when an eclipse occurs over any part of the current system, the change in the magnetic field below the currents (that is, at the ground) will be perpendicular to the local direction of current-flow, and of the said order of magnitude. However, if an eclipse occurs at the focus of a current circuit, the eclipse will mainly affect the vertical force—in the centre of the circuit we have vertical lines of force, and we should expect a reduction in the vertical force in a region like that. In the equatorial electrojet one would expect a corresponding reduction in the north horizontal force. At times of magnetic storms there are very much stronger electric currents, certainly partly flowing in the atmosphere, especially in high latitudes, and possibly partly flowing outside the atmosphere. One might apply the same reasoning to such cases, but the difficulty of detecting an eclipse effect in such cases is very much greater, because of the natural variability and irregularity of the field during what we call magnetic disturbance.

Referring to the simple model, when the conductivity is reduced over a certain area, the current-flow tends to be reduced by partly avoiding this region. It does this by producing an electric charge at the boundary of this discontinuity. This electric charge will, of course, have an electric field outside this plane as well as inside it, and this is of interest in connection with the movements in the F layer, because these movements are supposed to be affected very much by the electric fields in the ionosphere (as pointed out especially by MARTYN). It is certain that the actual eclipse effect on the electric flow is more complicated than in my simple treatment, and needs much more mathematical attention, though I think that apart from the effect of this polarization field on the $F2$ layer, that is, from the magnetic point of view, the labour required to construct an accurate theory could better be spent otherwise.

Actually, the atmosphere is not isotropically conducting, as for simplicity I assumed. It is anisotropic, and there are equations such as

$$i_x = k_{xx}E_x + k_{yx}E_y$$
$$i_y = k_{xy}E_x + k_{yy}E_y$$

expressing the relation between the components of the electric-current intensity i_x, i_y and the electric field E_x, E_y. All the conductivity coefficients k are proportional to the electron density n_e, and so they will all be reduced *pari passu* during an eclipse. But the effect of such a reduction in conductivity, even in the simple model case that I took, where n_e was supposed uniform outside the eclipse circle and of reduced amount inside it, will not be so simple as in my treatment; nor will it be so simple to calculate the polarization effects in this case. I think it is worth doing, however, for the sake of the effects on the *F2* layer. Problems of this kind have, indeed, already been discussed by MARTYN (1953).

REFERENCES

APPLETON, E. V., and CHAPMAN, S.	1935	*Proc. Inst. Radio Engr.* (New York), **23,** 658.
CHAPMAN, S.	1932	*Monthly Not. Roy. Astron. Soc.*, **92,** 413.
	1933	*Terr. Mag. Atmos. Elec.*, **38,** 175.
MARTYN, D. F.	1953	*Phil. Trans. Roy. Soc.*, **A246,** 306.
Netherlands P.T.T.	1955	Bibliography, The effects of solar eclipses upon the ionosphere, 2nd ed., No. 768a.

Solar Eclipses and the Ionosphere (Special Suppl. Vol. 6 to J. Atmosph. Terr. Phys.), Pergamon Press, London, 1956

On the Effect on Geomagnetism of Solar Eclipses

J. EGEDAL
Danish Meteorological Institute, Copenhagen

Abstract—While earlier observations of the effect on geomagnetism of solar eclipses have given more or less doubtful results, some recent observations seem to be more promising. L. A. BAUER in an early paper outlined a theory of the phenomenon, and S. CHAPMAN in 1933 worked out a theory, according to which the effect manifests itself as a diminution of the departure of the daily variation. The maximum magnetic effect occurs simultaneously with maximum obscuration, and the diminution of the departure of the daily variation is of the order given by the theory. The necessary international co-operation in this work has already been started.

1. INTRODUCTION

SINCE 1870 the study of the effect on geomagnetism of solar eclipses has been undertaken by several investigators, and most actively by L. A. BAUER. The results were in many cases negative, but BAUER, NORDMANN, and others in some cases obtained promising results. This is also the case for recent studies.

2. THE PHENOMENON

The magnetic effect of solar eclipses is so small that ordinary magnetic disturbances often mask it and make the study of it difficult. L. A. BAUER (1900) expected the effect of a solar eclipse on geomagnetism to be similar to the effect which would be produced if night hours were suddenly interposed among the day hours. In other words, the effect should be a diminution of departures from the mean value for the night hours. S. CHAPMAN (1933) based an investigation on the same assumption and on observations of the reduction of the ionization in the E layer during the solar eclipse of 31 August 1932, and found that a diminution of 28 per cent of the departure from night values of horizontal components should be expected if the currents producing the daily variation are flowing in the E layer.

3. RECENT OBSERVATIONS

During the solar eclipse of 25 February 1952, the Committee to promote observations of daily magnetic variations in low latitudes arranged magnetic observations at Kubbo, Sudan, a place situated in the track of totality and at the magnetic equator (MADWAR, 1953–4). The observations were examined by D. F. MARTYN. A great diminution of the range of the daily variation of the horizontal force H was found, but as only a few instantaneous values of H through the day were at hand, no conclusive result could be obtained.

During the solar eclipse of 30 June 1954, the track of totality passed several magnetic observatories in Europe, and N. AMBOLT and the author compiled magnetograms from eighteen magnetic observatories; but magnetograms from two of these observatories situated in the Arctic region were not used, the magnetic disturbances being too great. From the following observatories, magnetograms were examined: Reykjavik (Iceland); Lerwick, Eskdalemuir, and Abinger (United Kingdom); Dombås (Norway); Manhay (Belgium); Wingst (Germany W); Rude Skov (Denmark); Edh (Värnamo) and Lovö (Sweden); Nurmijärvi (Finland);

Niemegk (Germany E); and Leningrad (Seltsy), Moscow (Pakhra Krasnaya), Odessa (Stepanovka), and Tiflis (Dusheti) (U.S.S.R.). Only records with normal speed have been used, and only the magnetic declination D has been examined.

4. The Treatment of Observations

The effect being rather small, special methods have to be used in order to detect its existence. In most cases the magnetogram for the day of the solar eclipse has been compared with those from the preceding and succeeding days. However, using this method may lead to difficulties, first from the day-to-day change of the daily variation, and secondly from lunar effects, especially when other days than the consecutive ones are applied. Gama (1948) has used the method mentioned in a modified form, so that the influence of lunar effects is reduced. If the classic method had been used in the treatment of the observations during the solar eclipse of 30 June 1954, no useful result would have been obtained, as the range of the daily variation of D on 30 June was 30 per cent larger than the mean range of all days of June 1954. A new procedure has therefore to be applied.

V. H. Ryd (1917) has worked out a method appropriate for an adjustment of daily magnetic variations given by hourly values, and, using this method, it has been possible by interpolation to derive normal hourly values of the magnetic element considered during the solar eclipse, and to draw a smooth curve, based on the adjusted hourly values, representing the normal daily variation of the day. All deviations are reckoned from this curve; thus it is used in the determination of the time of the maximum deviation as well as of the magnitude of the deviation during the solar eclipse. The formula used in the adjustment is found in an article by Egedal and Ambolt (1955).

5. The Difference in Time between Maximum Obscuration and the Occurrence of the Maximum Deviation

The difference in time has been determined for D for the eclipse of 30 June 1954, the very favourable conditions making it possible to carry through such a determination. In the determination of the time of maximum obscuration at the magnetic observatories the altitude and the azimuth of the sun have to be considered. The moment of time of the maximum deviation is taken from the records for the period of the solar eclipse, and if therefore even small magnetic disturbances are present, the time determined may be the time of a magnetic disturbance and not of a deviation caused by the solar eclipse. In Table 1 are given the names of the magnetic observatories considered, their latitude and longitude, and the occurrence of maximum obscuration and of maximum deviation.

For the observatories Reykjavik and Tiflis (Dusheti) it has not been possible to trace the effect. The mean difference between the time of maximum obscuration and the time of maximum deviation for the other observatories is:

$$+1{\cdot}6 \pm 2{\cdot}5 \text{ min}$$

and excluding the value for Manhay, possibly affected by magnetic disturbances, the difference is:

$$-0{\cdot}4 \pm 1{\cdot}6 \text{ min}$$

The maximum obscuration and the maximum deviation in the present case seem to occur nearly simultaneously.

Table 1. Occurrence of maximum obscuration and of maximum deviation of the declination during the solar eclipse of 30 June 1954

Station	ϕ	λ	Maximum obscuration G.M.T.	Maximum deviation G.M.T.	Difference (min)
Reykjavik	64° 08′	338° 03′	1204	—	—
Lerwick	60° 08′	358° 49′	1226	1219	+7
Eskdalemuir	55° 19′	356° 12′	1226	1233	−7
Abinger	51° 11′	359° 37′	1235	1231	+4
Dombås	62° 04′	9° 07′	1236	1236	0
Manhay	50° 18′	5° 41′	1244	1216	+28
Wingst	53° 45′	9° 04′	1245	1243	+2
Rude Skov	55° 51′	12° 27′	1246	1243	+3
Edh (Värnamo)	57° 10′	14° 01′	1246	1246	0
Lovö	59° 21′	17° 50′	1248	1247	+1
Niemegk	52° 04′	12° 40′	1251	1245	+6
Nurmijärvi	60° 30′	24° 39′	1252	1303	−11
Leningrad (Seltsy)	59° 57′	30° 42′	1255	1305	−10
Moscow (Krasnaya Pakhra)	55° 28′	37° 19′	1308	1311	−3
Odessa (Stepanovka)	47° 47′	30° 53′	1318	1315	+3
Tiflis (Dusheti)	42° 05′	44° 42′	1336	—	—

6. The Maximum Diminution in per cent of a Derived Departure

It is assumed that the effect of the solar eclipse is proportional to the departure of the daily variation which would have occurred if no solar eclipse had taken place, and that it varies with the obscured part of the sun's disk. The ionization will be reduced on account of the eclipse, and the distribution of existing electric currents in the E layer may be changed; but it is estimated that the effect in the central part of the eclipsed area may be considered proportional to the obscured part of the sun's disk. In order to calculate the maximum diminution of the magnetic departure at totality, it is necessary to calculate the ratio between the maximum effect at the place of observation and the maximum effect at the central line of totality. This calculation has been made in two different ways. In the first method the ratio was determined simply by summing up the effects at different equidistant points on a line perpendicular to the magnetic north–south direction through the considered station as well as through the centre of totality and for the moment of time when the totality is nearest to the station. This method has been applied to observations from the eleven first-mentioned observatories of Table 2 (Egedal and Ambolt, 1955).

The second method has been based on the following considerations: supposing the daily variation to be due to a sheet of electric currents at a height of 100 km above the earth flowing in the magnetic north–south direction, the effect on the magnetic needle originates from all parts of the sheet; but during a solar eclipse the contributions from the different parts of the sheet will be changed, the effect of the eclipse being greatest in the zone of totality. If the effect at the place of observation is determined for all parts of the sheet, the resulting effect can be computed. The computation can be made in the following way: a map of the surroundings of the observatory in question is drawn, and curves are drawn representing places where 100, 98, 96, . . . per cent of the sun's disk is covered at the moment of time when the totality is nearest to the observatory. These curves approximate to ellipses. On a sheet of transparent paper a network of points—at a distance apart equal to 100 km on the map—are entered and furnished with a number giving the magnetic effect on the magnetic declination at the considered observatory of a certain electric current flowing in the direction of the declination needle at a height of 100 km above the point. The network comprises an area of 500 km in radius, the points outside this limit showing values below $\frac{1}{2}$ per cent of the effect at the centre (see Fig. 1).

			0·007	0·008	0·009	0·008	0·007			
		0·008	0·011	0·014	0·016	0·014	0·011	0·008		
		0·012	0·019	0·028	0·033	0·028	0·019	0·012		
	0·009	0·018	0·037	0·068	0·091	0·068	0·037	0·018	0·009	
	0·012	0·026	0·065	0·189	0·352	0·189	0·065	0·026	0·012	
0·006	0·012	0·028	0·085	0·347	1·000	0·347	0·085	0·028	0·012	0·006
	0·012	0·026	0·065	0·189	0·352	0·189	0·065	0·026	0·012	
	0·009	0·018	0·037	0·068	0·091	0·068	0·037	0·018	0·009	
		0·012	0·019	0·028	0·033	0·028	0·019	0·012		
		0·008	0·011	0·014	0·016	0·014	0·011	0·008		
			0·007	0·008	0·009	0·008	0·007			

Fig. 1. Magnetic effect from electric currents at a height of 100 km. Distance between the points: 100 km.

Placing the transparent paper on the map with the centre on the observatory and orientated in the magnetic north-south direction, corresponding values on the map and the transparent paper are noted down for all points. The two corresponding values are multiplied for all points, and the products are summed up.

Next the centre of the transparent paper is placed on the centre of totality on the map, and the same operation is carried through. From the two sums obtained the ratio between the maximum effect at the place of observation and at the central line of totality is found. This method has been used for the East German and Russian stations (EGEDAL and AMBOLT, 1956). The difference between ratios found by the two methods is slight.

In Table 2 are given: station; distance from central line of totality (to the north +, to the south −) (column 1); 0·9° cot h_s cos d (column 2), where 0·9° is the height of the electric currents considered (100 km) expressed in latitude degrees, h_s the altitude of the sun at totality as observed from the point of the central line nearest to the place of observation in question (the magnetic observatory), and d the difference between the azimuth of the line from the point mentioned to the place of observation and the azimuth of the sun at the above-mentioned moment of time; the distance from the projection upon the earth's surface of the ionospheric central line (E layer) (column 3); derived departure (column 4); effect at maximum obscuration (column 5); per cent of maximum effect (column 6); and the maximum effect in per cent of the derived departure (column 7).

Table 2. Position of stations in relation to the track of totality, maximum effect expressed in per cent of the derived departure.

Station	Distance from central line	0·9° . cot h_s . cos d	Distance from projection of ionospheric central line	Derived departure	Deviations at maximum obscuration	Per cent of maximum effect	Maximum effect in per cent of the derived departure
Reykjavik	+1·2°	+0·7°	+1·9°	−0·8′	untraceable	96	—
Lerwick	−1·1°	+0·7°	−0·4°	−7·4′	2·1′	99	29
Eskdalemuir	−6·3°	+0·6°	−5·7°	−5·0′	0·8′	84	19
Abinger	−9·6°	+0·5°	−9·1°	−4·9′	0·8′	71	23
Dombås	+2·4°	+0·7°	+3·1°	−7·8′	2·85′	93	39
Manhay	−9·0°	+0·5°	−8·5°	−6·3′	0·9′	74	19
Wingst	−5·0°	+0·5°	−4·5°	−6·8′	1·9′	88	32
Rude Skov	−2·3°	+0·6°	−1·7°	−7·4′	2·2′	97	31
Edh (Värnamo)	−0·7°	+0·6°	−0·1°	−7·5′	2·2′	100	29
Lovö	+2·2°	+0·7°	+2·9°	−7·8′	2·3′	93	30
Nurmijärvi	+5·1°	+0·7°	+5·8°	−7·5′	1·5′	85	24
Niemegk	−5·4°	+0·5°	−4·9°	−7·0′	2·1′	89	34
Leningrad (Seltsy)	+6·7°	+0·7°	+7·4°	−7·1′	2·0′	83	34
Moscow (Krasnaya Pakhra)	+5·9°	+0·8°	+6·7°	−6·0′	1·8′	87	34
Odessa (Stepanovka)	−2·5°	+0·8°	−1·7°	−4·8′	0·7′	98	15
Tiflis (Dusheti)	−0·6°	+1·1°	+0·5°	−2·7′	0·0′	100	0
				Weighted mean: 29 ± 2 per cent			

In the above-mentioned paper by CHAPMAN (1933) the diminution was predicted to be 28 per cent, of the same order of magnitude as the weighted mean found.

The effect of the solar eclipse of 30 June 1954 has been examined for times where the departures of the daily variation of D were to the west, and the effect itself showed a variation to the east. As it was of interest to examine the case where the departures of D during a solar eclipse were to the east, solar eclipses occurring near Rude Skov in the morning were sought for. It was found that a total solar eclipse took place near Rude Skov on 29 June 1927 during the time when the departures of D were to the east, and the effect did show a variation to the west, but, unfortunately, the day was disturbed, so that the result was rather uncertain. However, BAUER (1900) in a figure facing page 154, gives a reproduction of a record of D showing clearly a diminution of the departure to the east in connection with the occurrence of a total solar eclipse.

7. The Variation of the Effect

At this stage of the examination it was found of value also to examine the variation of the effect with time. Deviations from the normal daily variation have been determined from the magnetograms of the following ten magnetic observatories for every 10 min in a period from 100 min before to 100 min after maximum obscuration: Lerwick, Eskdalemuir, Abinger, Dombås, Manhay, Wingst, Rude Skov, Edh (Värnamo), Lovö, and Nurmijärvi. Means of the deviations for each 10 min group have been computed, and the result is represented in Fig. 2. The zero line represents the normal daily variation.

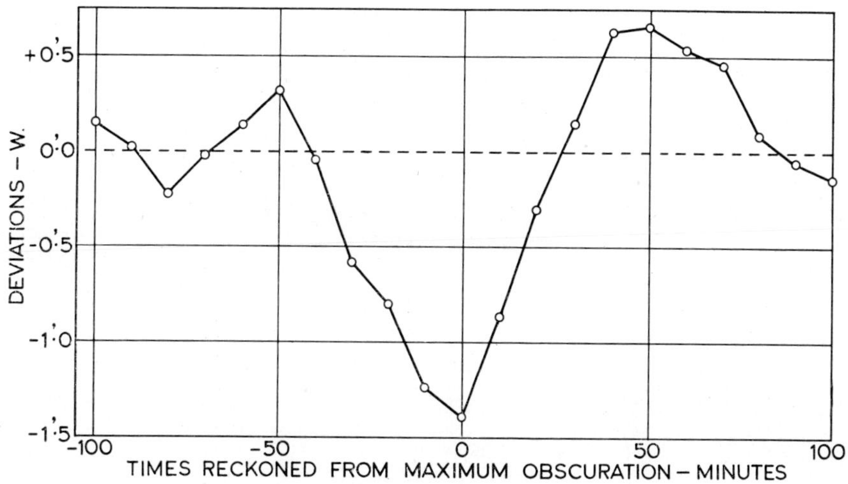

Fig. 2.

For the observatories considered the solar eclipse began about 75 min before and ended about 75 min after the maximum obscuration. The variation shown by the departures given in the figure is a sum of the magnetic disturbances in the period and the effect of the solar eclipse. As the deviation given in the figure is a mean of deviations at ten observatories for different moments of time, the influence of magnetic disturbances of short duration will be reduced, but any effect which occurs simultaneously with the maximum obscuration will suffer no reduction.

The diminution of westerly declination reaching its greatest easterly value just at maximum obscuration undoubtedly is an effect of the solar eclipse, because similar effects in the same sense are found at all ten observatories, although the time of maximum obscuration is different at the different places.

There seems to be an augmentation of the departure just before and just after the diminution. In the above-mentioned paper by CHAPMAN (p. 179) a figure is given representing the current-flow in the upper atmosphere, and it is diminished in the central part of the eclipse-shadow, while an augmentation of the current-flow takes place at a distance of 800 km from the centre. An augmentation of the daily *magnetic* variation certainly would be expected to take place at a greater distance from the centre than that for the current-flow. From Fig. 2 it has been found that the distance from the centre to the augmentations is about 1700 km, but the value is uncertain. Even this radius of 1700 km is too small if our estimate that the effect should be nearly proportional to the obscured part of the sun's disk is correct. If the effect is diminishing faster than the obscured part of the sun's disk, then calculated ratios should be smaller than those found, and consequently the maximum diminution should be greater than the found value of 29 per cent. However, the curve of Fig. 2 is not symmetrical in relation to the point of totality, possibly on account of magnetic disturbances. Therefore the radius of the area for which a diminution takes place may be greater than 1700 km, but even so it seems as if the observations will give a maximum diminution slightly higher than that predicted by CHAPMAN, but with an uncertainty so great that the predicted and found values nearly agree with each other.

It should be remembered that the maximum diminution depends on the duration of the totality. The effect will be more than 4 per cent higher if the duration of totality is maximum (nearly 8 min) as compared with the case where the duration is zero.

8. CONCLUDING REMARKS

As the magnetograms obtained during the solar eclipse of 30 June 1954 clearly show, there should be no doubt as to the existence of an effect on geomagnetism of solar eclipses.

The simultaneous occurrence of maximum obscuration and maximum magnetic effect is remarkable and does not fully agree with the results of RYDBECK (1946) from the eclipse of 9 July 1945, where the minimum of electron density of the E layer took place about 5 min after the occurrence of maximum obscuration; but possibly magnetic disturbances may have changed the time found for the maximum effect.

A study of the magnitude of the area where a diminution of the departures representing the daily variation takes place might give some information concerning the distribution of the electric currents of the E layer during solar eclipses.

The magnitude found for the diminution, using the results of CHAPMAN and BARTELS (1940), shows that the currents producing the daily variation are flowing at a height of about 100 km (E layer), and that they are not flowing in the F_2 layer (where the diminution would be 22 per cent).

The nature of the large daily magnetic variation at the magnetic equator ought to be investigated by means of magnetic registrations during solar eclipses.

Finally, an investigation of the effect on earth-currents is desirable.

It is of considerable value that the UGGI Committee No. 10 will take care that the necessary observations are carried out, in order to procure data for the study of effects of solar eclipses on different geophysical phenomena.

References

Bauer, L. A. 1900 *Terr. Mag. Atmos. Elect.*, **5**, 143.

Chapman, S. 1933 *Terr. Mag. Atmos. Elect.*, **38**, 175.

Chapman, S., and Bartels, J. 1940 *Geomagnetism*, Oxford Univ. Press, p. 796.

Egedal, J., and Ambolt, N. 1955 *J. Atmosph. Terr. Phys.*, **7**, 40.

1956 *J. Atmosph. Terr. Phys.*, **8**, 105.

Gama, L. I. 1948 *Terr. Mag. Atmos. Elect.*, **53**, 405.

Madwar, M. R. 1953–54 *Bul. de l'Inst. d'Égypte*, XXXVI.

Ryd, V. H. 1917 *Publ. Danish Met. Inst., Medd.*, No. 3.

Rydbeck, O. E. H. 1946 *Trans. Chalmers Univ. Tech., Gothenburg*, No. 53.

Solar Eclipses and the Ionosphere (Special Supp. Vol. 6 to J. Atmosph. Terr. Phys.), Pergamon Press, London, 1956

The Effect of a Solar Eclipse on the Lower Part of the Ionosphere and on the Geomagnetic Field

T. NAGATA, T. RIKITAKE, and Y. NAKATA

Geophysical Institute, University of Tokyo

Abstract—The changes in the electron density of the E layer for the solar eclipse of 9 May 1948 have been investigated on the basis of many observations in Japan and China. The distribution of the change can be shown on a chart in which local time is conventionally taken in place of longitude, and it is evident that the electron density decreased over a large area, the boundary of which reached far beyond the area of the optical eclipse.

The changes in the electrical conductivity, integrated throughout the layer, have been obtained, using the electron-density distribution determined and making suitable assumptions.

The modification of the S_q current-system by the presence of the non-uniformly conducting area due to the eclipse was thus determined, and the three elements of the magnetic field caused by the "additional current-system" (opposite to the original current at the centre of the eclipsed area) were obtained. The geomagnetic effect of the eclipse can be obtained for different observation points by moving the distribution of the magnetic field on the earth's surface along the path of the eclipse, making due allowance for the change in the direction of the original S_q current during the eclipse. With this convention, the changes in the magnetic elements at a given station were obtained and were found to agree well with the observed changes.

INTRODUCTION

MANY observations have hitherto been carried out in order to detect the effect of solar eclipses on both the ionosphere and the geomagnetic field. On the basis of those observations, we are quite sure that both the electron density of the ionosphere and the amplitude of the S_q variation of the geomagnetic field decrease to some extent during eclipses. The relation between the ionospheric and geomagnetic changes can be approximately explained by the theory proposed by S. CHAPMAN (1933). He estimated the modification of the electric current flowing in the ionosphere (the origin of S_q) by the presence of the eclipsed area in which the electrical conductivity becomes smaller than that of the surrounding area. The theory suggests that the reduction of S_q is nearly one-quarter of the value of S_q at that time, if we assume suitable values for the decrease of the conductivity, the radius of the eclipsed area, and so on.

Since, however, almost all the observations have been reported individually, no investigation has been attempted to obtain a unified view concerning both the ionospheric and geomagnetic effects. The writers would here like to obtain the distribution of the changes in the electron density of the E layer during a particular eclipse. Starting from the distribution, the electrical conductivity integrated throughout the layer will be presumed in and around the shadow area. Then the electric current in the layer as well as the magnetic field at the earth's surface will be calculated, following CHAPMAN's theory. Finally, the geomagnetic field thus calculated will be compared with the changes actually observed. The process stated here would be the best way to investigate the effect of a solar eclipse both qualitatively and quantitatively.

THE CHANGES IN THE ELECTRON DENSITY AND THE ELECTRICAL CONDUCTIVITY OF THE E LAYER

One example of the horizontal distribution of the changes in the values of f_0E will be examined below in relation to the solar eclipse that occurred in China and

Japan on 9 May 1948. The writers were able to collect the $h'f$ curves during the eclipse from ten ionospheric observatories. Since it is sometimes difficult to determine the critical frequencies of the E layer by direct reading of the $h'f$ curves, because of active Es and absorption in the layer, the writers first obtained the empirical relation between $(f_0E)_{eclipse}/(f_0E)_{mean}$, and the sun's area that is not masked by the moon, where $(f_0E)_{eclipse}$ and $(f_0E)_{mean}$ denote respectively the critical frequency at the maximum eclipse and that of the normal day. Next, the above ratio for various local times during the eclipse was obtained at Wakkanai, Kokubunji, and Nanking. Values of f_0E can be obtained throughout the eclipse

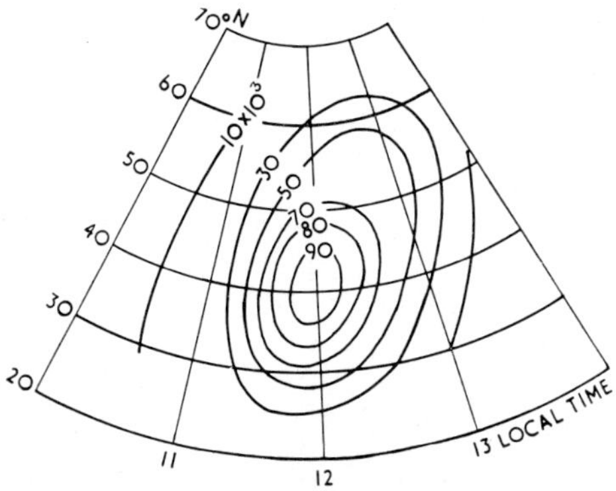

Fig. 1.

only at these three stations on account of unfavourable ionospheric conditions. It is noticeable that we find a different rate of decrease and increase of the ratio before and after the maximum of the eclipse, this asymmetry being attributed to the heterogeneous distribution of the concentrated active region on the surface of the sun.

Taking into account both the empirical relation between $(f_0E)_{eclipse}/(f_0E)_{mean}$ at the maximum eclipse and the uncovered area of the sun's surface and the above asymmetry, the decrease of the electron density in the E layer at the respective local times was finally determined at Wakkanai, Fukaura, Shibata, and Kokubunji, which are situated approximately on the same meridian. Hence we obtained the distribution of the decrease of electron density in the E layer, as shown in Fig. 1, though we use the local time in place of the longitude. The same procedure failed to give similar distributions for the other eclipses owing to unfavourable ionospheric conditions.

The eclipse moved along an approximate straight line from SW to NE with a fairly constant speed. If we assume that the distribution of the decrease of electron density also moved along the path without changing its shape, the distribution pattern is readily obtained from the one which is shown in Fig. 1, by use of a linear transformation. The pattern thus obtained is nearly circular. Under the assumption that the change in the electrical conductivity integrated throughout the layer is proportional to that in the electron density, we can obtain the distribution of the decrease of the conductivity from the above pattern.

If we ignore the azimuthal change in the distribution, the decrease of the conductivity varies with the distance from the centre, as shown in Fig. 2, in which an approximation of the distribution with a parabolic curve is also shown. The maximum decrease is about one-half of the normal conductivity appropriate

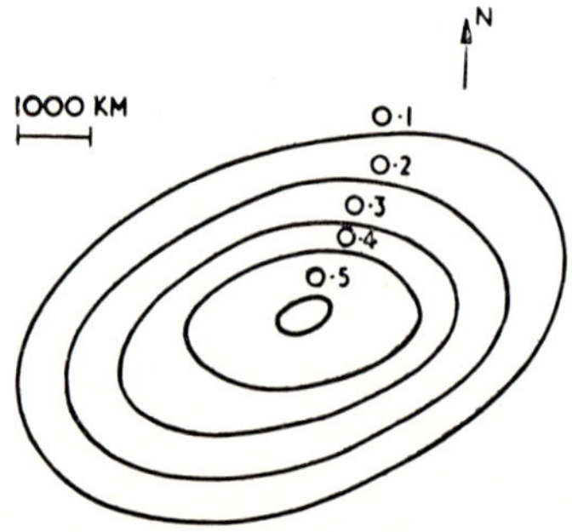

Fig. 2

to no-eclipse conditions, while the area in which we observe a conspicuous decrease is so large that its diameter exceeds 7000 km, this value being surprisingly larger than that suggested by CHAPMAN.

THE ADDITIONAL CURRENT-SYSTEM AND ITS MAGNETIC FIELD

If there were no eclipse, there would be a flow of electric current which produces the S_q variation. We shall now investigate the distribution of the electric current

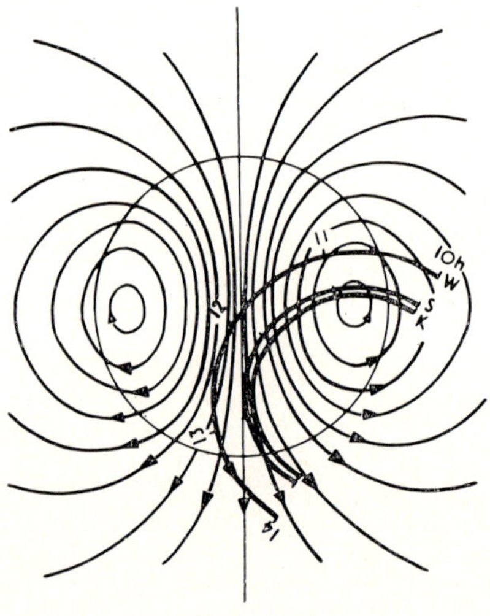

Fig. 3

when there is a particular depression of the conductivity as obtained in the last section as a parabolic curve, the electromotive force being taken as that of the S_q current which is uniform at great distances from the centre of the eclipsed area. After some complicated calculations, we obtain an electric current-system superposed upon the original uniform current. The additional current-system, which may be regarded as the effect of the solar eclipse, is shown in Fig. 3, the shape

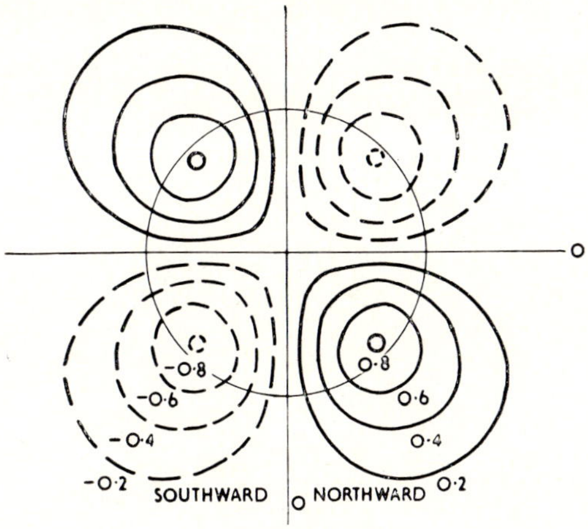

Fig. 4a

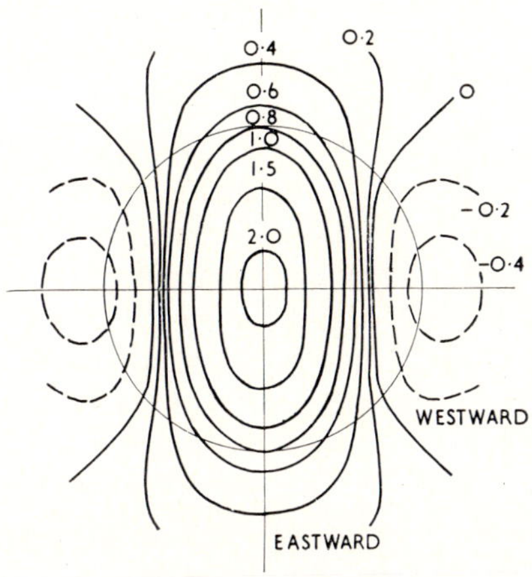

Fig. 4b

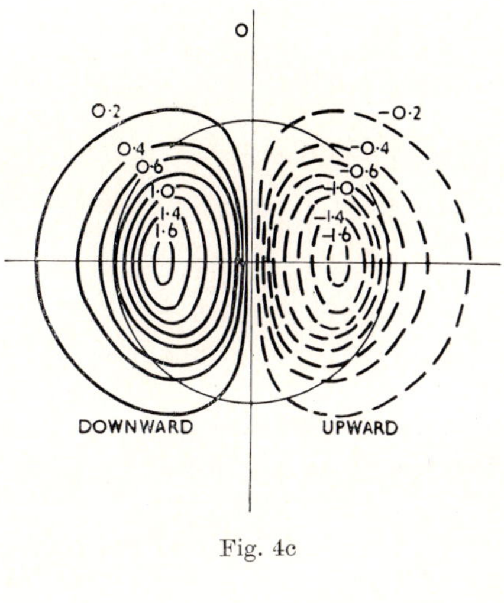

Fig. 4c

of the system being not greatly different from that obtained by CHAPMAN. The direction of the current is opposite to that of the original current at the centre of the eclipsed area.

The magnetic field at the earth's surface produced by the additional current is also calculated by the use of potential theory. The height of the layer is taken to be 100 km. The three components of the magnetic field thus calculated are shown in Figs. 4a, 4b, and 4c. Therefore we may consider that the patterns of the magnetic field proceed along the path of the eclipse.

Strictly speaking, however, the parallel current model of the S_q variation is not applicable because the eclipsed area is quite wide. Although it is no easy matter to take into account the non-uniformity of the S_q current, this may be done, to a rough degree of approximation, by rotating the direction of the parallel current with passage of time. Since it is likely that the eclipse passed through the northern part of the S_q current vortex on that day, the direction of the parallel current at the beginning of the eclipse should be nearly opposite to that at the end of the eclipse. If we take into consideration the above-mentioned influence of the S_q current, the relative movements between the eclipsed area and the observation-points Wakkanai (45° 24′ N, 141° 41′ E), Sado (38° 01′ N, 138° 22′ E), and Kakioka (36° 14′ N, 140° 11′ E) are given as is also shown in Fig. 3. By tracing the changes in the magnetic field along the paths in Fig. 3, the time-variations of the three geomagnetic components at the respective stations become as shown in Fig. 5.

On the other hand, the difference-curves between the observation on the eclipse day and the mean observation before and after the eclipse are as shown in Fig. 6, the mean observation being defined as the variation averaged through several calm days. Comparing the difference curves of Fig. 6, which give approximately the effect of the solar eclipse on the geomagnetic field, with the calculated ones in Fig. 5, we find a good agreement between them not only in their magnitude but also in their time-variations. If we assume that the east component of the S_q field amounted to 50 gammas (as determined from the observations), the calculated magnitude of the changes in the east component are as given in Table 1, in which the observed changes are also given.

Table 1. The maximum changes in the east component

Station	Calculated	Observed
Wakkanai	20 gammas	23 gammas
Sado	15 gammas	17 gammas
Kakioka	14 gammas	13 gammas

In addition to the agreement between the calculated and observed values, we also find that the decrease of the east component of the S_q field due to the eclipse effect is nearly one-half of the original S_q field. This magnitude greatly exceeds that which has been suggested by CHAPMAN.

The writers also examined the results of the geomagnetic observations in connection with several solar eclipses that occurred in the vicinity of Japan. Although it is not always possible to detect the effect of the eclipse, on account

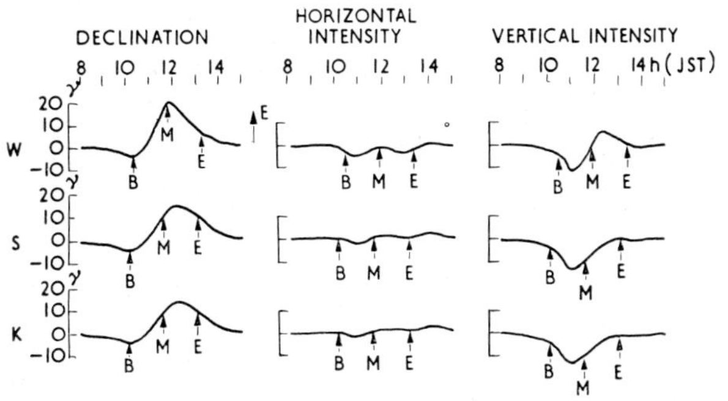

Fig. 5

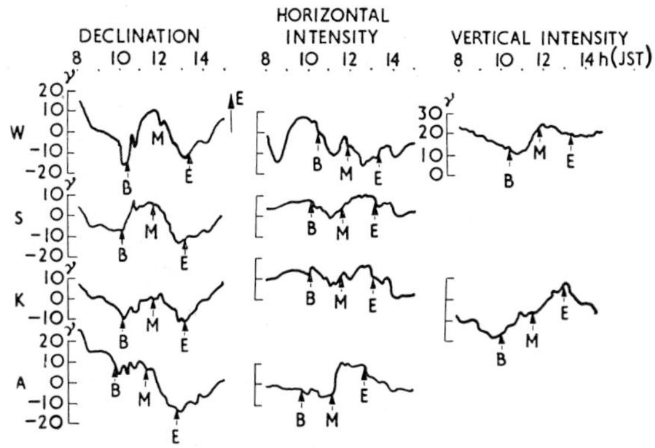

Fig. 6

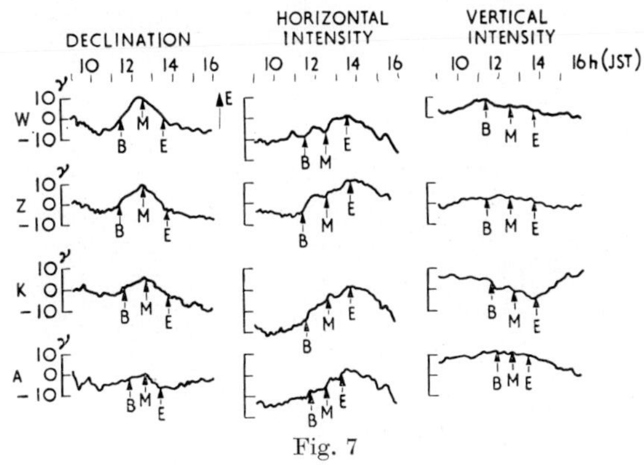

Fig. 7

of geomagnetic disturbances, it seems certain that the effect is of the type investigated above. A further example of the geomagnetic effect of a solar eclipse (that on 12 September 1950) is shown in Fig. 7, though the ionospheric conditions are not favourable for a detailed study.

CONCLUDING REMARKS

An extension of CHAPMAN's theory concerning the effect of a solar eclipse on the geomagnetic field is attempted in this paper with success in synthesizing the results of the ionospheric and geomagnetic observations. It is concluded that the electron density of the E layer decreases during the eclipse over a wide area whose extent much exceeds that of the optical eclipse, that the decrease of the electron density of the E layer amounts to nearly one-half of the normal value of the electron density when there is no eclipse, that the geomagnetic changes observed on the earth's surface are of the right magnitude and kind expected from the theory based on the actual distribution of the decrease of the electron density in the E layer, and that the decrease of a component of the geomagnetic field in the horizontal direction amounts to as much as one-half of the S_q field on the eclipse day.

REFERENCE

CHAPMAN, S. 1933 *Terr. Mag. Atmos. Elec.*, **38**, 175.

Solar Eclipses and the Ionosphere (Special Suppl. Vol. 6 to J. Atmosph. Terr. Phys.), Pergamon Press, London, 1956

Preliminary Report on the Effect of the Solar Eclipse of 20 June 1955 on the Earth's Magnetic Field

Y. KATO, J. OSSAKA, and A. SAKURAI
Geophysical Institute, Tohoku University, Sendai

Abstract—Changes of the earth's magnetic field were observed at Ceylon, which was situated in the zone of totality of the eclipse and also is near the geomagnetic equator. In addition to the conventional recording equipment, a specially designed variometer-type magnetometer was used which recorded the changes of the magnetic field visually on paper with a sensitivity of about 1 gamma per mm record.

Definite changes were observed which are attributed to eclipse effects: the horizontal intensity decreased at the time of maximum phase of the eclipse, the vertical intensity increased, and the declination deviated eastward for the first half of the eclipse and westward for the latter half. It is considered that these changes result from disturbances of the electric-current system due to a decrease of the electrical conductivity as a result of the eclipse.

THE total solar eclipse occurred in the south-eastern part of Asia and southern part of the Pacific Ocean on 20 June 1955. The duration of the time-interval between second and third contact of this total eclipse was long (about 7 min at Bangkok and 4 min in Ceylon).

In order to investigate the effect of the solar eclipse on the terrestrial magnetic field, we observed the changes of the earth's magnetic field in Ceylon, because Ceylon was situated on the total zone of this eclipse and also is near the geomagnetic equator.

The geomagnetic observations were made at the University of Ceylon, the situation of the observation site being as follows:

Geographic latitude	7° 15′ 34″ N
Geographic longitude	80° 35′ 36″ E
Geomagnetic latitude	−2° 41′
Geomagnetic longitude	149° 49′
Altitude	460 metres.

The absolute value of the magnetic elements was as follows:

Horizontal comp.	40142γ	1909, June 19
	40138γ	1839, June 21
Declination	3° 03·2′ W	1823, June 19
	3° 02·8′ W	1825, June 21
Dip. (North Pole upwards)	−4° 18·0′	1927, June 19
	−4° 18·4′	1851, June 21
Local mean time	82·5° E local mean time.	

The circumstances of the eclipse were:

1st contact	2nd contact	3rd contact	4th contact
0703	0811	0815	0930

In order to observe the changes of the earth's magnetic field, we used a special variometer-type magnetometer besides the ordinary unifilar suspension-type magnetometer and an induction magnetometer. The sensitivity of this special instrument was about 1 gamma per mm on the record, and the paper-speed was 6 cm per hour.

The first half of the eclipse day was geomagnetically quiet, but the latter half was slightly disturbed. However, the following day (21 June) was perfectly

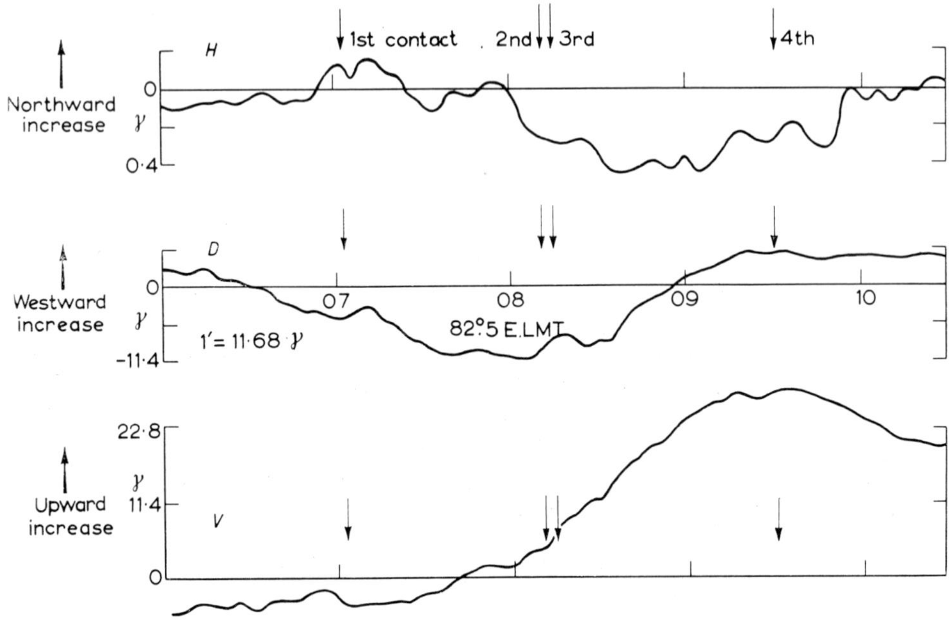

Fig. 1.

calm, and we can compare the records for these two days. The differences between measurements on the 20th and 21st over the eclipse period (0703 to 0930) for the three elements—horizontal component, vertical component, and declination are shown in Fig. 1. It will be seen that the horizontal intensity decreased by about 12 gammas at the time of maximum phase of the eclipse, the vertical intensity increased, and the declination deviated eastward in the first half of the eclipse and then westward. The beginning, maximum phase, and ending of the changes seem about half an hour later than the corresponding times of the solar eclipse. These changes in the earth's magnetic field indicate the disturbance of the electric-current system due to the decrease in the electric conductivity in the ionosphere at the time of the eclipse.

A fuller description of these results will be published in the near future.

DISCUSSION OF PAPERS IN SECTION VI

Chairman: Dr. J. BARTELS

Mr. RATCLIFFE I hope that Professor CHAPMAN or someone else will do a first-order calculation to show what changes of electric field in the F region might accompany an eclipse in the E region. Professor CHAPMAN's original calculations on the first-order magnetic effect were so interesting that I think we ought to have the first-order electric effect without any complications.

Prof. CHAPMAN I think that Mr. EGEDAL's analysis on the magnetic effect during an eclipse is one of the best yet made—a result of the eclipse coming over the dense network of observatories in Europe. Often one complains about the poor distribution of magnetic observatories, with so many in Europe and so few in other parts of the world. It is a disadvantage on most occasions, but it was an advantage in this case. I should like to ask Mr. EGEDAL why he only observed the effect in declination. I should have thought that observations of horizontal force would be at least as necessary.

Dr. BARTELS As a matter of fact, equatorial eclipse observations would be most valuable, because effects would show in H.

Mr. EGEDAL Declination is the easiest magnetic element to measure. I think, too, that the effect will be less in H, and is unlikely to show in the eclipses we have studied.

Dr. BARTELS I think the declination effect would be biggest, but it really depends on the location of the observatory.

Mr. EGEDAL Observations near the magnetic equator were made for the eclipse of 25 February 1952 near Khartoum. These observations have been studied by Dr. MARTYN, but the results were not conclusive, since the observations consisted only of a limited number of isolated measurements in a day. It has been suggested that observations should be taken at two observatories lying near the magnetic equator, but I don't know whether this has been done.

Prof. NAGATA The observations described in the paper by Dr. KATO were made in geomagnetic latitude 2° 41′ S.

Solar Eclipses and the Ionosphere (Special Suppl. Vol. 6 to J. Atmosph. Terr. Phys.), Pergamon Press, London, 1956

SECTION VII

SOLAR RADIO NOISE

Radio Investigations of the Structure of the Solar Corona

M. Ryle

Cavendish Laboratory, Cambridge

Abstract—Observations of the solar emission at metre wavelengths may be used to derive information on the distribution of electron density and temperature in the undisturbed solar corona. It is necessary to determine at each of several wavelengths not only the total intensity but also the distribution of "radio brightness" across the solar disk. These measurements have been made by observations of the variation of intensity during an eclipse, and more extensively by using interferometric methods to provide a sufficiently high resolving power. Using the results of such observations preliminary models of the solar corona have been derived.

An entirely different method of observation has been used to study the outermost layers of the solar envelope. In this method, observations are made of a radio star at different angular distances from the sun. The results show that as this angle decreases the apparent angular size of the radio star increases, due to scattering by irregularities in the electron density. It has been possible to detect this effect out to distances of $20R_0$, and important deductions about the electron density in this region can be made.

The evidence obtained from recent ionospheric observations during solar eclipses has suggested that part of the ionizing radiation may originate in the solar corona; it is therefore important to collect all the available evidence concerning the density and temperature at different heights in the solar atmosphere. During the past five years a number of different radio methods have been applied, and in certain regions these new methods already seem capable of providing more accurate data than the optical methods.

The most important method of observation is based on measurements of the radio emission from the sun; in the absence of sunspots or other solar activity, it appears possible to explain the radio emission from the sun in terms of the thermal emission produced by a heated ionized gas. The absorption coefficient, and hence the emissivity of any region, increases with increasing wavelength, and thus we should expect that the outer layers of the sun's atmosphere, where the density is lower, might be a good emitter of long-wave radiation whilst being a relatively poor emitter at a shorter wavelength. A lower layer, on the other hand, might be an effective radiator at the short wavelength, but any radiation emitted at the longer wavelength would be reabsorbed by the outer layers. In this way we might expect different wavelengths to be emitted predominantly by different layers; the measurement of the intensity of the solar radiation over a wide range of wavelengths might therefore enable us to determine directly the temperatures at the corresponding layers where the density had a value capable of producing (for each wavelength) effectively complete absorption.

This simple interpretation of the results is, however, not valid, except at the shorter wavelengths (1–25 cm) which originate in the chromosphere. At wavelengths greater than 25 cm it is necessary to take account of the refractive effects which arise in the solar atmosphere. It is perhaps simpler to reverse the problem and consider the transmission of a wave into the sun: in the absence of refractive effects the wave would travel inwards until it became absorbed. In practice, however, it will be deviated as shown in Fig. 1; the total absorption throughout the whole ray-path may only be small, so that the sun would act as a partial reflector. Under these conditions the intensity of the radiation emitted by the sun will be less than the value corresponding to a "black body" radiator at the same temperature, and measurements of the total intensity will not provide us with information on the temperature in the region in which the radiation originated. The emission could be small either because the temperature was low or because the reflection coefficient was high.

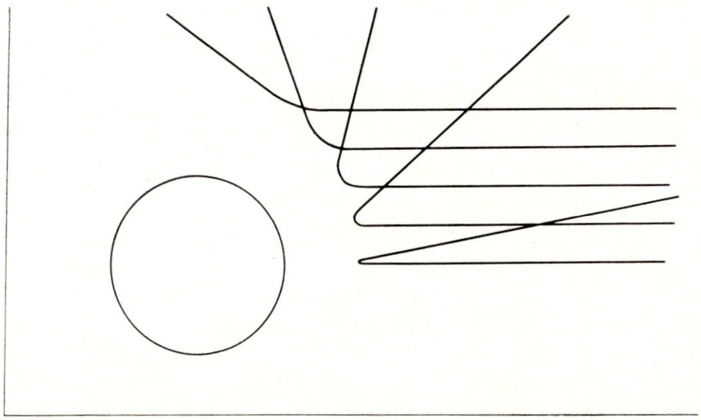

Fig. 1. Diagram to illustrate the refraction of radio waves in the solar atmosphere.

By considering the integrated emission along different ray-paths, however, it is possible to distinguish between these two effects, and a measurement of the distribution of radio "brightness" across the solar disk at a number of different wavelengths can in principle be used to find the variation of both temperature and density with height.

The first problem is to determine experimentally the distribution of radio brightness across the solar disk; for the investigation of the coronal regions it is necessary to use wavelengths in the range 50 cm to 10 m, and in order to get a resolving power sufficiently great to distinguish between different parts of the sun, apertures of several hundred metres are involved.

A number of observations have been made to determine the distribution by measuring the variation of intensity during an eclipse, and some of these, notably those of the French workers (LAFFINEUR *et al.*, 1952; BLUM *et al.*, 1952) have provided important results, in particular, evidence for the ellipticity of the distribution at metre wavelengths. However, unlike the similar observations at centimetre wavelengths, there are serious practical difficulties in deriving an unambiguous distribution from such measurements. The main difficulty arises from the greater extent of the sun at metre wavelengths; the relatively small size of the lunar disk leads to eclipse curves which are insensitive to variations of the distribution. A further difficulty, which has upset many of these observations, is due to the very great intensity of the non-thermal sources associated with sunspots; unless the sun is very free from activity on the day of the eclipse, large and irregular fluctuations of the intensity may be produced which mask the variation caused by the passage of the moon's disk across the corona.

In 1949 a new method was devised at Cambridge by STANIER (1950), which has since been extended by O'BRIEN (1953) and used by him on a wide range of wavelengths. The method makes use of two movable aerials as an interferometer of variable spacing d. If these are used with a phase-switching receiver (RYLE, 1952) which measures the difference in received power when the aerials are connected alternately in and out of phase, a sinusoidal reception pattern is produced, having maxima separated by λ/d; a point source will then produce a sinusoidal record as the pattern is moved across the source by the earth's rotation. If an extended source is observed, the amplitude of the record is reduced and may be written $\int P(\theta) \cos(2\pi d\theta/\lambda)\, d\theta$ where $P(\theta)$ represents the variation of intensity across the source. It can be seen that the amplitude of this record is proportional to one term of the Fourier transform of the distribution across the source. By making observations with a number of different spacings d, the complete Fourier transform can then be derived. In this way it is possible to investigate the fine structure in a source with a detail equivalent to what could be obtained by using an aerial system as large as the maximum spacing d used. In the general case it is necessary to repeat the observation with interferometer axes which cross the sources at different angles.

Observations with this system were first made by STANIER in 1949 at a wavelength of 60 cm, and subsequently at longer wavelengths by O'BRIEN. Although all of these early measurements were made on days when the sunspot activity was very small, they were made near the maximum phase of the sunspot cycle, and may therefore have been affected by the presence of active areas remaining from old spot groups.

During the last two years of minimum sunspot activity a new series of observations has therefore been made in order to get the most reliable data on which to base theoretical work of the density and temperature distribution in the corona (O'BRIEN and CONWAY, 1956). Observations have been made at wavelengths of 60 cm, 1·4 m, 3·7 m, and 7·9 m. In addition, advantage was taken of the occurrence of the eclipse in June 1954 to confirm and supplement the interferometric measurements. The most complete interferometric observations were made at 60 cm (O'BRIEN and TANDBERG-HANSSEN, 1955), where a complete range of axes was used. The distribution derived from these observations is shown in Fig. 2, which shows two interesting features. Firstly the sun becomes markedly elliptical at large radial distances, and secondly limb-brightening occurs, especially in equatorial regions; unlike the shorter wavelengths, this occurs at a radius considerably smaller than the optical limb.

Most of the observations at the longer wavelengths (1·4 m, 3·7 m, and 7·9 m) were restricted to an east–west axis owing to the limited space available in the north–south direction. The radial distribution in this direction, computed on the assumption of elliptical symmetry, is shown in Fig. 3. Slight limb-brightening is visible at 1·4 m, and it is displaced inwards still further. The results obtained at a wavelength of 21 cm by CHRISTIANSEN and WARBURTON (1955) in Australia are also shown, indicating the important transition to the centimetre waves.

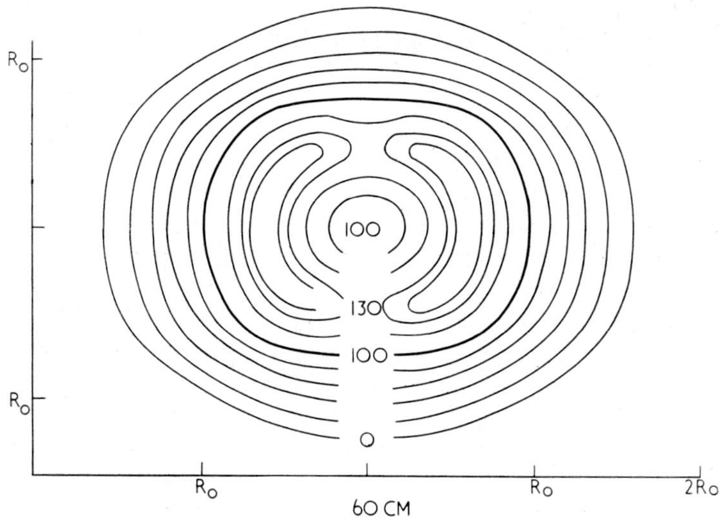

Fig. 2. The distribution of radio brightness at a wavelength of 60 cm. R_0 represents the radius of the optical disk.

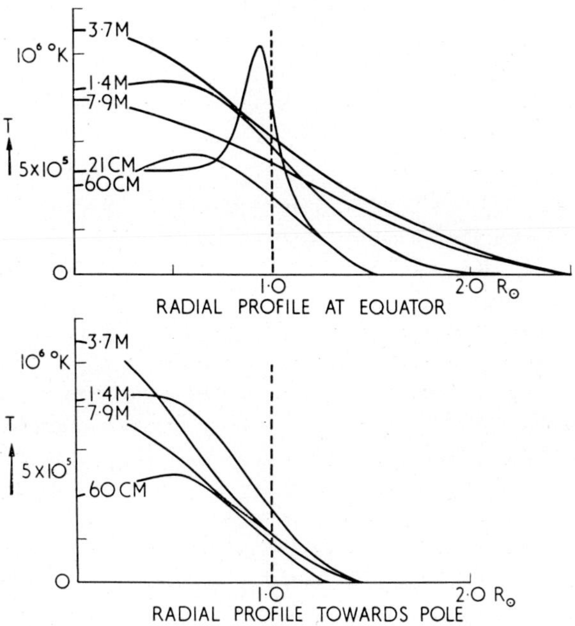

Fig. 3. The radial distribution of radio brightness at wavelengths of 60 cm, 1·4 m, 3·7 m, and 7·9 m. Upper curve—equator; lower curve—pole.

In these experiments at longer wavelengths no information was obtained on the extent in polar directions, but advantage was taken of the partial eclipse of June 1954, firstly to provide confirmation of the 60 cm observations, and also to determine the ellipticity of the contours at the longer wavelengths (O'Brien and Conway, 1956). The results obtained made it possible to derive the lower curves shown in Fig. 3, which represent the variation with radial distance in the polar directions; it can be seen here that there is a progressive increase in size and ellipticity with wavelength.

A theoretical study of these results is now being made in order to derive a model of the solar corona at the minimum phase of the sunspot cycle. The results of this analysis are not yet complete, but some of the preliminary results based on earlier, less extensive observations indicated that, although the inner regions of the corona agree well with previous determinations, the density in the outer corona ($>2R_0$) must be considerably greater than that of previous models (O'Brien and Bell, 1954).

The possibility of a considerable extension of the corona has led to a further investigation of these outer regions by an entirely new method. Although the outer layers of the solar corona in the region 5–$20R_0$ are probably of little significance in connection with the origin of the radiation responsible for ionospheric ionization, they may have a direct relationship with the electron density in the neighbourhood of the earth, as derived from observation of the zodiacal light and studies of whistling atmospherics. For this reason, a brief account will be given of a new method of investigation which has been developed by Hewish (1955).

The method is based on the observation of a radio star as its position is brought close to that of the sun by the earth's annual motion. So far, detailed observations have been restricted to one intense radio star which passes within a radial distance of $4{\cdot}6R_0$ of the sun; observations of this source have been made every June since 1952. On its passage through the outer layers of the solar corona, the radiation from the radio star is subject to multiple scattering by irregularities in the electron density of the corona; this scattering increases the apparent angular size of the radio star, and by making observations with interferometers it is possible to measure this scattering at different wavelengths and as a function of the angular distance from the sun. It is then possible to make some deductions about the electron density in the outermost layers of the solar atmosphere. Some typical results are shown in Fig. 4, which gives the scattering (for a wavelength of 3·7 m) as a function of angular separation between the radio star and the sun. It can be seen that it is possible to detect the effect of the solar atmosphere at radial distances up to $20R_0$; that is, about one-third of the radius of the orbit of Mercury.

An interesting feature of all the observations 1953–55 has been a significant asymmetry in the curve; the scattering is less when the star approaches than when it recedes from the sun. Since the axis of the sun is not perpendicular to the track of the radio star, the explanation of this effect may again be related to elliptical contours of the density in these outer regions.

The magnitude of the scattering depends both on the variation of electron density (ΔN) in the irregularities, and also on their physical scale; we cannot therefore derive the variation of density with radial distance directly from the scattering curve. However, by considering possible mechanisms for producing the irregular structure, it seems likely that the scale lies in the range 10^4–10^5 km. In this case the density ΔN (which must, of course, represent a minimum value for the total density N) has the values shown below:—

Radial distance	ΔN
$5R_0$	2×10^4
$10R_0$	2×10^3
$15R_0$	$1{\cdot}5 \times 10^3$

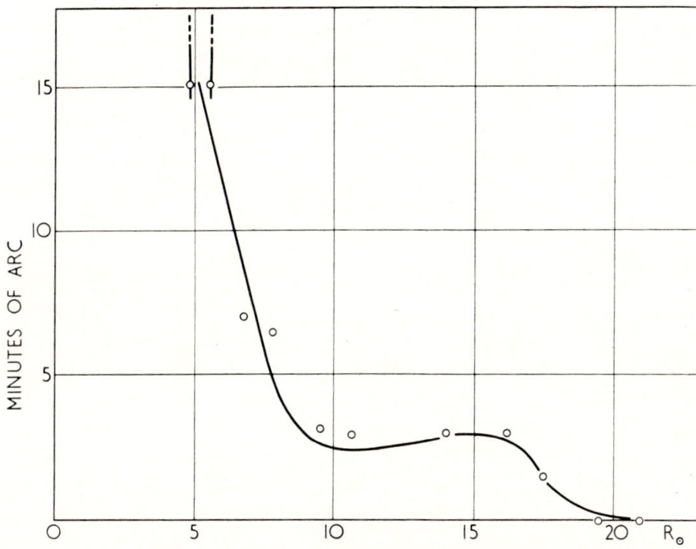

Fig. 4. Scattering produced by the irregular structure of the solar corona for a 3·7-m wave passing at different distances from the sun.

Although the scattering curve (Fig. 4) follows a steep law for radial distances less than $7R_0$, it is very much more gradual for greater distances. It even seems possible that the outermost layers of the solar corona extend indistinguishably into the interplanetary region; measurements of the zodiacal light and studies of whistling atmospherics have indicated electron densities of about 500 near the earth.

It is important to remember that these observations correspond to a normal state of the solar atmosphere, which has remained virtually unaltered over the past four years; they are not related to the corpuscular streams associated with prominence activity and magnetic storms. It would indeed be of great interest if similar observations could be repeated during the coming years of growing solar activity. Unfortunately, at these times we must expect very greatly enhanced solar radio emission, which would probably prevent the observation of the very weak emission from the radio star.

REFERENCES

BLUM, E. J., DENISSE, J. F., and STEINBERG, J. L.	1952	*Compt. Rend.*, **234,** 1597.
CHRISTIANSEN, W. N., and WARBURTON, J. A.	1955	*Observatory*, **75,** 9.
HEWISH, A.	1955	*Proc. Roy. Soc.*, **A228,** 238.
LAFFINEUR, M., *et al.*	1952	*Compt. Rend.*, **234,** 1528.
O'BRIEN, P. A.	1953	*Monthly Not. Roy. Astron. Soc.*, **113,** 597.
O'BRIEN, P. A., and BELL, C. J.	1954	*Nature*, **173,** 219.
O'BRIEN, P. A., and CONWAY, R. G.	1956	In preparation.
O'BRIEN, P. A., and TANDBERG-HANSSEN, E.	1955	*Observatory*, **75, 11**.
RYLE, M.	1952	*Proc. Roy. Soc.*, **A211,** 351.
STANIER, H. M.	1950	*Nature*, **165,** 354.

Solar Eclipses and the Ionosphere (Special Suppl. Vol. 6 to J. Atmosph. Terr. Phys.), Pergamon Press, London 1956

Radial Brightness Distribution of the Sun at 8 mm

J. P. HAGEN
Naval Research Laboratory, Washington

Abstract—The principal solar radiation in the centimetre and millimetre regions comes from the chromosphere, and measurements of the variation in flux at these wavelengths during the course of an eclipse may be used to give a model of the solar atmosphere. Measurements of distribution of radio brightness at 8 mm made in Sweden during the 1954 eclipse show limb-brightening and a marked amount of brightness at the centre of the disk, which may be explained in terms of a two-fluid theory of the chromosphere: cool, dense spicules extending into a hotter, sparser gas of the lower corona.

HIGH-RESOLUTION measurements at 8·5-mm wavelength made with the 50-ft diameter antenna in Washington and at the total solar eclipses of 1952 and 1954 in Khartoum (Sudan) and Oskarshamn (Sweden) have shown that radio measurements of solar radio radiation at these wavelengths can best be explained by assuming a two-fluid chromosphere—one in which cool, dense spicules extend into the hotter, sparser gas of the lower corona.

The atmosphere of the sun is transparent to visible radiation, is nearly transparent to millimetre and centimetre radio radiation, and becomes opaque to the metre wave radiation and longer. Information about the chromosphere can then be given by observing the radiation from the sun at short radio wavelengths. In its outer part the atmosphere of the sun is highly ionized. Absorption in any region varies with the density n, the wavelength λ, and the temperature T, according to the relation.

$$K \propto \frac{n^2\lambda^2}{T^{3/2}}.$$

This is the familiar equation for the absorption of radio waves in an ionized medium. By consequence of this, the longer-wave radiation is absorbed in the outer layers of the sun's atmosphere and can escape only from these outer regions. The shorter-wavelength radiation is absorbed very little in the outer part of the solar atmosphere, where the density is quite low, and hence, radiation from the chromosphere escapes as centimetre and millimetre radio waves. In fact, the principal radiation from the sun in the centimetre and millimetre region comes from the chromosphere.

Since the sun subtends an angle which is small compared with most radio antenna beam-widths, it is necessary to resort to unusual methods to resolve the sun to obtain the variations in brightness across its surface. At 8-mm wavelength, this can be done with the NRL 50-ft antenna, which has a beam-width of about 3 min of arc at this wavelength. A better way, however, is to observe the sun at the time of a total solar eclipse when the advancing edge of the moon provides the necessary resolution at any wavelength. Eclipses are particularly useful in the centimetre region, where the radio diameter of the sun is not too much larger than the visual, and where the moon will still effectively eclipse the sun. At metre wavelengths the radio size of the sun is so large that at totality more than half of the sun is seen around the moon.

A consideration of early measurements of the radio flux from the sun at many wavelengths revealed that the radio measurements were an effective means of defining the temperature gradient in the sun's atmosphere. The curve marked (1) in Fig. 1 shows the results of such an analysis. Note that there is a steep temperature rise in the chromosphere. It is one of the consequences of the eclipse measurements that the location of the steep rise in the chromosphere may be better defined. This model of the sun requires limb-brightening at all short radio wavelengths, and recent measurements here and abroad have now amply confirmed that limb-brightening does exist.

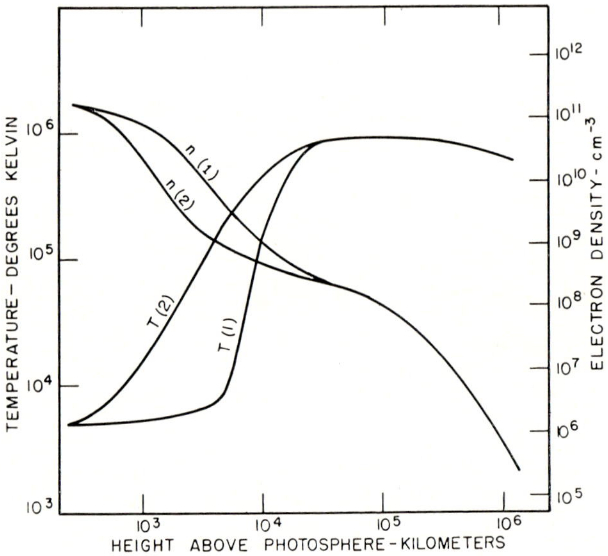

Fig. 1.

The most effective means in the centimetre region for measuring the distribution of radio brightness has been to measure the changing flux through the course of the total eclipse. Our most recent measurement was at Oskarshamn, Sweden, in 1954, where measurements at 8-mm and 10-cm wavelength were made. Both measurements improved and confirmed measurements made in Khartoum in 1952. The 10-cm results show a clear case of limb-brightening, and reduce to a model atmosphere similar to those in Fig. 1, with a temperature rise occurring near that of model (1) of that figure.

The 8-mm results will now be considered in somewhat more detail. The antenna used at 8-mm had an aperture of 16 ft by 2 ft, producing a fan-beam 1 deg wide in one dimension and 8 min wide in the other. The antenna beam was so oriented that the elliptically shaped cross-section was centred on the sun and aligned so that the path of the moon was along the major axis of the pattern. In this way we watched a central slice of the sun being eclipsed, and derived a radio-brightness curve for the equatorial region. The antenna consists of a cylindrical parabola, 16×2 ft, which illuminates along a line-focus a second smaller cylindrical parabola 2 ft $\times \frac{1}{4}$ in. in cross-section; this in turn brings the radiation to a focus at a point. This antenna was mounted equatorially, and driven by a synchronous motor to follow the motion of the sun.

With this equipment, the flux of the sun was measured during the entire course of the eclipse. The record was obtained on a recording milliammeter, and the raw data are shown in Fig. 2. In this curve there are many breaks for calibrations of sky temperature, the temperature of a calibrating hot load, and ambient

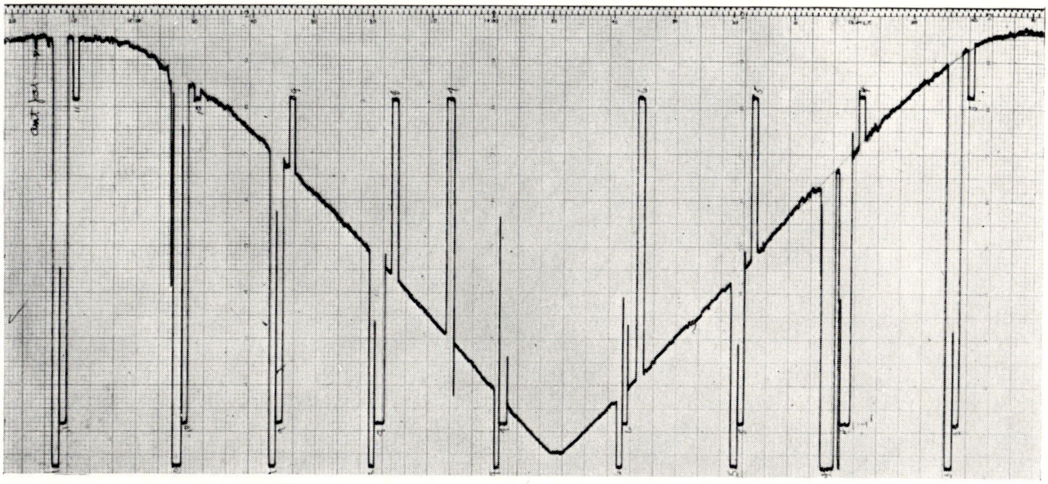

Fig. 2.

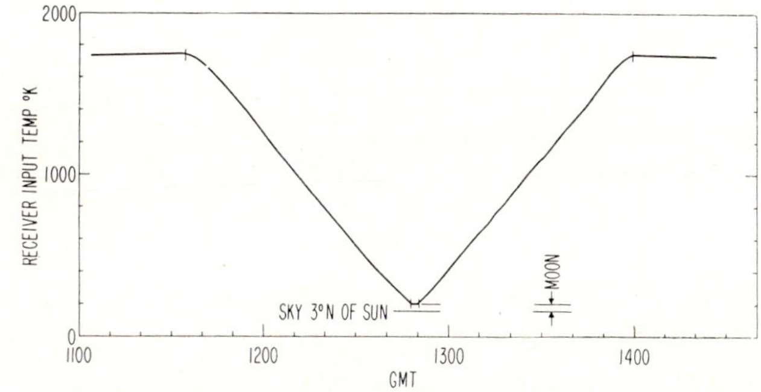

Eclipse curve corrected for gain and atmospheric variations, $\lambda = 8{\cdot}65$ mm, Oskarshamn, 30-6-54

Fig. 3.

temperature. While the weather was quite cloudy at Oskarshamn, it is apparent even from this curve that there was little effect on the radio data by the clouds.

This curve has been corrected for variations in receiver gain, for deviations from linearity, for effects due to antenna beam shape, and for atmospheric absorption. The corrected form of the curve is shown in Fig. 3. It is to be noted that there is a very small residual radiation at totality, less than $\frac{1}{2}$ per cent of the total flux of the sun, and in addition that, at totality, the principal radiation entering the receiver appears to be thermal radiation from the moon. Earlier measurements on the moon have allowed this radiation to be evaluated.

The radio-brightness distribution derived from this 8-mm curve differs markedly from that predicted on the basis of the simpler model. The model shows a relatively flat distribution with a narrow bright ring at the edge of the disk. The measured

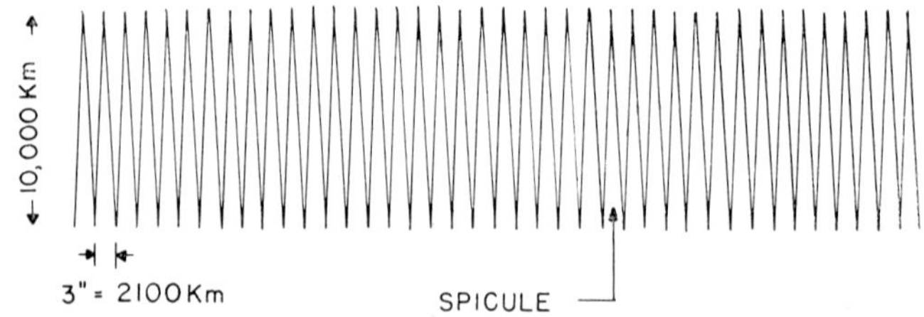

Fig. 4.

curve, Fig. 5, shows limb-brightening, but in addition shows a pronounced amount of brightness at the centre. A model of the chromosphere that could produce this effect has been constructed, and this will now be described.

It was earlier suggested by Giovanelli that the chromosphere may not be uniform, but may have a grass-like structure. Roberts and others have observed

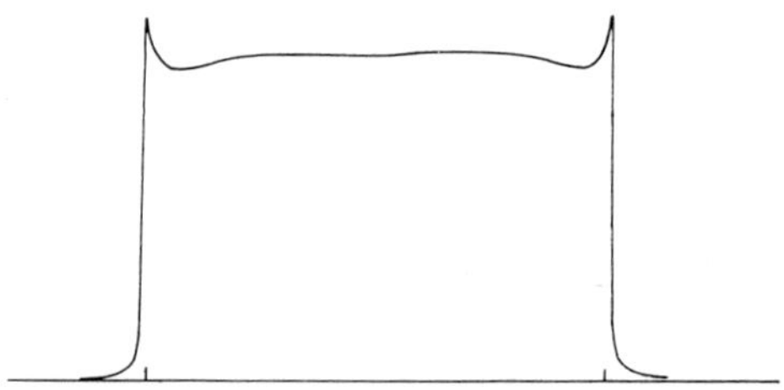

RADIAL BRIGHTNESS DISTRIBUTION OF THE SUN AT λ 8.5 mm

Fig. 5.

spicules extending through the chromosphere. I have assumed that the chromosphere is made up of spicules blending into a gas which smoothly joins the photosphere with the corona. The spicules would have a random or Gaussian distribution about a given size and would also be random in duration and occurrence. For simplicity of analysis, it is assumed that all spicules are pyramidal in shape, have the same base-width (3 sec of arc), and also the same height (10,000 km), which is assumed to be the average height of all spicules. These distributions were arrived at by trial and error to obtain a fit with the distribution curves of radio radial brightness. The derived temperatures and pressure gradients in the spicules (1) and in the inter-spicule material (2) are shown in Fig. 1. It is seen that the spicules are cooler, and hence more dense, than the ambient gas. Hydrostatic equilibrium across the face of the spicules was assumed. Schematically the chromosphere will appear as shown in Fig. 4.

Qualitatively, one can see that such a model will fit the radio results. At millimetre waves, where the penetration is deep, the central ray will emerge from the region high in the spicule and low in the inter-spicule material. The equivalent temperature will be the average of the two, and will be high. For the ray emerging nearer the limb, the spicules will be seen more nearly face on, and since their absorption coefficient is higher than that for the inter-spicule material, they will contribute more to the escaping radiation than will the inter-spicule material. Hence, the equivalent temperature of this ray will be lower. In this way, centre brightening is seen. At the extreme limb the escaping ray must pass tangentially through the atmosphere, and absorption will occur at greater heights, where temperatures are higher, and thus limb-brightening will be observed. In this way the radial-brightness distribution of Fig. 5 can be reconstructed from the spicule model. At the longer centimetre wavelengths, where absorption and radiation by the corona plays a larger part, the penetration is such that the spicules will play little part, and one would expect strong limb-brightening with only a mild centre brightening at 10-cm wavelength.

It is thus seen that radio measurements predict a two-fluid chromosphere with cooler spicules surrounded by a more tenous hotter gas.

Solar Eclipses and the Ionosphere (Special Suppl. Vol. 6 to J. Atmosph. Terr. Phys.), Pergamon Press, London, 1956

Solar Radio Noise and the Study of Corpuscular Streams from the Sun

J. P. WILD
Radiophysics Division, C.S.I.R.O., Sydney

Abstract—Observations in Australia over several years with a dynamic spectroscope covering the frequency range 40–240 Mc/s have shown that most sporadic bursts of radio emission from the sun fall into two distinct spectral classes, having quite discrete ranges of outward radial velocity. Both types of radio burst may occur at the time of solar flares. It is thought that the slower velocities may be associated with the outrush from the sun of geomagnetic storm particles, while the very fast range of velocities may be associated with the ejection from the sun of the agency responsible for the increase in cosmic rays sometimes observed about one hour after flares.

THE sporadic bursts and outbursts of radio emission received from the sun at times of solar activity are believed to be due to plasma oscillations excited by the passage of streams of ionized matter passing through the solar corona: as the streams move outwards through layers of successively decreasing electron density,

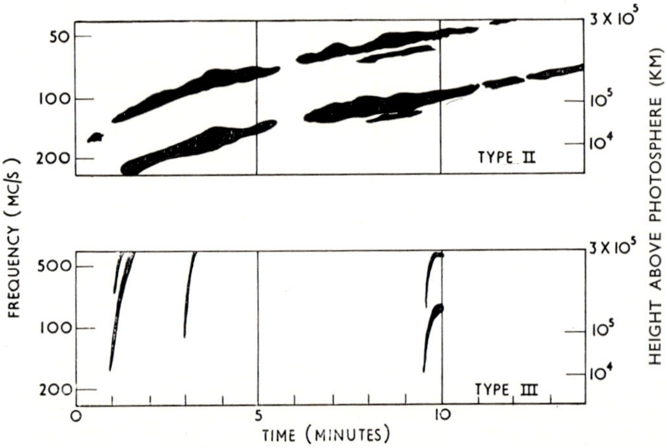

Fig. 1. Idealized examples of the dynamic spectrum of the two known types of sporadic burst.

so the emitted plasma oscillations decrease in frequency. This accounts for the characteristic spectral behaviour of sporadic bursts, which show narrow emission bands that gradually decrease in frequency. Plasma oscillations are inherently nonlinear, and it is significant that the bursts are radiated strongly at second-harmonic as well as fundamental frequencies. The height-variation of electron density in the solar atmosphere is known from optical eclipse observations, and so it is possible to deduce a height-time plot of the ejections from the observed patterns of the dynamic radio-spectrum.

Observations of solar activity with a dynamic spectrograph covering the frequency range 40–240 Mc/s have been in progress at the Radiophysics Laboratory, Australia, since 1952. The great majority of sporadic bursts have been found to fall in two distinct spectral classes, which are illustrated by the idealized examples of these as shown in Fig. 1. In these figures intense emission is represented by black areas in the time-frequency plane. In both types, the fundamental emission band is seen to be duplicated at the second-harmonic frequency. The height-scale on the right-hand side refers only to the fundamental frequency.

The type III bursts may occur sporadically throughout days of solar activity; also intense clusters of them are sometimes observed at the time of solar flares. The type II bursts are much less common, and are highly correlated with flares.

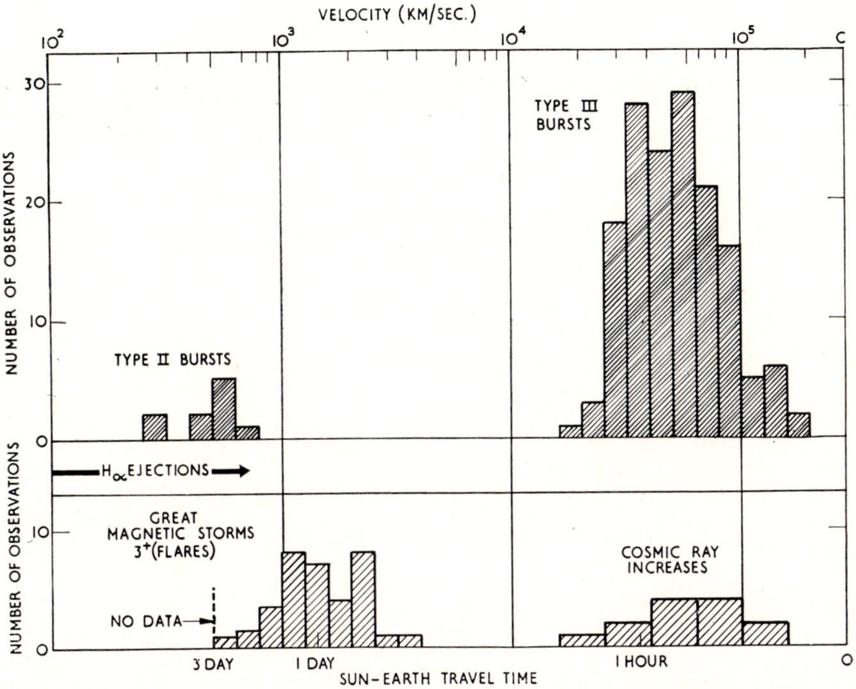

Fig. 2. Histogram of velocities derived from solar radio bursts compared with histograms of velocities derived from sun-earth delays of terrestrial effects following solar flares.

Fig. 2 (upper) shows a histogram of velocities derived from the radio observations. The two types of burst are seen to correspond to two quite discrete ranges of velocity, the mean type II velocity being about 500 km/sec, and the type III 5×10^4 km/sec.

It is thought that in the case of the slower velocities we may be observing the outrush from the sun of geomagnetic storm particles which are known to reach the earth one to three days after outstanding flares. Similarly, the very fast range of velocities correspond to sun-earth delays of the order of one hour, and in these cases we may be observing the ejection from the sun of the agency responsible for the increases in cosmic rays sometimes observed about one hour after flares.

These hypotheses are examined in Fig. 2, in which the velocities corresponding to published sun-earth time-delays of magnetic storms and cosmic-ray increases are compared with those derived from the radio bursts. There is seen to be a close correspondence in velocity between type III bursts and cosmic-ray increases. The type II velocities appear to be rather lower than those of the magnetic-storm particles, but it should be borne in mind that data for the latter refer only to the very large flares of importance 3+, while the former refer to smaller and presumably less energetic flares.

It is suggested that the very fast corpuscles inferred from this work may well produce observable effects in the earth's atmosphere, and that eclipses at suitable times of solar activity may provide the means of recognizing them.

REFERENCES

WILD, J. P., MURRAY, J. D., and ROWE, W. C. 1954 *Aust. J. Phys.*, **7**, 439.
WILD, J. P., ROBERTS, J. A., and MURRAY, J. D. 1954 *Nature*, **173**, 532.

Solar Eclipses and the Ionosphere (Special Suppl. Vol. 6 to J. Atmosph. Terr. Phys.), Pergamon Press, London, 1956

Observations Radioélectriques de l'Eclipse du 30 juin 1954

M. LAFFINEUR, P. COUPIAC, and B. VAUQUOIS
Laboratoire de Radio-Astronomie, Institut d'Astrophysique, Paris

Abstract—The authors have observed the eclipse with the aid of the radio telescope already used by one of them in 1952 at Khartoum. The curve of the eclipse appears very regular, more so than in 1952, while the emission at the instant of optical totality is less intense than in 1952 (14·7 per cent against 19·5 per cent on 545 Mc/s).

LE laboratoire de Radioastronomie de l'Institut d'Astrophysique a observé l'éclipse du 30 juin 1954 à l'île d'Öland (Suède) et à l'Observatoire de Meudon (LAFFINEUR *et al.*, 1954). Le matériel employé fut analogue à celui utilisé à Khartoum pour l'éclipse du 25 février 1952 (LAFFINEUR, 1954).

A Högby, le collecteur d'onde était un miroir parabolique de 6 m de diamètre et de 3 m 40 de distance focale au foyer duquel se trouvaient un dipole demi

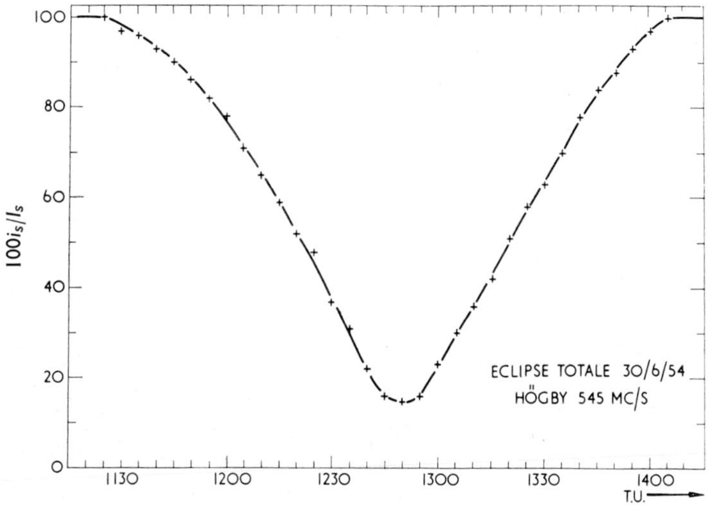

Fig. 1.

onde et son réflecteur; le lobe de l'aérien avait une largeur de 22° entre les premiers zéros. La monture équatoriale, entraînée par un moteur synchrone, possédait un système de débrayage à 2 positions permettant de pointer soit le soleil, soit une région bien déterminée du ciel. Le récepteur du type superhétérodyne était accordé sur 545 Mc/s et possédait une bande hors tout de 1,5 Mc/s; il comprenait un étage H.F. coaxial à tube 2 C 40, suivi d'un mélangeur à cristaux de germanium; l'amplificateur, moyenne fréquence, accordé sur 30 Mc/s, avait une bande passante de 1,5 Mc/s, et un gain de 120 db. A la sortie de cet amplificateur, le signal était détecté par 1 diode et enregistré sur un millivoltmètre potentiométrique Philips PR 2000. La constante de temps était de l'ordre de 1 seconde.

Toutes les 5 minutes, on comparait l'intensité solaire à celle d'une région du ciel prise comme étalon.

Tableau 1. Observations faites à Högby (Suède) $\varphi = 57° 9'$; $L = -17° 2'$.
Fréquenec 545 Mc/s

T.U.	$100 i_s/I_s$	T.U.	$100 i_s/I_s$	T.U.	$100 i_s/I_s$
1125	100	1220	52	1315	42
1130	97	1225	48	1320	51
1135	96	1230	37	1325	58
1140	93	1235	31	1330	63
1145	90	1240	22	1335	70
1150	86	1245	16	1340	78
1155	82	1250	14,7	1345	84
1200	78	1255	16	1350	88
1205	71	1300	23	1355	94
1210	65	1305	30	1400	97
1215	59	1310	36	1405	100

Tableau 2. Observations faites à Meudon (éclipse partielle) $\varphi = 48° 48'$; $L = 2° 14'$.
Fréquence 545 Mc/s

T.U.	$100 i_s/I_s$	T.U.	$100 i_s/I_s$	T.U.	$100 i_s/I_s$
1117	100	1221	51	1341	84
1127	98	1228	46	1350	93
1137	88	1236	44	1358	100
1143	84	1244	42	1406	100
1148	79	1256	46		
1156	74	1306	51		
1202	69	1315	58		
1209	64	1324	65		
1216	56	1332	74		

Tableau 3. Observations faites à Meudon (éclipse partielle) $\varphi = 48° 48'$; $L = 2° 14'$.
Fréquence 255 Mc/s

T.U.	$100 i_s/I_s$	T.U.	$100 i_s/I_s$	T.U.	$100 i_s/I_s$
1106	100	1202	75	1306	54
1117	100	1208	71	1315	63
1126	99	1216	65	1324	71
1137	96	1221	58	1332	77
1144	91	1227	54	1342	87
1150	86	1236	49	1351	94
1157	83	1243	43	1359	99
		1256	42	1417	100

A Meudon, les mesures ont été faites par une méthode analogue; le matériel employé est celui qui sert aux mesures journalières de l'intensité solaire (LAFFINEUR, 1953). L'aérien est 1 paraboloïde de Würzburg de 7 m 50 de diamètre, au foyer duquel se trouvent 2 dipoles demi onde perpendiculaires. Le lobe d'antenne a une largeur de 16° en azimut et 24° en hauteur pour 545 Mc/s. La monture est azimutale et la poursuite du soleil se fait par asservissement du paraboloïde à un équatorial pilote.

Des dépointés de 15° furent faits en hauteur, vers le zénith, toutes les 5 minutes. Les 2 récepteurs sont des superhétérodynes du même type que celui décrit plus haut.

Les résultats bruts des mesures faites sur les enregistrements furent ensuite corrigés des erreurs dues:

(1) à la présence de lobes secondaires dans le diagramme de rayonnement de l'aérien: la correction à apporter fut jugée négligeable.

(2) à la présence de franges d'interférence entre l'onde incidente et l'onde réfléchie par le sol. Les enregistrements du 29 juin et du ler juillet nous ont permis de réduire cet effet.

(3) au rayonnement de la région du ciel prise comme source de comparaison; ce rayonnement fut comparé 6 mois aprés à celui d'une région froide du ciel et introduit une erreur inférieure à 1%.

(4) à la variation de sensibilité du récepteur (essentiellement d'origine thermique); l'étude des dépointés nous permit de faire une correction qui fut d'ailleurs partout inférieure à 2%.

On trouvera ce-dessous les résultats de nos observations (tableaux 1, 2, 3) et la courbe réduite d'éclipse observée en Suède.

BIBLIOGRAPHIE

LAFFINEUR, M. 1953 *Bull. Astron.* (Paris), **18,** 1.

LAFFINEUR, M. et al. 1954 *Ann. Astrophys.*, **17,** 358.

LAFFINEUR, M., VAUQUOIS, B., COUPIAC, P., and CHRISTIANSEN, W. 1954 *Compt. Rend.*, **239,** 1589.

Solar Eclipses and the Ionosphere (Special Suppl. Vol. 6 to J. Atmosph. Terr. Phys.), Pergamon Press, London, 1956

A Model for the Solar Enhanced Region at Centimetre Range Derived from Partial Eclipse Observations

T. HATANAKA,* K. AKABANE,* F. MORIYAMA,* H. TANAKA,† and T. KAKINUMA†

Ionosphere Research Committee, Japan

Abstract—The partial eclipse of the sun on 20 June 1955 was observed at three stations in Japan (Kagoshima, Toyokawa, and Tokyo) at frequencies ranging from 3000 Mc/s to 4000 Mc/s. At Kagoshima and Toyokawa the moon covered a large sunspot-group in the southern hemisphere, and a considerable decrease in flux was observed, while at Mitaka the flux decreased by only a few per cent. Interferometric observations at Toyokawa isolated the flux variation of the enhanced region. By combining the observed curves, the location, size, and brightness distribution of the enhanced region were derived. The observed size of the region is approximately 4′. The brightness temperature is of the order of 10^6 degrees, but is not uniform over the region. Limb-brightening only at the equatorial zone is also suggested.

1. INTRODUCTION

THE partial solar eclipse of the sun on 20 June 1955 was observed at Kagoshima, Toyokawa, and Tokyo. The location, the frequency used, and other information are given in Table 1. The path of the moon's disk for the three stations is illustrated in Fig. 1.

Table 1

Station (latitude and longitude)	Magnitude of the eclipse	Frequency (Mc/s)	Type of aerial	Organization
Kagoshima 31° 37′ N 130° 32′ E	0·38	3000	2-m dish	Tokyo Astron. Obs.
Toyokawa 34° 50′ N 137° 22′ E	0·21	3750 4000	2·5-m dish Eight 1·5-m dish interferometer	Res. Int. Atmospherics
Tokyo (Mitaka) 35° 40′ N 139° 33′ E	0·16	3000	10-m dish	Tokyo Astron. Obs.

In this paper the observational results are presented first. The reductions of the observations by two institutions are described separately, and then the results are compared. It is very remarkable that the independent analyses produced almost identical results. By combining the results, a model for the enhanced region and also the brightness distribution for the quiet sun are discussed in Section 4.

* Tokyo Astronomical Observatory, University of Tokyo.

† Research Institute of Atmospherics, Nagoya University.

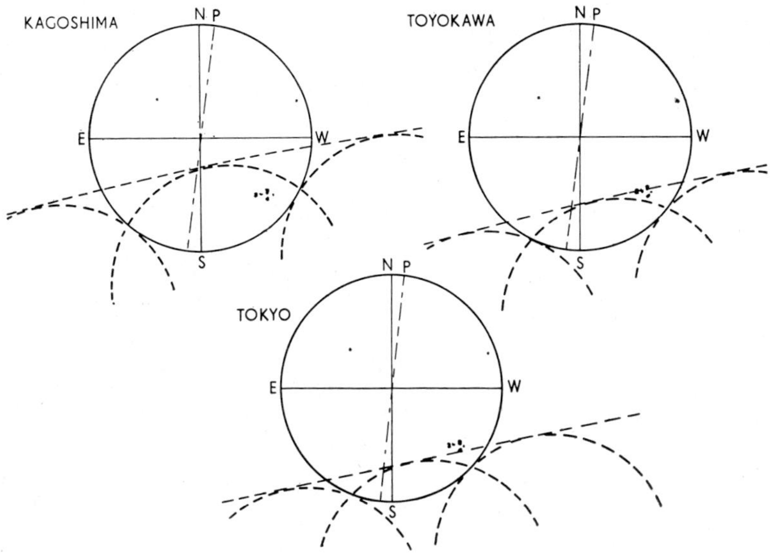

Fig. 1. Path of the moon's disk.

2. Observational Results

The variations of the flux at Kagoshima (3000 Mc/s), Toyokawa (3750 Mc/s), and Tokyo (3000 Mc/s) are shown as curves marked A in Fig. 2(a, b, c) respectively. The flux (for two polarizations) is expressed in units of 10^{-22} watts $m^{-2}(c/s)^{-1}$, which will be referred to as "unit" in what follows. The instants for the first and

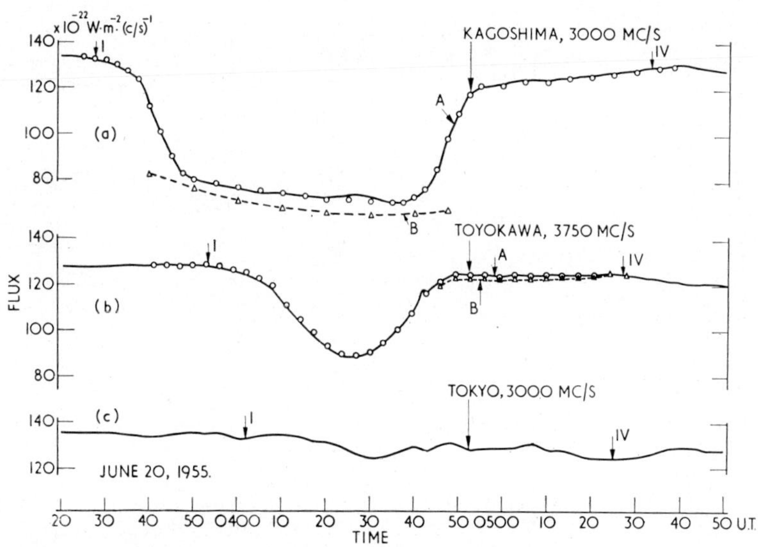

Fig. 2. Eclipse curves. (a) Kagoshima, (b) Toyokawa, (c) Tokyo.

the last contacts for the optical sun are also indicated by the arrows marked I and IV respectively. In Fig. 3 curves A illustrate examples of the drift determined by interferometric observations at Toyokawa. Other curves in the figure will be described later.

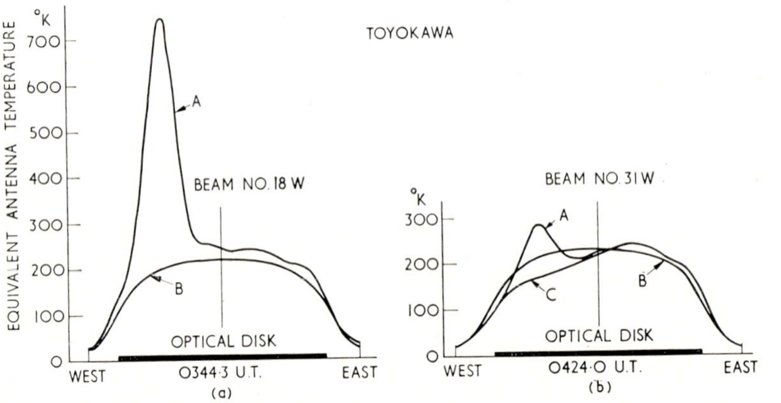

Fig. 3. Examples of the drift curves observed with the eight-element interferometer.

3. Reductions

(a) *Reduction of the observations made at Toyokawa*

Two kinds of observations are available at Toyokawa. One is the flux measurement at 3750 Mc/s with the 2·5-m dish, and the other the interferometric observation at 4000 Mc/s with eight dishes. In the following discussions the flux at 4000 Mc/s is all reduced to the scale at 3750 Mc/s. It is apparent that a sharp peak is observed on the drift curves shown in Fig. 3 in the west part of the sun, which decreased very much during the eclipse. The position line of the peak agrees with that of a large sunspot group in the southern hemisphere seen on the day. It is possible to calculate the flux which corresponds to the peak if the correction due to the quiet sun is applied to a drift curve observed outside the eclipse. It is also possible to calculate the flux for the peak from a drift curve during the eclipse if the correction for the flux due to the covered solar disk is also applied. The last correction depends upon the model assumed for the quiet sun. If the correction due to the change in the activity of this peak, i.e., the "strong radio spot," is applied to the theoretical eclipse curve based on the assumed model, the reconstructed curve must agree with the flux observation made with the 2·5-m dish. This procedure is applied to the part of the eclipse curve after the "spot" has reappeared.

The total flux just before the eclipse is 128 units. The "lower envelope" obtained by superposing the drift curves throughout the month of June 1955 is shown by the curves B in Fig. 3, and the integration under the curve B gives 83 units. There is some extra radiation over the curve B in the eastern part of the disk in Fig. 3(a). Since the flux for this part, or the "eastern part," does not change during the eclipse, we ascribe this to some less-enhanced region in the area not occulted at Toyokawa. The flux for the eastern part is 4 units. Therefore the flux due to the strong radio spot is 128 − 83 − 4 = 41 units at the time just before the eclipse.

In the first trial, the quiet sun is assumed to have the total flux of 83 units and to have a radially symmetrical brightness distribution which gives the drift curve corresponding to the "lower envelope" as shown by curve B in Fig. 3. A solution (KAKINUMA, 1955) gives a model for the quiet sun which shows the limb-brightening with the maximum brightness 1·6 times that at the centre of the disk. By adopting this model for the quiet sun, the corrected eclipse curve was constructed, which is shown as curve B in Fig. 2(b). The agreement is not satisfactory.

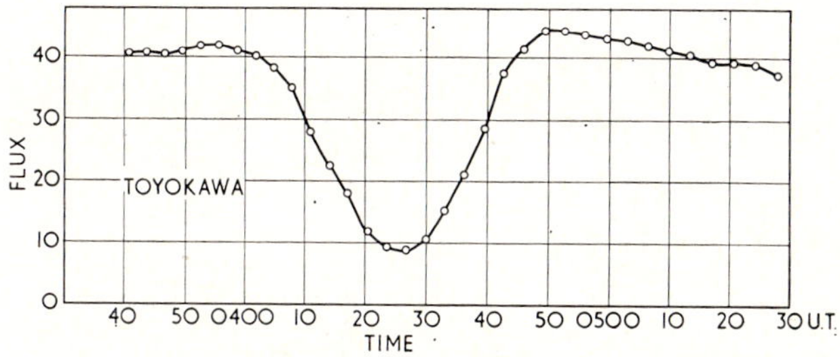

Fig. 4. Variation of the flux due to the strong radio spot alone deduced from the observation with the interferometer.

In the second model assumed, the flux due to the "lower envelope" is divided into two parts: 77 units to the uniform disk with the radius of $1{\cdot}1 R_0$ (R_0 is the radius of the optical sun), and the remaining 6 units to the "other area" not eclipsed at Toyokawa. The calculation is again not consistent with the flux observation with 2·5-m dish.

In the final model adopted, the flux due to the lower envelope is divided into a uniform disk of $1{\cdot}1 R_0$ with the flux of 62 units, and the other area with the flux of 21 units. The flux observation after 0450 U.T. is consistent with this model, as shown by open circles in Fig. 2(b). The other part of the eclipse curve, i.e., before 0450, is also consistent with the calculation based upon this model, as shown in Fig. 2(b).

The variation in the flux of the strong radio spot alone in this case is shown in Fig. 4. An example for the procedure in getting the plots in Fig. 4 is shown by curve C in Fig. 3(b), which is the calculated drift curve for the eclipsed quiet sun when the final model is assumed.

The analysis made so far gives the contribution of each source as shown in column 2 (3750 Mc/s) of Table 2.

The position and the brightness distribution of the "strong radio spot" are analysed as follows: The drift curves obtained by the interferometer are available from 0130 to 0530 (U.T.). Because the sweeping direction of the interferometer varies relative to the solar disk as time advances, it is possible to locate the position of the strong radio spot in a two-dimensional sense by finding the intersection of the position lines of the spot computed from drift curves which are separated by considerable intervals of time. The result is shown in Fig. 5. The centre of activity of the strong radio spot is situated in the spot area. Then the brightness distribution of the strong radio spot is estimated by trial and error by assuming three concentric circles with different brightness temperatures within each circle, until the variation in the flux of the strong radio spot as shown in Fig. 3 is explained. The final model for the brightness distribution is shown in Fig. 5. It is also found that to explain the observations, a "small patch" with the flux of 3 units is to be assumed to the west side of the main part of the strong radio spot.

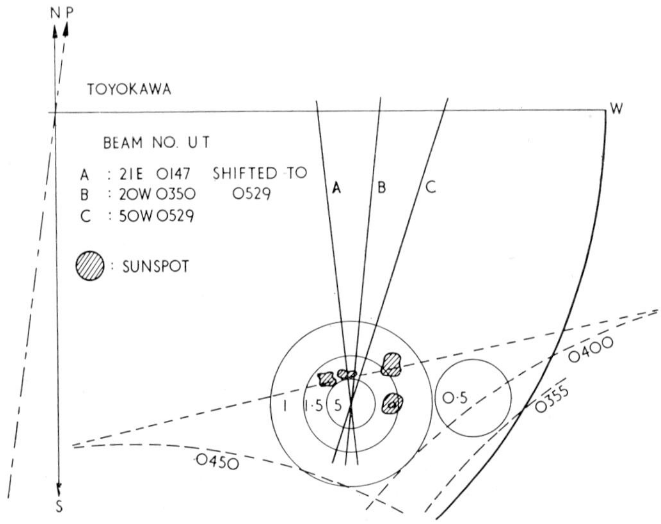

Fig. 5. Isophotes of the brightness for the strong radio spot (Toyokawa).

(b) *Reduction of the observations made at Kagoshima and Tokyo*

The eclipse curve at Kagoshima shows a flat minimum with a sharp fall and rise, as shown in Fig. 2(a). It is reasonable to assume that the strong radio spot has been completely covered during this time-interval at Kagoshima. The total flux just before the eclipse is 135 units at 3000 Mc/s, and the amount of the drop when the spot is covered is 45 units. Therefore the flux due to the strong radio spot is 45 units, and there remain 90 units of the flux.

Under the assumption, which seems reasonable, that the strong radio spot has been completely covered at the time of the flat minimum, it is possible to calculate the variation of the flux due to the covered area other than the spot if a model for the brightness distribution for the covered area is assumed, and hence to check the validity of the model by comparing with the observed variation during the flat minimum.

As the first trial, the brightness distribution of the type described by COVINGTON (1954) is assumed for the covered area with the total flux of 90 units over the whole disk. The computed curve is shown as B in Fig. 2(a), and it is very much in disagreement with the observed curve A.

After many trials, the best fit is obtained when a "uniform disk" of $1{\cdot}1 R_0$

with the total flux of 66 units is assumed to calculate the variation of the flux due to the covered area other than the spot. Open circles in Fig. 2(a) show the calculated values for such a model, and the agreement is satisfactory. The last model is checked also for the part of the curve after the whole of this spot has reappeared. The observed curve at Kagoshima after 0500 (U.T.) is corrected for the possible variation of the spot by using the observation at Tokyo, where the eclipse is already over. The agreement is satisfactory. It is to be mentioned,

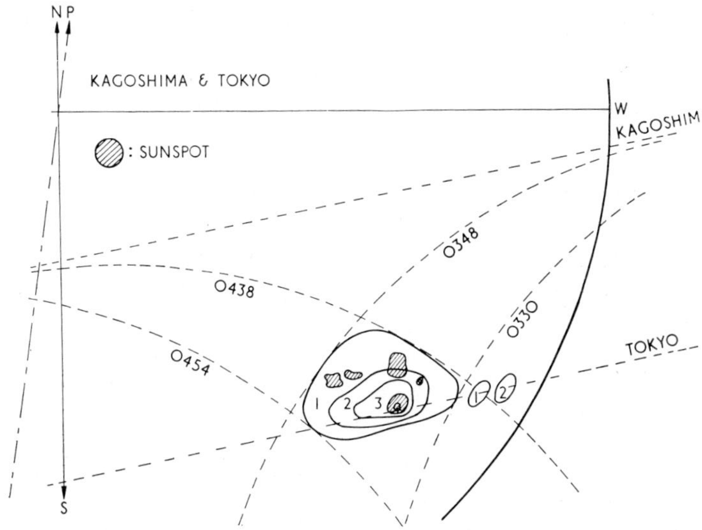

Fig. 6. Isophotes of the brightness for the strong radio spot (Kagoshima and Tokyo).

however, that if the flux of 66 units is used for the total flux of the radially symmetric disk to compute the eclipse curve, it is difficult to distinguish the difference between the uniform disk and Covington-type distribution of limb-brightening.

If we adopt the flux of 45 units for the strong radio spot and that of 66 units for the uniform disk, the remaining 24 units must be in the portion of the solar disk which was not occulted by the moon at Kagoshima. By using the observation at Toyokawa, the contribution of the eastern part is assumed to be 4 units. Therefore the contribution to the total flux by each source becomes as shown in the third column (3000 Mc/s) of Table 2. It should also be remarked here that the eclipse curve between 0350 and 0435 (U.T.) is better explained if a bright zone with the total flux of 3 units is assumed near the western limb just south of the equator (see Fig. 7). This point will be discussed later.

We now turn to the discussion of the strong radio spot. The sharp fall and rise in the eclipse curve well define the possible boundaries of the spot. The flux at Tokyo has dropped a few per cent just at the time when the moon covered the area at the south of the spot. This suggests that only a small portion of the spot is covered at Tokyo, and so it defines the southern extent of the spot. It is rather fortunate that the edge of the moon at the disappearance of the spot at Kagoshima is approximately perpendicular to that at the reappearance. Therefore, if we differentiate with respect to time the gradient of the observed curve at the time of sharp fall and rise, we can get the brightness distribution over the spot. The solution is not unique, but there is a condition that the attributed brightness must be a non-negative value. A smoothed solution for the brightness distribution thus obtained is shown in Fig. 6. It is to be noticed that small patches with the total flux of 3 units to the west of the main part of the strong radio spot are also found from this solution.

The final eclipse curve is reconstructed (open circles in Fig. 2a) by adopting (i) the "uniform disk" with the radius of $1{\cdot}1 R_0$ with the flux of 66 units, (ii) the strong radio spot with the flux of 45 units, (iii) the bright region near the limb at the equator, and (iv) the flux coming from the other area and the eastern part. The agreement with the observed curve A is satisfactory.

4. DISCUSSIONS

The reductions in the preceeding section have been worked out almost independently by two institutions. The results for the contribution to the total flux by each source are summarized in Table 2.

Table 2. Contribution to the total flux by each source (unit: 10^{-22} watt/m^{-2} (c/s)$^{-1}$)

Sources	Res. Inst. Atmospherics (3750 Mc/s)		Tokyo Astron. Obs. (3000 Mc/s)	
Radially symmetric disk	62		66	
Strong radio spot	41		45	
Main part		38		42
Small patch		3		3
Eastern part	4		(4)	
Other area	21		20	
Total	128		135	

It is very remarkable that two sets of reductions give almost identical results. It is seen from Figs. 5 and 6 that the position of the strong radio spot is also in agreement for the two determinations (one by interferometric measurements and the other by the eclipse curve) and that the brightness distribution over the spot is similar. Even the small patch to the west of the spot is deduced from the two reductions. The size of the main part of the strong radio spot is approximately 4′, and it seems to be slightly elongated in the east-west direction. The brightness temperature is: $0{\cdot}43 \times 10^6$ °K for 1 and $2{\cdot}2 \times 10^6$ °K for 5 in Fig. 5 (3750 Mc/s); $1{\cdot}2 \times 10^6$ °K for 1 and $3{\cdot}6 \times 10^6$ °K for 3 in Fig. 6 (3000 Mc/s). These values are subject to errors in the absolute values and also to errors in the estimated area, but the order of 10^6 °K for the brightness temperature of the enhanced region is established. It should also be remarked that the brightness temperature over the spot is not at all uniform.*

* Comparison with the optical observations kindly sent to us by Dr. S. B. NICHOLSON of Mt. Wilson and Palomer Observatories reveals that the location and the form of the strong radio spot as shown in Fig. 6 are in close agreement with those of Ca-plage areas around the sunspot group.

Finally, a remark on a possible source for the flux attributed here to the other area will be given. The flux to be attributed to this area is 21 units (3750 Mc/s) or 20 units (3000 Mc/s). As was discussed in Section 3(b), the observation at Kagoshima suggests a bright region near the western limb just south of the equator. On the other hand, by subtracting the theoretical curve (curve B in Fig. 7b) corresponding to the adopted model for Toyokawa (uniform disk with the radius of $1{\cdot}1R_0$ and total flux of 62 units) from the observed lower envelope for the month of June 1955 (curve A in Fig. 7b), we have a drift curve which has peaks at both limbs (curve C in Fig. 7b). It is tempting to assume a model for the

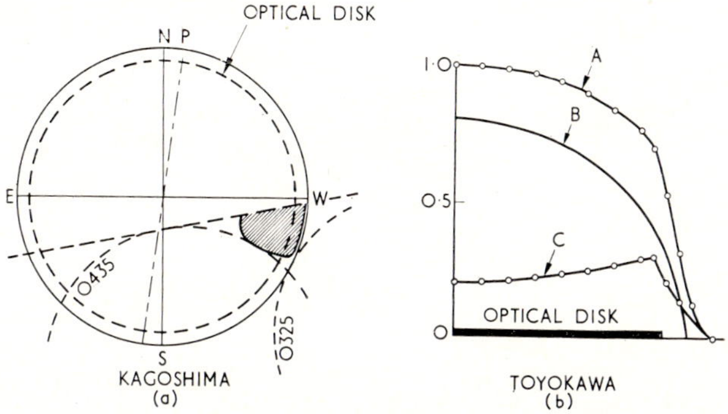

Fig. 7. Illustrations for the bright zone along the equator.

"quiet sun" which consists of a uniform disk of $1{\cdot}1R_0$ and a bright zone along the equator. Then the bright region suggested by the Kagoshima observation (see Fig. 7a) corresponds to the one of limb-brightening which appears only at the equatorial zone with this model.*

5. Acknowledgements

The authors wish to express their thanks to Prof. Y. Hagihara, Director of the Tokyo Astronomical Observatory, and Prof. A. Kimpara, Director of the Research Institute of Atmospherics, for their encouragement, and to Messrs. S. Suzuki and T. Takakura of the Tokyo Astronomical Observatory for their valuable discussions and suggestions. Our thanks are also due to the members of the Ionosphere Research Committee in Japan for discussions.

The detailed description for the observation and reduction will be published elsewhere.

References

Covington, A. E., and Broten, N. W. 1954 *Astrophys. J.*, **119**, 569.

Kakinuma, T. 1955 *Proc. Res. Inst. Atmospherics* (to be published).

* After the present paper was written, Dr. A. E. Covington kindly sent to one of the authors (T. H. a pre-publication copy of a paper: "Radio Brightness Distribution of the Sun at a Wavelength of 10·7 Centimetres, June 30, 1954." Our model for the "quiet sun" is essentially in agreement with the one therein described. The authors are very grateful to Dr. Covington.

Solar Eclipses and the Ionosphere (Special Suppl. Vol. 6 to J. Atmosph. Terr. Phys.), Pergamon Press, London, 1956

Eclipse Observations of Solar Radio-Frequency Radiation on 200, 545, 3000, and 9100 Mc/s

A. D. Fokker, J. C. de Munck, and L. D. de Feiter

Netherlands Postal and Telecommunications Services, The Hague

Abstract—A brief summary is presented of solar radio-noise observations on four frequencies during the eclipse of 30 June 1954. In the case of the frequencies 545 and 200 Mc/s, the duration of the radio eclipse exceeded that of the optical eclipse by about 10 and 13 min. respectively. For the higher frequencies the durations of the radio and optical eclipses are approximately equal. In most cases, maximum phase of the radio eclipse occurred some minutes later than that of the visual eclipse.

Observations were made of the eclipse of 30 June 1954, by the "Ionosphere and Radio Astronomy" section of the Netherlands P.T.T., at Nederhorst den

SOLAR ECLIPSE JUNE 30,1954

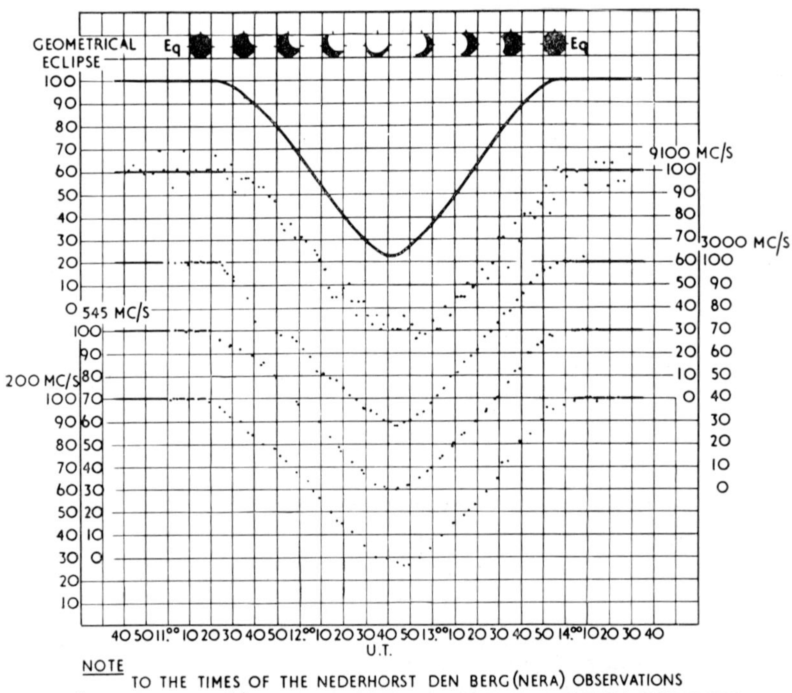

NOTE
TO THE TIMES OF THE NEDERHORST DEN BERG (NERA) OBSERVATIONS (200 AND 545 MC/S) ONE MINUTE HAS BEEN ADDED TO MAKE THEM COMPARABLE TO THE KOOTWIJK ECLIPSE CONDITIONS (KOOTWIJK IS SITUATED SOME 50 KM EAST FROM NERA). ALSO, THE CURVE OF THE GEOMETRICAL ECLIPSE, AS COMPUTED FOR NERA, HAS BEEN SHIFTED ONE MINUTE TO THE RIGHT.

Fig. 1.

Berg-Radio (eastern longitude: 0 h 20 m 18 s; latitude: $+52°\ 14{\cdot}5'$) on 200 Mc/s and on 545 Mc/s, and at Kootwijk-Radio (eastern longitude: 0 h 23 m 23 s; latitude: $+52°\ 09{\cdot}8'$) on 3000 Mc/s and on 9100 Mc/s. The results are given in Fig. 1. Some information on the equipment, the method of observation, and the reliability of the data, is given on following pages.

200 Mc/s. The observations were made with a 7·5-metre paraboidal reflector, mounted alti-azimuthally, the position being adjusted every five minutes. Before and after the eclipse the antenna was pointed due south, to the sun's declination, at which position the galactic background radiation was recorded. From the known brightness-distribution of galactic radiation in this region the background-radiation at the position of the sun could be derived. During the month of October, when the sun was situated in another part of the sky, additional measurements of the background-radiation were made. These new measurements gave a somewhat larger value for the background correction. Since the galactic background noise-level was only very little below the level corresponding to the dummy load, from which the calibration by means of a noise diode was made, no uncertainty was introduced by any extrapolation of the calibration curve to levels below the dummy-load noise-level.

545 Mc/s. Observations were made with the same reflector that was used for 200 Mc/s, the dipole being oriented orthogonal to that for 200 Mc/s. The eclipse curve was read in two ways: one set of readings was made at the times for which the positions were computed beforehand, the other set was made at the times where the inner envelope to the little arcs, described by the sun during its 5-min passage through the antenna pattern, contacts the eclipse curve. The two sets of readings corresponded closely, and are combined in Fig. 1. The error of the minimum residual intensity of the radio eclipse is estimated to be smaller than 0·5 per cent.

3000 Mc/s. Observations were made with a 1·2-metre paraboidal reflector. The comparison level corresponds to the temperature of the sky in the neighbourhood of the sun's position. This level was taken as zero point for the sun's radiation. Due to instrumental difficulties, the measurements between 1130 and 1155 were not considered reliable.

9100 Mc/s. A 60-cm reflecting mirror was mounted alti-azimuthally and measurements were made every two or three minutes, the mirror being given a position corresponding to maximum deflection. In between the measurements the mirror was pointed to an adjacent part of the sky, the recorder thus indicating the zero point to be adopted. Between 1207 and 1307 the pointings were made by observing the sun visually through a hollow tube attached to the mounting and approximately parallel to the radio axis. The deflections obtained by visual pointing turned out later to be 18 per cent less than those obtained by the "maximum deflection" method. This has been allowed for in the reduction. Owing to the imperfections of the receiver, the scatter of the individual readings is large. From a calibration of the receiver, it proved possible to make a linear reduction of the deflections to corresponding intensities.

Principal Results

200 Mc/s. The beginning of the radio eclipse can be estimated to have taken place between 1115 and 1118, the end between 1401 and 1404 (compare visual eclipse: 1123 and 1356 respectively. The radio eclipse thus exceeded the visual eclipse by some ten to sixteen minutes. The maximum obscuration (73 per cent) of the radio eclipse occurred at approximately 1144; that is, some three minutes later than the maximum phase of the visual eclipse.

545 Mc/s. The beginning probably took place between 1117 and 1120, the end probably between 1400 and 1403. The total duration thus exceeded the duration of the visual eclipse by seven to thirteen minutes. Maximum obscuration (70 per cent) occurred at 1241 or 1242, which is about simultaneous with the maximum phase of the visual eclipse.

3000 Mc/s. The beginning took place between 1123 and 1125, the ending between 1357 and 1400 (compare visual eclipse: 1124 and 1357 respectively). The radio eclipse thus seems to have been of almost exactly the same duration as the visual one. Maximum obscuration (72 per cent) occurred at 1243 or 1244, that is, one or two minutes later than the visual eclipse.

9100 Mc/s. The beginning and the ending of the radio eclipse cannot be determined very accurately from the measurements, but it is probable that its duration did not exceed the duration of the visual eclipse. Neither can the moment of maximum obscuration (69 or 70 per cent) be determined with great precision, but it seems to have occurred after the maximum phase of the visual eclipse—some three to nine minutes later. The minimum of the eclipse curve for this frequency is flatter than that found for the other frequencies.

Note: The difference of maximum obscuration of the geometrical eclipse corresponding to the different locations of Nederhorst den Berg and Kootwijk amounts to +0·3 per cent (Kootwijk minus Ned. d. Berg).

Solar Eclipses and the Ionosphere (Special Suppl. Vol. 6 to J. Atmosph. Terr. Phys.), Pergamon Press, London, 1956

SECTION VIII

MISCELLANEOUS

The Ionospheric Eclipse of 23 October 1957

SIMONE D. GOSSNER
U.S. Naval Observatory, Washington

Abstract—Unusual geometric conditions, arising from the fact that the axis of umbra will pass very near the surface of the earth without actually touching it, are discussed for the eclipse of 23 October 1957. The axis of umbra will intersect each ionospheric layer twice at every instant, and the central lines so generated are computed for various heights.

THE solar eclipse of 23 October 1957 will present unusual geometric conditions owing to the fact that the axis of umbra will pass very near the surface of the earth without actually touching it. At the surface there will be a total eclipse of brief duration in a small area off Bruce Coast (Weddell Sea) in Antarctica, because the cone of umbra will touch the earth even though its axis will not. The nearest approach of the axis of umbra to the surface will be 33 km, and the eclipse will thus be central in the ionosphere.

If central lines are computed in the conventional manner (observer above the fundamental plane, i.e., $\zeta > 0$), for heights of 100, 200, and 300 km, it is found that the ionospheric eclipse would be observable only from relatively inaccessible regions of Queen Maud Land and Coats Land. But because the axis of umbra does not intersect the earth at any time, a second intersection with each ionospheric layer will occur at every instant. Geometrically, this is the case of the observer located below the fundamental plane, i.e., $\zeta < 0$, and the computation is similar to that of eclipse phenomena below the horizon.

The central lines generated by this second intersection were computed for heights of 100, 200, and 300 km, and it was found that, for vertical sounding, they will be observable in the general area of the Falkland Islands Dependency (Palmer Peninsula). The positions of the central lines at five-minute intervals are listed in Tables 1(a), 1(b), and 1(c), together with the half-width of path and duration of totality. They are plotted in Fig. 1. The shaded area on the figure shows the region where the eclipse will be total at the surface. The sections of the central lines corresponding to $\zeta > 0$ are located east of the curve marked "Middle at Sunrise," whereas those corresponding to $\zeta < 0$ are located west of that curve.

At those times at which the eclipse is total at the surface, the earth cuts off a portion of the cone of umbra, thereby reducing the half-width of the path on the westernmost side of the central lines generated by the second intersection. These modified values of the half-width are given in a note at the bottom of Tables 1(a), 1(b), and 1(c).

The possibility of a deflection of the central lines owing to refraction was investigated. Since only the order of magnitude of the displacement was sought, the conditions of the International Standard Atmosphere were assumed. It was found that the maximum displacement would be of the order of 2 km.

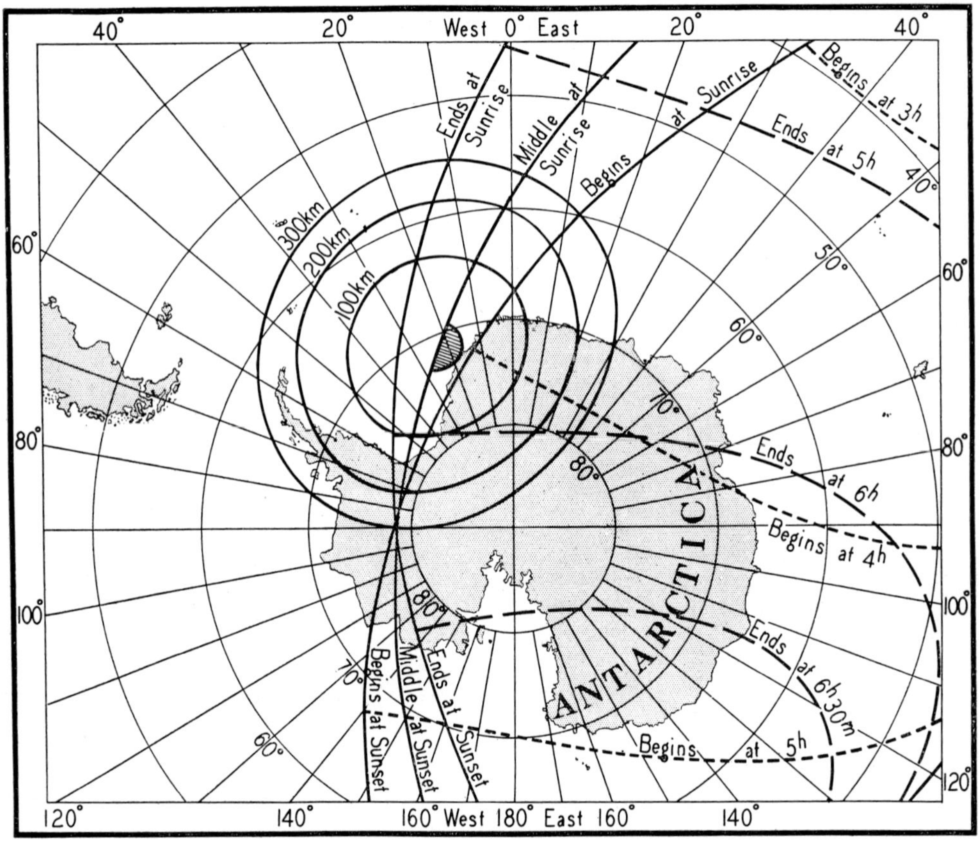

Fig. 1.

Table 1(a). Central line at 100 km

U.T.	ζ > 0				ζ < 0			
	Lat.	Long.	Half-width	Duration	Lat.	Long.	Half-width	Duration
h m	° ′	° ′	km	s	° ′	° ′	km	s
4 38·0	−64 12	+7 49	279	68	−64 12	+7 49	279	68
40	66 44	+0 25	304	74	63 02	17 15	246	62
45	70 08	−3 41	303	78	63 19	27 11	227	59
50	73 00	3 56	293	80	64 23	34 35	224*	57
55	75 39	−0 55	283	80	65 57	41 01	230†	57
5 00	77 59	+7 07	278	79	68 07	46 51	244	58
05	79 22	24 22	278	75	71 18	51 44	266	60
5 08·5	−76 45	+49 09	284	68	−76 45	+49 09	284	68

* 217 km on western side.
† 209 km on western side.

Table 1(b). Central line at 200 km

U.T.	$\zeta > 0$				$\zeta < 0$			
	Lat.	Long.	Half-width	Duration	Lat.	Long.	Half-width	Duration
h m	° ′	° ′	km	s	° ′	° ′	km	s
4 29·1	−59 56	+1 21	177	68	−59 56	+1 21	177	68
30	61 35	−4 45	191	73	58 51	8 07	162	63
35	65 15	12 54	203	81	58 18	20 01	142	57
40	68 06	16 50	202	85	58 44	27 57	134	54
45	70 47	19 12	198	87	59 33	34 45	131	52
50	73 24	20 23	194	88	60 39	41 00	131*	51
55	76 02	20 10	190	88	62 01	46 57	133†	51
5 00	78 42	17 39	186	87	63 43	52 47	137	51
05	81 22	−10 18	183	85	65 48	58 38	144	52
10	83 37	+9 30	182	82	68 32	64 35	154	55
15	83 19	50 52	181	76	72 43	70 45	167	60
5 17·5	−78 28	+71 51	178	67	−78 28	+71 51	178	67

* 127 km on western side.
† 121 km on western side.

Table 1(c). Central line at 300 km

U.T.	$\zeta > 0$				$\zeta < 0$			
	Lat.	Long.	Half-width	Duration	Lat.	Long.	Half-width	Duration
h m	° ′	° ′	km	s	° ′	° ′	km	s
4 22·6	−56 52	−2 39	140	68	−56 52	−2 39	140	68
25	59 49	12 49	158	78	55 15	+9 00	120	59
30	62 56	19 35	165	85	54 59	18 53	108	54
35	65 33	23 51	166	89	55 19	26 20	102	51
40	68 02	27 05	164	92	55 58	32 51	99	49
45	70 27	29 41	161	94	56 49	38 53	97	47
50	72 53	31 47	159	94	57 53	44 39	97*	47
55	75 23	33 22	156	94	59 08	50 17	98†	47
5 00	77 57	34 15	153	94	60 36	55 52	101	47
05	80 41	33 44	151	92	62 20	61 33	104	48
10	83 36	29 30	149	89	64 25	67 25	109	49
15	86 33	−8 37	147	86	67 03	73 44	116	52
20	86 27	+68 26	145	80	70 46	81 01	126	56
5 24·0	−78 38	+90 12	140	67	−78 38	+90 12	140	67

* 94 km on western side.
† 89 km on western side.

For the purpose of interpreting ionospheric observations of the second intersections, it will be of value to know what parts of the atmosphere are intersected at every instant by the cone of umbra. Pertinent data at five-minute intervals are listed in Table 2. They may be interpolated linearly for intermediate times. In using these values, it should be noted that the westernmost edge of the path corresponds to the side of the umbral cone nearest to the surface of the earth.

Table 2. Least distance of umbra to earth's surface

U.T.	Least distance of axis	Radius of umbra	Least distance of nearest edge of umbra
h m	km	km	km
4 25	259	38	221
30	187	38	149
35	129	38	91
40	84	38	46
45	53	38	15
50	36	38	—
55	34	37	—
5 00	46	37	9
05	73	37	36
10	113	37	76
15	168	37	131
20	236	37	199

Solar Eclipses and the Ionosphere (*Special Supp. Vol. 6 to J. Atmosph. Terr. Phys.*), *Pergamon Press, London*, 1956

The E Layer of the Ionosphere

I. Physical Theory: D. H. MENZEL

II. Statistical Analysis: D. H. MENZEL, J. G. WOLBACH, and H. GLAZER

Harvard College Observatory, Cambridge, Mass.

Abstract—Part I considers an extended physical theory of a stratified static ionospheric layer. The equations derived are used in Part II, which gives a statistical analysis of combined data for the E region for the sunspot cycle 1944–53 derived from a large number of stations. Results are compared with those obtained from a similar but more limited study by HARNISCHMACHER.

I. Physical Theory

AN atmosphere in hydrostatic equilibrium obeys the law

$$\frac{\mathrm{d}p}{\mathrm{d}h} = -n\overline{m}g = -p\,\frac{\overline{m}g}{kT}, \qquad p = nkT, \tag{1}$$

where p is the pressure, h the height, n the number of molecules per cm^3, $\overline{m}$ their mean molecular mass, g the acceleration of gravity, k Boltzmann's constant, and T the absolute temperature. To integrate this equation, we must know the variation of $\overline{m}g/kT$ with h. We adopt a linear variation, such that

$$kT/\overline{m}g = b^{-1} + h/q, \tag{2}$$

where b and q are constants

$$b = \overline{m}_0 g/kT_0 \qquad \text{and} \qquad 1/q = d(kT/\overline{m}g)/\mathrm{d}h, \tag{3}$$

where the subscript zero refers to values at the ground or other convenient reference level. Equation (1) integrates to

$$p = p_0\left(\frac{1 + bh_0/q}{1 + bh/q}\right)^q = p_0\left(\frac{T_0}{T}\right) . \tag{4}$$

For an isothermal atmosphere, $q \to \infty$, and

$$p = p_0 e^{-b(h-h_0)} . \tag{5}$$

CHAPMAN (1931) has developed the theory of ionospheric stratification for monochromatic solar radiation and an isothermal atmosphere. I shall here extend the theory.

Let α be the atomic absorption coefficient, n the number of "active" neutral atoms per cm^3, τ the optical depth at height h, and χ the zenith angle of the sun. Then, for an assumed flat-earth model we have

$$\tau = \int_h^\infty n\alpha \,\mathrm{d}h \sec \chi = N\alpha \sec \chi, \tag{6}$$

where N is the number of active atoms per cm^2 vertical column above height h.

In general, α and τ will depend on the frequency ν. Let

I = specific intensity of solar radiation at h,
I_0 = specific intensity of solar radiation outside the atmosphere,
Ω = solid angle subtended by the sun,
$h\nu_0$ = energy of ionization,
$h\nu$ = energy of incident quantum,
n_ε = number of electrons per unit volume,
n_i = number of ions per unit volume,
β = recombination coefficient,
τ_0 = $N\alpha$, optical depth for zenith sun.

$$I = I_0 e^{-\tau} = I_0 e^{-\tau_0 \sec \chi} . \tag{7}$$

In equilibrium, we must have

$$\text{number of ionizations per cm}^3 \text{ per sec} = n\int_{\nu_0}^{\infty} \frac{I\alpha\, d\nu}{h\nu}\Omega$$
$$= \text{number of captures per cm}^3 \text{ per sec} = \beta n_i n_\varepsilon = \beta n_\varepsilon^{\,2}, \tag{8}$$

since $n_i \sim n_\varepsilon$ as long as we can disregard the formation of negative ions. Substituting equations (6) and (7) into equation (8) and differentiating with respect to h, we find that

$$2n_\varepsilon \beta \frac{dn_\varepsilon}{dh} = \int_{\nu_0}^{\infty} \frac{I_0 \alpha\, d\nu}{h\nu} \Omega e^{-\int_h^\infty n\alpha\, dh \sec \chi} \left(\frac{dn}{dh} + n^2 \alpha \sec \chi\right) = 0 \tag{9}$$

for some height, h_m, where n_ε reaches a maximum. Thus:

$$\left[\frac{\cos \chi}{n^2} \frac{dn}{dh}\right]_{h = h_m} = - \frac{\displaystyle\int_{\nu_0}^{\infty} I_0\, \alpha^2 \frac{d\nu}{\nu} e^{-\int_{h_m}^\infty n\alpha\, dh \sec \chi}}{\displaystyle\int_{\nu_0}^{\infty} I_0\, \alpha \frac{d\nu}{\nu} e^{-\int_{h_m}^\infty n\alpha\, dh \sec \chi}} = - \bar{\alpha}_2, \tag{10}$$

where $\bar{\alpha}_2$ is a mean value, taken with respect to the weighting function in the denominator of equation (10).

Now, define another mean $\bar{\alpha}$, such that

$$e^{-\int_{h_m}^\infty n\bar{\alpha}_1\, dh \sec \chi} = \frac{\displaystyle\int_{\nu_0}^{\infty} \frac{I_0 \alpha\, d\nu}{h\nu} \Omega e^{-\int_{h_m}^\infty n\alpha\, dh \sec \chi}}{\displaystyle\int_{\nu_0}^{\infty} \frac{I_0 \alpha\, d\nu}{h\nu} \Omega}, \tag{11}$$

and set

$$S = \int_{\nu_0}^{\infty} \frac{I_0 \alpha\, d\nu}{h\nu} \Omega , \tag{12}$$

the number of solar ionizing quanta absorbed per atom outside the earth's atmosphere. We then have, from equations (10, 8, 11, and 12)

$$\left[\frac{1}{n}\frac{dn}{dh}\right]_{h=h_m} = \left[\frac{d}{dh}\ln n\right]_{h=h_m} = -n_m\bar{\alpha}_2 \sec\chi$$

$$= \frac{\beta}{S}[n_\varepsilon^2]_{\max}\,\bar{\alpha}_2 \sec\chi e^{\int_{h_m}^{\infty} n\bar{\alpha}_1\, dh \sec\chi}. \tag{13}$$

The value of $[n_\varepsilon]_{\max}$ determines the critical frequency f (in Mc/s) by the equation

$$f = 81 \times 10^{-6}[n_\varepsilon]_{\max}. \tag{14}$$

Let us now adopt, from equation (4), the law for density distribution in a polytropic atmosphere. Then:

$$n = n_0/(1 + bh/q)^{q+1}, \tag{15}$$

so that

$$\left[\frac{d}{dh}\ln n\right]_{h=h_m} = -\frac{q+1}{q}\frac{b}{(1+bh_m/q)} = -\frac{q+1}{q}\left(\frac{n_m}{n_0}\right)^{\frac{1}{q+1}} b$$

$$= -n_m\bar{\alpha}_2 \sec\chi. \tag{16}$$

Also,

$$\int_{h_m}^{\infty} n\, dh = \frac{n_0}{b}\frac{1}{(1+bh_m/q)}\,q = \frac{n_m^{\frac{q}{q+1}}\,n_0^{\frac{1}{q+1}}}{b} = \frac{q+1}{q}\frac{1}{\bar{\alpha}_2\sec\chi}, \tag{17}$$

and

$$N = n_0/b, \tag{18}$$

the total number of active atoms per cm² vertical column.

Thus, from equation (16), we have

$$n_m = n_0\left(\frac{q+1}{q}\frac{b\cos\chi}{\bar{\alpha}_2 n_0}\right)^{\frac{q+1}{q}} \tag{19}$$

From equations (13) and (14) we obtain

$$f^4 = 6{\cdot}56 \times 10^{-9}\frac{S}{\beta}\,n_m e^{-\int_{h_m}^{\infty} n\bar{\alpha}_1\, dh \sec\chi}$$

$$= 6{\cdot}56 \times 10^{-9}\frac{S}{\beta}\,n_m e^{-\frac{q+1}{q}\frac{\bar{\alpha}_1}{\bar{\alpha}_2}}$$

$$= 6{\cdot}56 \times 10^{-9}\frac{SbN}{\beta}\left[\frac{q+1}{q}\frac{\cos\chi}{N\bar{\alpha}_2}e^{-\gamma}\right]^{\frac{q+1}{q}} \tag{20}$$

where we employ equations (17) and (19), in turn. We have set

$$\gamma = \bar{\alpha}_1/\bar{\alpha}_2. \tag{21}$$

Thus:

$$f = S^{1/4}\Psi \cos^p \chi, \tag{22}$$

where

$$p = (q + 1)/4q \tag{23}$$

and Ψ is a function of N, q, $\bar{\alpha}_1$, $\bar{\alpha}_2$, and b. Since these quantities may vary seasonally or with position on the earth's surface, Ψ will probably not be a constant. Since $\bar{\alpha}_1$ may show a dependence on χ, we cannot expect Ψ to be constant with χ. However, if Ψ is constant with time at a given location we expect the frequency to obey equation (22). As $q \to \infty$, the isothermal case, $p \to \frac{1}{4}$, in agreement with CHAPMAN's result. We may also note that a positive temperature gradient, $q > 0$, requires $p > \frac{1}{4}$, whereas a negative gradient gives $p < \frac{1}{4}$.

An analysis of observed data appears in the following paper.

This research was supported in part by the Air Force Cambridge Research Center, Geophysics Research Directorate, through Contract AF 19(604)–146 with Harvard University.

REFERENCE

CHAPMAN, S. 1931 *Proc. Phys. Soc.*, **43**, 26 and 483.

II. Statistical Analysis

IN the previous paper, one of us has shown that E-layer critical frequencies should follow the law

$$f = E = S^{1/4}\Psi \cos^p \chi = E_0 \cos^p \chi, \tag{1}$$

where S is a measure of the solar ionizing radiation, Ψ is an unknown function dependent on geographical position and other physical parameters of the ionosphere, χ is the zenith angle of the sun, and p is a constant that depends primarily on the temperature gradient in the ionosphere. Since we cannot separate S and Ψ in our preliminary analysis, we combined them as in equation (1), into a single parameter E_0. The logarithmic form,

$$\log E = \log E_0 + p \log \cos \chi, \tag{2}$$

is the equation of a straight line with slope p and intercept $\log E_0$. We therefore plotted log (median monthly f_0E) against the corresponding $\log \cos \chi$ for the years 1944–1953, with a full year's observations from a given ionospheric station on a single graph. Fitting straight lines to the data by eye, we measured the intercept and slope of the line. Fig. 1 provides a representative sample of these plots to show differences in the amount of scatter of the data among stations, and to give some idea of the reliability of the values ($\log E_0$ and p) entering into the calculations of the regression equations. We represent $\log E_0$ and p as a function of the location of the station ϕ_i and phase of the spot cycle θ. Thus $\log E$ itself becomes a function of location, phase of spot cycle, and time of day. We adopted the following empirical equations, in an attempt to find a satisfactory representation for statistical study.

$$\log E_0 = A + B_1 \sin \theta + B_2 \cos \theta + B_3 \cos^2 \theta + B_4 \cos \varphi_i \quad (3)$$

$$p = A' + B_1' \sin \theta + B_2' \cos \theta + B_3' \cos^2 \theta + B_4' \cos \phi_i + B_5' \log E_0 \quad (4)$$

In equations (3) and (4), the time-factor $\theta = 360k/11$, $k = 0, 1, 2, \ldots, 10$; for 1944 (sunspot minimum) $k = \theta = 0$; for 1945, $k = 1$, $\theta = 32{\cdot}7°$, etc. We

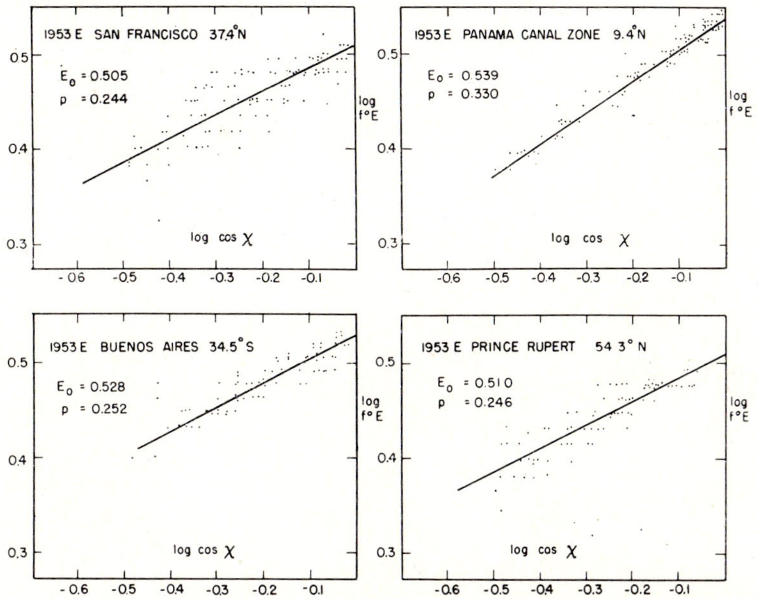

Fig. 1

denote geographic latitude as ϕ_1, geomagnetic latitude as ϕ_2, and magnetic dip as ϕ_3. With models such as these, we were able to investigate which of these three location parameters exhibits the closest relation with $\log E_0$ and p.

To handle the large amount of data in a statistical study, we employed punch cards and an IBM calculator, type CPC. Least-squares solutions for equations (3) and (4) for northern and southern stations separately led to nearly identical results for the two hemispheres (see Fig. 2). Since appropriate statistical tests indicated that the two sets did not differ significantly, we assumed hemisphere location had no effect on the determination of the values of $\log E_0$ and p, and that the slight differences merely arose from sampling and other variations independent of hemisphere.

Therefore we combined all the data in a single sample and, using the least-squares solution to the multiple regression model, we derived the constants that appear in Table 1.

Table 2 tabulates values for $\log E_0$ and p for geographic latitude, geomagnetic latitude, and magnetic dip. The differences between the three sets of tabulated

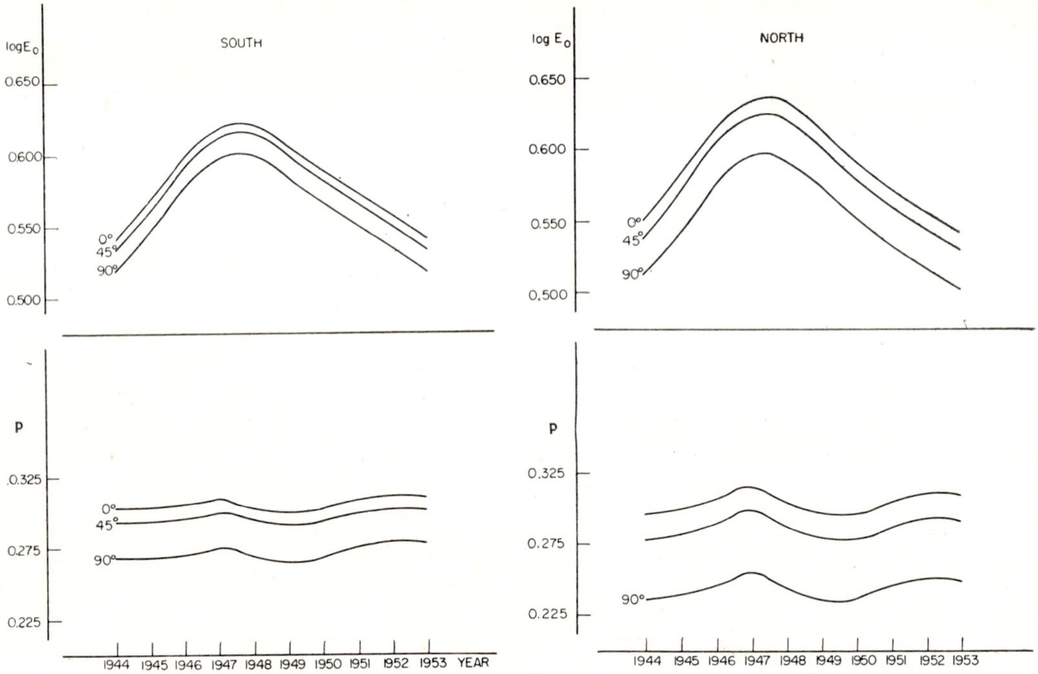
log E₀
SOUTH
0.650
0.600
0.550
0.500
0°
45°
90°
p
0.325
0.275
0.225
1944 1945 1946 1947 1948 1949 1950 1951 1952 1953 YEAR
log E₀
NORTH
0.650
0.600
0.550
0.500
0.325
0.275
0.225
1944 1945 1946 1947 1948 1949 1950 1951 1952 1953

Fig. 2.

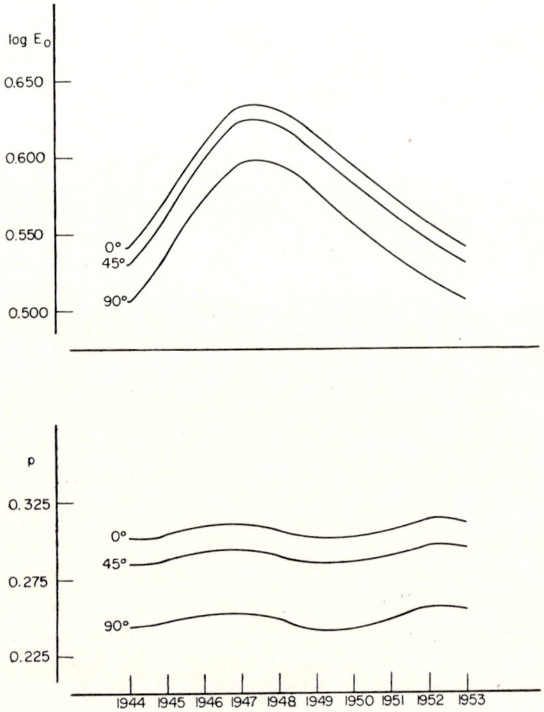
log E₀
0.650
0.600
0.550
0.500
0°
45°
90°
p
0.325
0.275
0.225
1944 1945 1946 1947 1948 1949 1950 1951 1952 1953

Fig. 3.

values arise from two facts: (a) systematic differences between three location parameters for any given station, and (b) rounding errors in the calculations. In our sample, the mean cos (geographic latitude) is 0·766, the mean cos (geomagnetic latitude) is 0·737, and the mean cos (magnetic dip) is 0·524. Hence, for any given station, the variation of the location parameters causes the values in Table 2 to alter. When we substitute these mean cos values into the equations, we obtain the values of log E_0 and p shown in the last three columns of the table. Here the discrepancies come mainly from rounding error. In Fig. 3 we give sample curves for log E_0 and p against time for three different values of magnetic dip (0°, 45°, 90°).

In Table 3 we list the multiple correlation coefficients for log E_0 and p with the independent variables. The square of the multiple correlation coefficient indicates the fraction of scatter in the dependent variable (log E_0 or p) removed by equations (3) and (4) and the constants of Table 1. In other words, Table 3 measures the fraction of variability of log E_0 or p that depends on time or time-location. In terms of the calculated variations (ANDERSON and BANCROFT, 1952), we derive the following facts concerning log E_0:

1. Although a significant reduction in the scatter of log E_0 results from use of a station-location parameter, trigonometric functions representing time account for 94 per cent of the regression.

2. Significant and roughly equivalent reduction occurs in terms of geographic latitude, geomagnetic latitude, or magnetic dip in the estimating equations. When the hemispheres are considered separately, we find that the equation involving magnetic dip reduces the scatter for the north slightly more than either of the other two. However, this reduction does not occur for southern stations.

3. The confidence interval for log E_0, estimated from a single station for any year, depends upon the values of the independent variables used, being smallest near the means and largest for the extreme values. If we substitute values of the independent variables near the means (calculated from the sample), we obtain, as 95 per cent confidence limits, the estimated log $E_0 \pm 0{\cdot}030$; when we use extreme values, we obtain log $E_0 \pm 0{\cdot}035$. If we employ the average value of the location parameter for any given year instead of the actual values, we obtain much smaller confidence regions for the average log E_0. These limits apply only to the years 1944–1953, the single cycle considered.

Our analysis of p contains an extra parameter involving log E_0, which we now may estimate from (3) and the constants of Table 1. However, when we calculated the equations for p, we used the original observed log E_0 values instead of estimated values for log E_0. Even so, the values of p derived from the use of equation (3) for log E_0 should be almost as reliable as those derived from use of the observed log E_0. Nevertheless, in setting up confidence intervals for p, we can no longer assume log E_0 to be an errorless (i.e., fixed) variable. We must take into consideration that log E_0 actually will assume, with a 95 per cent probability, some value in the interval log $E_0 \pm 0{\cdot}035$. Because of this fact, and also because the standard error of an estimated p is twice that of log E_0, even with an assigned log E_0, we do not give any confidence intervals for p.

Table 1. Constants for equations (3) and (4)

	Case 1	Case 2	Case 3
A	$+0{\cdot}5343$	$+0{\cdot}5388$	$+0{\cdot}5529$
B_1	$+0{\cdot}0359$	$+0{\cdot}0361$	$+0{\cdot}0370$
B_2	$-0{\cdot}0313$	$-0{\cdot}0306$	$-0{\cdot}0307$
B_3	$-0{\cdot}0187$	$-0{\cdot}0178$	$-0{\cdot}0169$
B_4	$+0{\cdot}0501$	$+0{\cdot}0455$	$+0{\cdot}0368$
A'	$-0{\cdot}4791$	$-0{\cdot}4693$	$-0{\cdot}4500$
B_1'	$-0{\cdot}0503$	$-0{\cdot}0496$	$-0{\cdot}0483$
B_2'	$+0{\cdot}0413$	$+0{\cdot}0408$	$+0{\cdot}0398$
B_3'	$+0{\cdot}0102$	$+0{\cdot}0099$	$+0{\cdot}0095$
B_4'	$+0{\cdot}0044$	$+0{\cdot}0073$	$+0{\cdot}0106$
B_5'	$+1{\cdot}3294$	$+1{\cdot}3086$	$+1{\cdot}2745$

Table 2. Estimated values for log E_0, p, north and south combined, 1944–1953

Year	Geographic lat.			Geomagnetic lat.			Magnetic dip			Geog. lat.	Geo-mag. lat.	Mag. dip
	0°	45°	90°	0°	45°	90°	0°	45°	90°	Sample mean cosine	Sample mean cosine	Sample mean cosine
log E_0												
1944	0·534	0·520	0·484	0·536	0·523	0·490	0·542	0·531	0·505	0·523	0·524	0·525
1945	0·564	0·550	0·514	0·566	0·552	0·520	0·572	0·561	0·535	0·553	0·554	0·554
1946	0·601	0·586	0·551	0·601	0·588	0·556	0·608	0·597	0·571	0·589	0·589	0·590
1947	0·624	0·609	0·574	0·624	0·611	0·578	0·630	0·620	0·594	0·612	0·612	0·613
1948	0·624	0·609	0·574	0·624	0·611	0·578	0·630	0·620	0·594	0·612	0·612	0·613
1949	0·607	0·592	0·557	0·608	0·594	0·562	0·614	0·603	0·577	0·596	0·596	0·597
1950	0·587	0·572	0·537	0·587	0·574	0·542	0·593	0·582	0·556	0·575	0·575	0·576
1951	0·570	0·550	0·520	0·569	0·556	0·524	0·575	0·564	0·538	0·558	0·557	0·557
1952	0·553	0·538	0·503	0·553	0·539	0·507	0·557	0·546	0·520	0·541	0·541	0·540
1953	0·535	0·521	0·485	0·536	0·522	0·490	0·540	0·530	0·504	0·524	0·524	0·523
p												
1944	0·284	0·266	0·216	0·285	0·271	0·223	0·301	0·284	0·243	0·271	0·272	0·273
1945	0·287	0·269	0·219	0·287	0·273	0·226	0·304	0·287	0·246	0·274	0·275	0·276
1946	0·294	0·276	0·226	0·293	0·279	0·232	0·309	0·292	0·252	0·281	0·281	0·282
1947	0·296	0·278	0·228	0·295	0·280	0·233	0·311	0·294	0·253	0·283	0·282	0·283
1948	0·291	0·273	0·223	0·289	0·275	0·228	0·306	0·289	0·248	0·278	0·277	0·278
1949	0·285	0·267	0·217	0·284	0·270	0·222	0·300	0·283	0·242	0·272	0·272	0·273
1950	0·287	0·269	0·219	0·285	0·271	0·223	0·301	0·284	0·243	0·273	0·273	0·273
1951	0·295	0·277	0·227	0·293	0·279	0·231	0·307	0·291	0·250	0·281	0·281	0·280
1952	0·301	0·284	0·233	0·299	0·285	0·238	0·313	0·296	0·256	0·288	0·287	0·286
1953	0·299	0·281	0·231	0·298	0·283	0·236	0·311	0·294	0·254	0·285	0·285	0·284

Table 3. Multiple correlation coefficient

	Geog. lat.	Geomag. lat.	Mag. dip
log E_0	0·920	0·921	0·921
p	0·616	0·616	0·618

As Table 3 indicates, the variation in p cannot be explained as readily as that of log E_0. Approximately two-thirds of the regression of p is accounted for by its relation with log E_0, one-third by its relation with the location parameter, and only one-thirtieth by time.

When we adopt a model given by equation (3), we are assuming that log E_0 varies with the phase of the spot cycle. However, we may alternatively assume that log E_0 depends upon the magnitude of the sunspot numbers. Fig. 4 shows

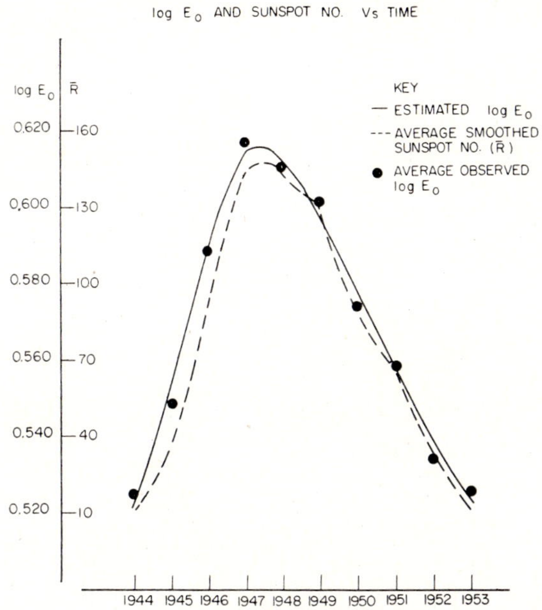

Fig. 4

the relationship that exists between log E_0 and time and also the mean annual sunspot number $\bar{R}$. The curves, which have been adjusted to fit as far as maximum and range are concerned, agree closely on the descending branch, but the values of $\bar{R}$ lie below the curve for the ascending branch. A replot of log E_0 (estimated) against $\bar{R}$, as shown in Fig. 5, shows the discrepancy. This curve, the ascending and descending branches (indicated by the arrows) of which do not coincide, suggests that a given spot number in the rising part of the cycle corresponds to greater ionization (larger log E_0) than does a similar number on the descending branch. With data for log E_0 available for only one spot cycle, we cannot decide which parameter is more fundamental. We can, however, employ the annual spot number instead of the phase, and compare the closeness of fit between log E_0 estimated by the two methods. Significant statistical differences in fit between the two methods for the cycle under consideration, determined by the closeness between estimated log E_0 and observed log E_0, will not necessarily establish the physical superiority of one model over the other.

We calculate the coefficients in the equation

$$\log E_0 = A + B_1\bar{R} + B_2 \cos(\text{mag. dip}) + B_3\, \mathrm{d}\bar{R}/\mathrm{d}t, \tag{5}$$

where $\bar{R}$ = average annual smoothed spot number, and $\mathrm{d}\bar{R}/\mathrm{d}t$ equals one-half the difference between average spot numbers for the prior and following half-year. We define $\mathrm{d}\bar{R}/\mathrm{d}t$ only for the ascending phase of the spot cycle, and set $\mathrm{d}\bar{R}/\mathrm{d}t = 0$ for the descending branch. We had two reasons for adopting this procedure. First, as noted earlier, examination of Figs. 4 and 5 suggested that values of $\log E_0$

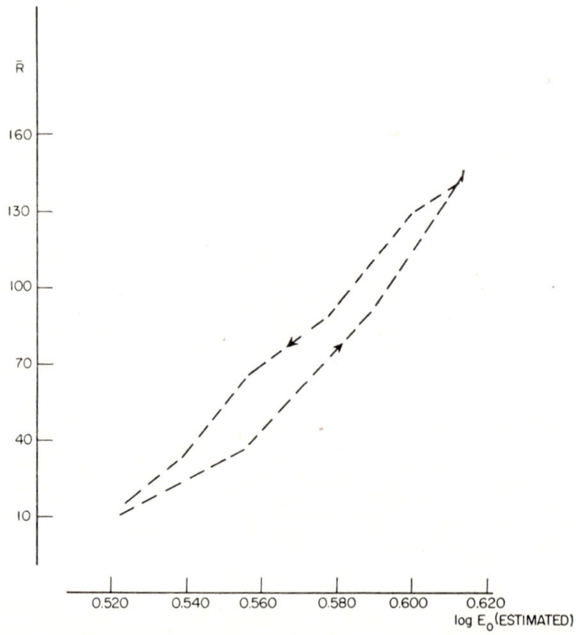

Fig. 5.

may be larger on the rising phase of the cycle for a given $\bar{R}$. The second reason stems from the mathematics of the calculations. Since many stations did not begin to produce records until the late 1940's or early 1950's, the four years 1944–1947 contribute less than one-third of the values entering into the analysis, while the last three years, 1951–1953, contribute almost half. Since each value received equal weight, the minimizing process automatically forced the curve closer to the values of the latter years. Hence, as partial compensation for these two effects, we introduced the $\mathrm{d}\bar{R}/\mathrm{d}t$ term as described above.

Solving equation (5), we obtain

$$\log E_0 = 0{\cdot}4960 + 0{\cdot}0667X_1 + 0{\cdot}0366 \cos(\text{mag. dip}) + 0{\cdot}0799X_3$$

$$X_1 = \bar{R}/100,\ X_3 = \frac{\mathrm{d}\bar{R}}{\mathrm{d}t}/100. \tag{6}$$

The data give a multiple correlation coefficient 0·925, with confidence intervals for $\log E_0$ slightly smaller than those obtained from equation (3), case 3. In

Fig. 6 we plot log E_0 estimated by equation (3) case 3 and equation (6) against observed annual values. If the correlation were perfect, both sets of values would lie on the indicated line with slope equal to 1. Note that for each of the ten points, the value determined from (6) tends to lie slightly closer to the observed value than does the one derived from phase alone. The *a priori* probability of obtaining all the points of one curve closer to the observed values is so small (less than one in one thousand), we may conclude that the spot numbers with

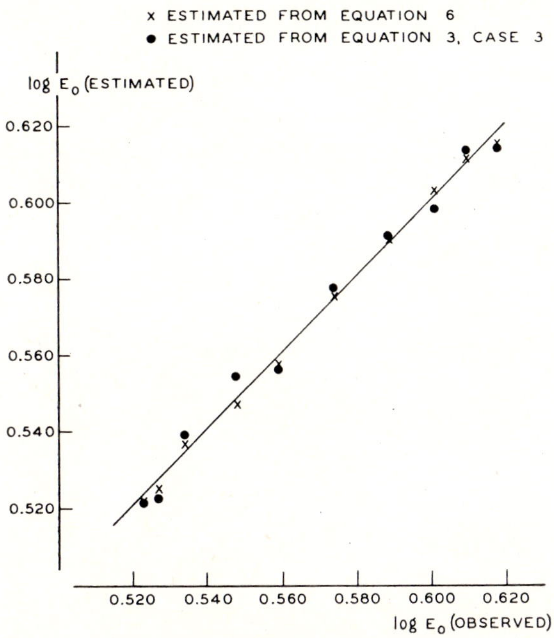

Fig. 6.

the term in $d\bar{R}/dt$ give a significantly closer fit than the phase. A test based on difference in the standard errors, however, does not indicate a significant difference.

We may compare (3) case 3 and (6) in a different manner. At some specified location, we ask at what phase of the spot cycle or for what value of $\bar{R}$ would we expect to find a given value of log E_0? Rather than determine the inverse of a sin and cos function, we set up the cubic in time, and determine the constants by least-squares, as before, with $t = -4, -3, \ldots, +4$, for 1945–1953.

$$\log E_0 = 0{\cdot}4949 + 0{\cdot}05800t - 0{\cdot}01042t^2 + 0{\cdot}00047t^3 + 0{\cdot}03652 \cos (\text{mag. dip}). \quad (7)$$

We solved this equation for t, for different given values of log E_0 and magnetic dip. Once again the latter years received greater weight because they contributed most of the observations. Values of $\bar{R}$ derived by direct solution of (6) in terms of given log E_0, cos (magnetic dip), and $d\bar{R}/dt$. Results are given in Table 4. With some computations we can obtain a confidence interval for R, $\pm(42 - 45)$

for an R estimated by any log E_0 and dip, or not usually more than ± 15 for the average R. This solution provides a means for evaluating solar activity from measures of the E layer, as discussed by the Mixed Commission on the Ionosphere. (1955). One of us (D. H. M.) has been a member of a sub-committee appointed to make such a study.

Table 4. Values of R

log E_0	$d\bar{R}/dt = 0$					$d\bar{R}/dt = 5$				
	Magnetic dip					Magnetic dip				
	0°	45°	60°	75°	90°	0°	45°	60°	75°	90°
0·500					6·0					0
0·520			8·5	21·8	36·0			2·5	15·8	30·0
0·540	11·1	27·2	38·5	51·8	66·0	5·1	21·2	32·5	45·8	60·0
0·560	41·1	57·2	68·5	81·8	96·0	35·1	51·2	62·5	75·8	90·0
0·580	71·1	87·2	98·5	111·8	126·0	65·1	81·2	92·5	105·8	120·0
0·600	101·1	117·2	128·5	141·8		95·1	111·2	122·5	135·8	150·0
0·620	131·1	147·2				125·1	141·2			
0·640	161·1									
	$d\bar{R}/dt = 10$					$d\bar{R}/dt = 15$				
0·500										
0·520				9·8	24·0				3·8	18·0
0·540		15·2	26·5	39·8	54·0		9·2	20·5	33·8	48·0
0·560	29·1	45·2	56·5	69·8	84·0	23·1	39·2	50·5	63·8	78·0
0·580	59·1	75·2	86·5	99·8	114·0	53·1	69·2	80·5	93·8	108·0
0·600	89·1	105·2	116·5	129·8	144·0	83·1	99·2	110·5	123·8	138·0
0·620	119·1	135·2	146·5	159·8		113·1	129·2	140·5	153·8	
0·640	149·1	165·2				143·1	159·2			

If we try to interpret these observational results in terms of the theory of Paper I, we find that the factor $4p^{4p}$ is in itself considerably larger than required to account for the change of E_0 with latitude. The remaining factors in Ψ must tend to compensate, but an exact theory is not available for determining all of the significant atmospheric parameters. The relative constancy of Ψ indicates, however, that the ionosphere is surprisingly uniform with latitude. One can show that γ must be very close to unity.

The mean of p for the magnetic pole is nearly 0·250, the value appropriate for an isothermal layer. For lower latitudes $p > 0{\cdot}25$, so that the temperature would appear to increase upwards.

We are indebted to Dr. RAWER for calling our attention to another statistical analysis, not unlike our own, by M. E. HARNISCHMACHER (1950). His study covered the years 1943–1953, and should, therefore, be comparable with our own. Using data from only six stations, four northern and two southern, he determined the formulae:

$$E_0 = 2{\cdot}25 + 1{\cdot}5 \cos \varphi + (0{\cdot}01 - 0{\cdot}007 \cos \varphi)R.$$

$$p = 0{\cdot}21 + 0{\cdot}12 \cos \varphi + 0{\cdot}0002R. \quad (8)$$

These formulae are said to be unreliable for $\varphi > 60°$ and spot number $R > 150$. To compare our formulae with his, we have calculated from equation (3), case 1, values of log E_0 and p for latitudes 0° and 60°. The results appear in Table 5.

Table 5. Comparison of present results (MWG) with those of Harnischmacher (H)

Year	$\varphi = 0°$				$\varphi = 60°$			
	log E_0		p		log E_0		p	
	MWG	H	MWG	H	MWG	H	MWG	H
1943	—	0·580	—	0·333	—	0·492	—	0·273
1944	0·534	0·578	0·284	0·332	0·509	0·486	0·250	0·272
1945	0·564	0·585	0·287	0·337	0·539	0·507	0·253	0·277
1946	0·601	0·605	0·294	0·348	0·576	0·556	0·260	0·288
1947	0·624	0·624	0·296	0·360	0·599	0·600	0·262	0·300
1948	0·624	0·619	0·291	0·357	0·599	0·589	0·257	0·297
1949	0·607	0·619	0·285	0·357	0·582	0·589	0·251	0·297
1950	0·587	0·602	0·287	0·347	0·562	0·549	0·253	0·287
1951	0·570	0·597	0·295	0·344	0·545	0·538	0·261	0·284
1952	0·553	0·585	0·301	0·336	0·528	0·505	0·267	0·276
1953	0·535	0·578	0·299	0·333	0·510	0·489	0·265	0·273

The agreement is better for latitude 60° than for the equator, and the discrepancy for p is greater than for log E_0. Equation (8) employs a product function connecting φ and R, whereas we have assumed, in accord with the physical theory of Paper I, that the effects of latitude and sunspot cycle are independent, to the order of our calculations here.

From equations (1) and (2) we have

$$\log \frac{S_{max}}{S_{min}} = 4 \log \frac{E_{0max}}{E_{0min}} = 0{\cdot}35 \quad (9)$$

Thus the ratio of ionizing radiation at sunspot maximum to that at minimum varies by a factor of 2.

This research was supported in part by the Air Force Cambridge Research Center, Geophysics Research Directorate, though Contract AF19(604)–146 with Harvard University.

References

Anderson and Bancroft 1952 *Statistical Theory in Research*, Chapter 15.
Harnischmacher, M. E. 1950 *Compt. Rend.*, **230**, 1301.
Mixed Commission on the Ionosphere 1955 *Proc. 4th M.C.I Meeting, Brussels*, p. 78.

APPENDIX

When we start with the equation

$$Y = a + \sum_{i=1}^{k} b_i X_i, \tag{A.1}$$

where $Y = \log E_0$, and the X_i are either sin and cos terms or sunspot functions, and station location terms, then for given values of the independent variables, which we shall denote by X'_i, the estimated values and confidence interval for *any* Y' is given by

$$y' = \sum_{i=1}^{k} b_i x_i \pm t_\alpha s \sqrt{1 + 1/N + \sum_{i=1}^{k} c_{ii} x_i'^2 + 2 \sum_{i<j}^{k} c_{ij} x_i x_j'} \tag{A.2}$$

where $y' = Y' - \bar{Y}$,

$x_i' = X_i' - \bar{X}_i$,

$t_\alpha = 1{\cdot}96$, the critical value of the standardized normal variate corresponding to a 95 per cent probability level,

$s =$ the standard error of estimate,

$N =$ sample size of observations from which the regression equation (A.1) was determined,

$c_{ij} =$ the inverse elements of the covariance matrix,

and $\bar{Y}$, $\bar{X}_i$ are the means determined from the sample.

To estimate the confidence interval for X_i (or for any other X_i) when Y and the remaining X'_is are given, we must solve (A.2) for X_i.

$$x_i' = r_1/r_0 \pm t_\alpha s \sqrt{Q(y', x_i')}, \quad i = 2, 3, \ldots, k,$$

$$r_0 = b_1^2 - (t_\alpha s)^2 c_{11} = 0{\cdot}004436,$$

$$r_1 = b_1 \left(y' - \sum_{i=2}^{k} b_i x_i'\right) + (t_\alpha s)^2 \left(\sum_{i=2}^{k} c_{1i} x_i\right). \tag{A.3}$$

The size of $Q(y', x'_i)$ depends on the proximity of the fixed Y', X'_i's to their means.

The contributions of many of the terms are negligible, so we may simplify (A.3). Returning to the variables $\bar{R}$ and $\mathrm{d}\bar{R}/\mathrm{d}t$, we give the formulae (approximate) for 95 per cent confidence regions, the first with values near the mean, the second with deviations fairly large.

$$R' = -746 + 1503{\cdot}61(Y' - 0{\cdot}0366 \cos(\text{mag. dip}) - 0{\cdot}0008\, \mathrm{d}\bar{R}/\mathrm{d}t)$$

$$\pm\ 42 \text{ (near means)}$$

$$\pm\ 45 \text{ (each variable about } 1{\cdot}5 \text{ standard deviations away)}.$$

We determined the confidence regions for a single Y or X_i, and not for the average of the estimated values. The corresponding confidence region for the average estimated value would be much smaller. The equation for the solution is given by (A.2), with the one in the term containing the square root being removed.

Solar Eclipses and the Ionosphere (Special Suppl. Vol. 6 to J. Atmosph. Terr. Phys.), Pergamon Press, London, 1956

L'Evaluation de la Densité Électronique dans la Région *E*

D. Lepechinsky
Laboratoire National de Radioélectricité, Bagneux/Seine

Abstract—From the point of view of propagation it is important to distinguish two very different regions in the ionosphere: an upper region where the collisional frequency is small and propagation is "quasi-transverse" (Q.T.), and a lower region where the collisional frequency is greater than a certain critical value and propagation is "quasi-longitudinal" (Q.L.). The altitude of the limit between these two regions varies with the angle between the direction of propagation and that of the terrestrial magnetic field.

For latitudes of France, for vertical incidence, the limiting Q.T.-Q.L. level is situated at about 95 km. Since the vertical incidence reflection conditions for the ordinary ray in the Q.T. and Q.L. regions are different, these considerations have an important bearing on the evaluation of the electron density in the E layer from values of f_0E. The error involved in N using the "classical" Q.T. condition when Q.L. condition is appropriate can be 30 per cent or more.

1. Introduction

On sait que l'évaluation habituelle de la densité électronique d'une couche ionisée au niveau de réflexion d'une onde émise sous incidence verticale, repose sur l'hypothèse, que, pour le rayon "ordinaire" la condition de réflexion est:

$$x = \frac{4\pi Ne^2}{mp^2} = 1, \tag{1}$$

où N est la densité électronique,
p est la pulsation de l'onde $(2\pi f)$,
e et m est respectivement la charge et la masse de l'électron.

Lorsqu'on néglige l'influence des chocs entre électrons et molécules neutres ou ions, la condition (1) correspond à l'annulation de l'indice de réfraction "ordinaire": μ_0. Rappelons que les deux autres conditions de réflexion "classiques" qui correspondent elles aussi à l'annulation de l'indice de réfraction μ sont:

$$x = 1 - y \tag{2}$$

$$x = 1 + y \tag{3}$$

où $y = f_H/f$, f_H étant la gyrofréquence (environ 1·2 Mc/s) et f, la fréquence considérée.

La condition (2) s'applique au rayon extraordinaire et (3) au rayon ordinaire dans le cas où la direction de propagation coïncide avec celle du champ magnétique terrestre.

Rappelons aussi qu'aux conditions (1), (2), et (3), vient s'ajouter une condition très particulière, d'ailleurs controversée, dite "4e condition de réflexion," qui correspond à une discontinuité brusque de l'indice de réfraction μ (μ passe par l'infini). Cette 4e condition est:

$$x = \frac{1 - y^2}{1 - {y^2}_L}, \tag{4}$$

où $y_L = y \cos\,\theta$, où θ est l'angle aigu formé par la direction de propagation et celle du champ terrestre.

La prise en considération des chocs entre électrons et molécules ou ions conduit à la formule bien connue d'Appleton et Hartree qui définit la constante diélectrique complexe du milieu ionisé, ayant la forme:

$$\varepsilon = \left(\mu - j\frac{ck}{p}\right)^2, \tag{5}$$

où μ est l'indice de réfraction, définissant la vitesse de la phase,
k est l'indice d'absorption, définissant l'atténuation de l'amplitude,
c est la vitesse de la lumière dans le vide.

Or la détermination de μ et de k à partir de la formule générale d'Appleton et Hartree, montre que les indices μ et k ne s'annulent pour aucune valeur de la densité électronique N, et cela quel que soit l'angle entre la direction de propagation et celle du champ magnétique terrestre.

Mais elle montre aussi que le *genre* de variation de μ et de k en fonction de N est très différent suivant que:

$$\left|\frac{y^2{}_T}{2y_L}\right| \gtrless z, \tag{6}$$

où $z = \dfrac{\nu}{p}$, ν étant le nombre de chocs par sec et $y_T = y \sin \theta$.

Or, il est aisé de voir que (6) correspond à

$$\left|p_H \frac{\sin^2 \theta}{2 \cos \theta}\right| \gtrless \nu \tag{6'}$$

condition qui est indépendante de la fréquence considérée f.

Si le nombre de chocs ν est supérieur à ν_c défini par:

$$\nu_c = \left|p_H \frac{\sin^2 \theta}{2 \cos \theta}\right| \tag{7}$$

on se trouve dans le cas où le type de propagation est dit "quasi-longitudinal" (Q.L.).

Si au contraire ν est inférieur à ν_c, on est dans le cas où le type de propagation est "quasi-transversal" (Q.T.).

H. G. Booker (1934) a montré comment varient μ et k en fonction de x (c. à d. de N pour une fréquence donnée f) dans les cas Q.L. et Q.T. lorsque ν est relativement faible. Rappelons seulement que dans le cas "Q.T." l'indice de réfraction μ_0 devient très petit au voisinage de $x = 1$, tout comme dans le cas où $\nu = 0$ et que dans le cas "Q.L.", l'indice μ_0 devient très petit au voisinage de $x = 1 + y$, tout comme dans le cas où l'on a à la fois $\nu = 0$ et $\theta = 0$ (propagation le long du champ terrestre).

Nous sommes donc amenés à distinguer dans l'ionosphère deux régions très différentes du point de vue de la propagation: la région à nombre de chocs ν, important, région inférieure, ou région "quasi-longitudinale" et la région supérieure, à nombre de chocs faible, ou région "quasi-transversale," puisque le nombre de chocs ν entre électrons et molécules neutres ou ions est une fonction rapidement décroissante de l'altitude.

La limite entre les deux régions, c. à d. l'altitude à laquelle $\nu = \nu_c$, variera, d'après (7) suivant la valeur de l'angle θ, c. à d. pour un sondage vertical, suivant l'inclinaison $(\pi/2 - \theta)$ du champ magnétique terrestre sur l'horizon au lieu considéré.

2. Altitude de la limite des regions Q.T. et Q.L. pour les sondages verticaux

Les données concernant le nombre de chocs ν à diverses altitudes étant assez variables et d'autant moins certaines que l'altitude est plus grande, nous avons eu recours:

(a) à celles indiquées dans les C.R. du Symposium ionosphérique de Cambridge de Septembre 1955;

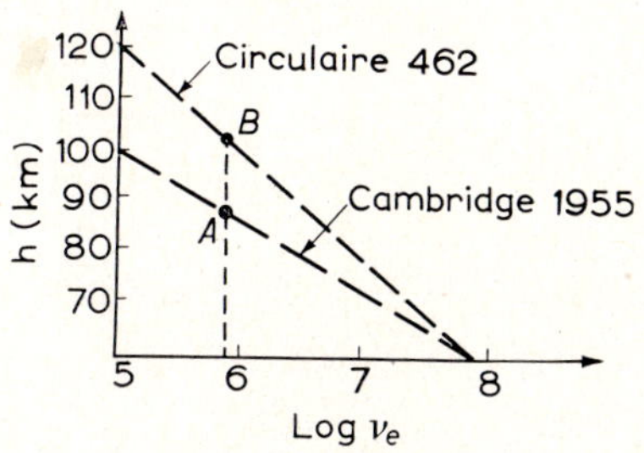

Fig. 1.

(b) à celles figurant dans la Circulaire 462 du National Bureau of Standards.

La fig. (1) reproduit ces données jusqu'à une altitude d'environ 120 km, en admettant une décroissance exponentielle de ν avec l'altitude.

Il est aisé de voir d'après (7) que pour la France (Chambon-la-Forêt, $I = \pi/2 - \theta = 64° \ 10'$) $\nu_c = 7{,}95 \cdot 10^5$ pour les sondages sous incidence verticale, ce qui correspond à une altitude d'environ 88 km (point A) avec les estimations de Cambridge et à une altitude d'environ 100 km (point B) avec celles de la Circulaire 462.

On peut donc admettre que pour la France cette altitude se situe, en moyenne, *à 95 km.*

Notons que pour une inclinaison du champ magnétique terrestre de 65°, $\nu_c = 7{,}42 \cdot 10^5$ et que pour une inclinaison de 70°, $\nu_c = 4{,}67 \cdot 10^5$ sec^{-1}.

Pour cette dernière inclinaison, d'après la moyenne des évaluations de ν, la région Q.L. s'étendrait donc *jusqu'a très près de* 100 *km d'altitude.*

3. Evaluation de la Densité Électronique dans la Couche E

On sait que la couche E est une couche mince dont la demi-épaisseur est de l'ordre de 10 km et dont le maximum d'ionisation se situe aux environs de 100 km d'altitude.

Son évolution diurne est très voisine de celle d'une couche de Chapman, puisque, se formant au lever du Soleil à une altitude de 120 km environ, cette couche descend régulièrement jusqu'à 95 et même 90 km d'altitude vers le milieu de la journée locale, pour remonter encore dans la soirée aux environs de 120 km avant de laisser son faible résidu nocturne.

Durant les éclipses, l'angle zénithal demeurant sensiblement le même, seule la densité électronique de E se modifie rapidement, son altitude restant pratiquement inchangée.

Considérons alors la couche E comme une couche parabolique par exemple dont le maximum est aux environs de 100 km d'altitude (fig. 2) et supposons que la limite des régions Q.L. et Q.T. se situe à 95 km, en XY.

Soient respectivement $N_{\max}$ et N_L les densités électroniques maximum de la couche et celle correspondant à l'altitude XY.

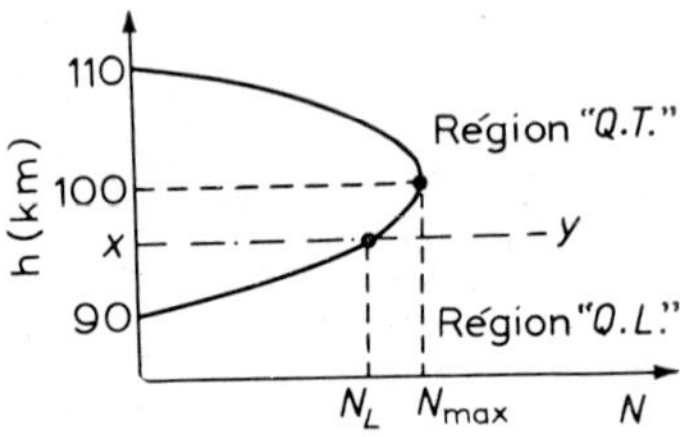

Fig. 2.

D'après ce que l'on a vu plus haut (§ 1), la condition de réflexion approximative d'une onde de fréquence $f = p/2\pi$ est:

$$x = \frac{4\pi N e^2}{m p^2} = 1,$$

dans la région Q.T. et

$$x = 1 + y = 1 + f_H/f$$

dans la région Q.L.

Pour mieux nous rendre compte de la manière dont varie l'indice de réfraction μ lorsqu'on passe de la région Q.L. à la région Q.T. nous avons calculé sa variation pour les rayons ordinaire et extraordinaire dans les 3 cas suivants pour l'inclinaison $(\pi/2 - \theta)$ du champ magnétique terrestre de 64° 10′ et la fréquence $f = 3f_H = 3{,}6$ Mc/s.

(1) $\nu = 1{,}5 \cdot 10^5 \text{ sec}^{-1}$ (cas Q.T.)

(2) $\nu = \nu_c = 7{,}95 \cdot 10^5 \text{ sec}^{-1}$ (cas limite)

(3) $\nu = 1{,}5 \cdot 10^6 \text{ sec}^{-1}$ (cas Q.L.)

(voir fig. 3a, 3b, 3c).

Il est bien évident que pour la densité électronique N_L correspondant à l'altitude limite XY, le type de variation est celui de la fig. 3b.

L'onde qui pénètre la couche E de la fig. 2, se propage évidemment d'abord suivant le mode Q.L. correspondant à la fig. 3c.

Plusieurs cas sont alors à considérer:

(a) si la densité électronique N_L correspond à une valeur x_L de x inférieure à l'unité, l'altitude limite XY correspond à la verticale A_1 B_1 de la fig. 3b; au delà de x_L, la propagation est du type Q.T. (de la fig. 3a) et tombe à une très petite valeur pour $x = 1$, ce qui revient à dire que l'approximation de réflexion classique: $x = 1$ est dans ce cas valable;

(b) si la densité électronique N_L correspond à une valeur x_L de x égale à l'unité, l'altitude limite XY correspond à la verticale A_2 B_2 de la fig. 3b; au delà de x_L la propagation est encore du type Q.T. et μ_0 tombe à une très petite valeur presqu'aussitôt; l'approximation de réflexion classique $x = x_L = 1$ reste valable également dans ce cas;

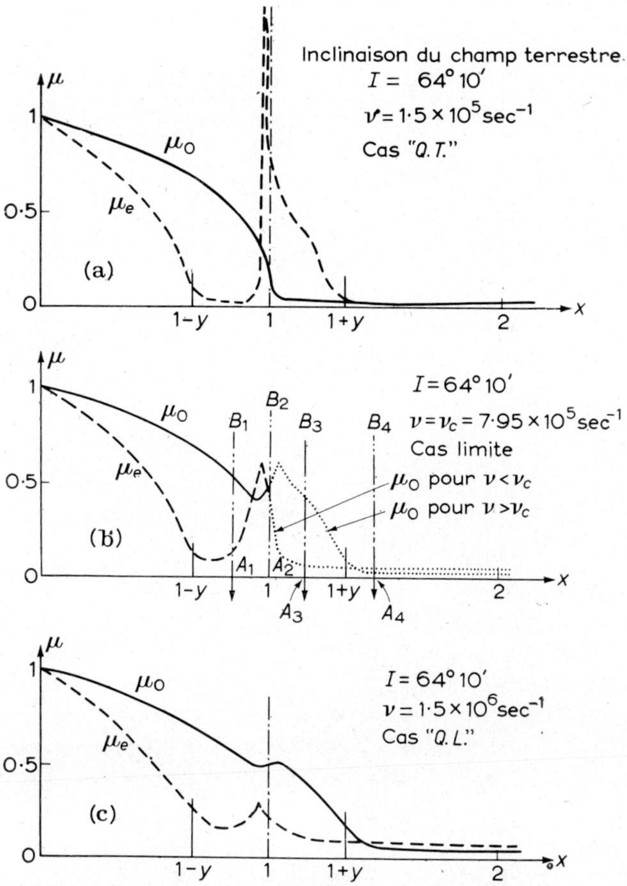

Fig. 3.

(c) si la densité électronique N_L correspond à une valeur x_L de x assez supérieure à l'unité, l'altitude XY, correspond à la verticale A_3 B_3 de la fig. 3b; en deçà de A_3 B_3, même pour $x = 1$ l'indice μ_0 varie suivant le type Q.L. (de la fig. 3c) et garde une valeur appréciable; il tombe toutefois rapidement le long de A_3 B_3 c. à d. pour une valeur de x égale à x_L, pour rejoindre par cette *discontinuité* le type de propagation Q.T. de la fig. 3a; on peut donc estimer que la condition de réflexion classique, $x = 1$ cesse dans ce cas d'être applicable à l'onde "ordinaire" et doit être remplacée par:

$$x = x_L; \tag{8}$$

avec $1 < x_L < 1 + y$;

(d) enfin si la densité électronique N_L correspond à une valeur x_L de x supérieure à $1 + y$, l'altitude XY, correspond à la verticale A_4 B_4 de la fig. 3b et μ_0 devient très petit pour $x = 1 + y$, en deçà de A_4 B_4 où le type de propagation est Q.L. comme dans la fig. 3c; il s'ensuit que dans ce cas, aussi la condition de réflexion classique $x = 1$ n'est pas applicable et doit être remplacée pour l'onde ordinaire par

$$x = 1 + y. \tag{3}$$

Il est important de noter que dans les trois cas: Q.L., Q.T. et limite QL — QT, l'indice μ_e de l'onde *extraordinaire* tombe à une très petite valeur pour

$$x = 1 - y. \tag{2}$$

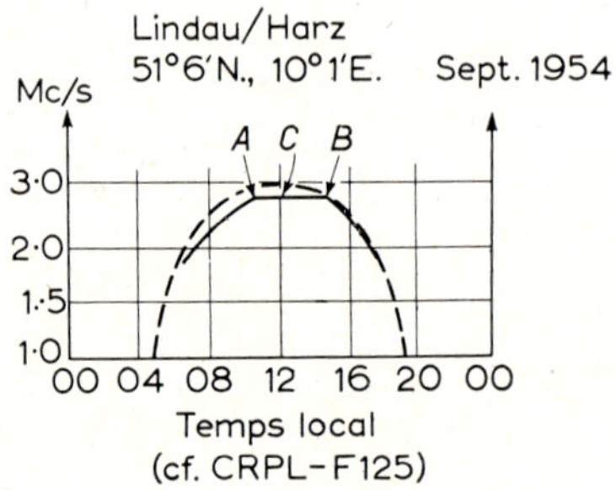

Fig. 4.

Une évaluation correcte de la densité électronique dans la couche E en fonction de l'altitude ne peut donc en général être faite qu'à partir du niveau de réflexion de la composante extraordinaire, au moyen de la condition (10).

En ce qui concerne les fréquences *critiques* "ordinaires" de la couche E, l'existence de l'altitude de transition XY (fig. 2) *au dessus* du niveau de densité électronique maximum, se traduit par un "*plafonnement*" apparent de ces fréquences critiques au milieu de la journée, lorsque la couche E se trouve à son niveau le plus bas.

Tel est fréquemment le cas de la f_0E à Poitiers (LEPECHINSKY et DURAND, 1955; LEPECHINSKY, 1955) et dans beaucoup de stations de sondage de latitude moyenne, comme p. ex. celle de Lindau en Allemagne (voir fig. 4).

A la lumière de ce qui précède, le plafond AB de f_0E de la fig. 4 correspond aux heures où il y a "anomalie Q.L." de la propagation, lorsque la couche E est le plus basse et se trouve "plongée" dans la région Q.L. par sa base. On peut dire en effet que les points A et B correspondent à la valeur $x = x_L = 1$, mais que les points tels que C correspondent à $x = x_L$ avec $1 < x_L < 1 + y$.

L'erreur d'évaluation de la densité électronique maximum peut évidemment atteindre dans ce cas la valeur $y = f_H/f$ c. à d. 30% si $f_H = 1{,}2$ Mc/s et $f = 3{,}6$ Mc/s.

4. CONCLUSION

De ce qui précède, il résulte le fait important que l'évaluation de la densité électronique de E à partir de la f_0E doit être faite avec prudence en raison de l'effet de la propagation Q.L. dans les régions du Globe où l'inclinaison du champ magnétique terrestre est de plus de 60°. La condition de réflexion y est approximativement pour le rayon "ordinaire"

$$x = x_L \qquad \text{avec } 1 < x < 1 + y,$$

durant les heures où la courbe de variation diurne de f_0E plafonne:
Elle est:

$$x = 1 + y \qquad \text{et non pas } x = 1,$$

durant toutes les heures de la journée dans les régions du Globe où l'inclinaison du champ magnétique terrestre est de plus de 80°.

Il est à noter par ailleurs que l'évaluation de la densité électronique de la couche E à partir de fxE, n'est pas entachée de cette anomalie en première approximation et peut donc toujours s'effectuer à partir de la condition

$$x = 1 - y.$$

Il apparait donc essentiel que dans tous les travaux de corrélation de la fréquence critique de E avec l'angle zénithal du Soleil, et en particulier dans ceux relatifs aux éclipses solaires et à la détermination d'un indice solaire ionosphérique, il soit strictement tenu compte de l'effet de la propagation Q.L. les erreurs résultant de l'approximation habituelle ($x = 1$) pour la reflexion de l'onde ordinaire, pouvant aller jusqu'à 30% et davantage.

Bibliographie

Booker, H. G. 1934 *Proc. Roy. Soc.*, **147**, 352.

Lepechinsky, D. 1955 La théorie magnéto-ionique et ses résultats; Méthode pratique de calcul des paramètres de la propagation ionosphérique. Note Prél. No. 185 du Laboratoire National de Radioélectricité.

Lepechinsky, D., et Durand, J. 1955 *Compt. Rend.*, **240**, 333

Solar Eclipses and the Ionosphere (Special Suppl. Vol. 6 to J. Atmosph. Terr. Phys.), Pergamon Press, London, 1956

Phenomena caused by Solar X-rays and Properties of the Solar Electric Ray Bundles producing the Aurorae

L. VEGARD
Universitetets Fysiske Institutt, Oslo

Abstract—Brief account of the effects on the terrestrial atmosphere of solar electric ray bundles. Electron rays appear to dominate auroral production, while the proton-flux influences the spectral composition of auroral luminescence. Disturbances in the ionosphere produced by the ray bundles are considered.

THERE is much evidence that solar X-rays can explain prominent solar and terrestrial phenomena, e.g., the constitution of the ionosphere, the E and $F2$ layers, the zodiacal light, the solar corona, and the limited bundles of solar electric rays which produce the aurorae. These bundles are kept concentrated to limited sources on the sun by electrostatic neutralization with positive ions, by the directing effect of local magnetic fields associated with sunspots, and by the mutual electromagnetic attraction (self-focusing effect) between the elements of the bundle.

Spectrographic results have shown that the electron radiation has a dominating influence on auroral excitation. The neutralization of the bundle alone yields no information concerning the proton flux (velocity) relative to that of the electrons. However, observations show that it varies upwards from very small values, and these variations lead to the following effects: (a) increase of proton flux increases the "magnetic stiffness" of the bundle, and this means that the relative intensity of the H lines increases towards lower latitudes; (b) when the relative proton flux is so small that the penetrating power of the protons is smaller than that of the electrons, the intensity of the H lines should be nearly zero near the lower border of an aurora, but might show up at greater altitudes. Both the latitude and the altitude effects have been found practically without exception from a great number of spectral observations.

A type of bundle essentially consisting of proton rays, as proposed by BENNETT and HULBURT (1954), cannot explain the observed variability effects. Furthermore, a proton bundle cannot explain the fact that aurorae can appear with all the usual forms and structures and yet without the appearance of the Hα and Hβ lines.

The great variability of the proton flux will influence the spectral composition of the auroral luminescence. A number of such variations caused by protons have been observed. Thus spectrograms of the red aurorae of type A indicate that the enhancement of the red O I($^3P - D$) doublet is caused by a great proton flux. Both proton flux and the intensity of the O I doublet show the same type of variability with latitude and altitude. Also the N I($S - D$) doublet is enhanced towards lower latitudes, and it seems as if proton excitation has a preference for transferring the O and N atoms from the normal to the metastable D state.

When a solar bundle enters the atmosphere and the protons and electrons in the bundle are separated on account of differences in absorption, the electric state of the ionosphere will be greatly disturbed. The positive protons will attract the electrons, and the negative surplus from the ray-bundle electrons further down will attract the positive ions and repel the electrons. This means that electric ionospheric currents will be set up, which will disturb the electric state of the *F2* layer. This disturbance of the conducting layers will be increased by the fact that the electron flux is much greater than that of the protons. The accumulation of negative electricity results in positive ionospheric ions being drawn in and electrons being driven away. The currents set up may be spread to great distances and may be accompanied by mass transport (ionospheric winds). In this way the ray bundles will produce sporadic layers and disturb the usual effect of solar eclipses on the *E* and *F1* layers in the auroral zone. The surplus of negative electricity transported by the bundle may partly account for the negative charge of the earth.

These changes in the state of the ionosphere caused by the solar bundles call for the greatest attention in connection with ionospheric studies and investigations of magnetic disturbances.

Reference

Bennett, W. H., and Hulburt, E. O. 1954 *Phys. Rev.*, **95**, 315.
J. Atmosph. Terr. Phys., **5**, 211.

Solar Eclipses and the Ionosphere (Special Suppl. Vol. 6 to J. Atmosph. Terr. Phys.), Pergamon Press, London, 1956

The Influence of the Inner Planets on the Earth's Magnetic Field

J. HOUTGAST

Sterrewacht Sonnenborgh der Rijksuniversiteit, Utrecht

Abstract—The international daily magnetic character figures indicate that Venus, during some days around inferior conjunctions, causes a decrease in the magnetic activity on earth. The explanation presumably lies in a dispersion of streams of solar corpuscles by the magnetic field of Venus. The fact that the influence of Venus is strongest in years of smaller solar activity is in accordance with the higher ratio of slower corpuscles against fast ones, and with less X-radiation during these years.

It seems interesting to go further into these problems and to find out something about the manner in which a planetary magnetic field affects the corpuscular streams from the sun in order to apply this to the theory of magnetic storms. For this purpose consideration should be given to the most effective way of getting quantitative data of the influences on magnetic activity of Venus, Mercury, and the moon.

THE author (HOUTGAST, 1955) has already drawn attention to a rather clearly indicated dip in the magnetic activity on earth as given by the character figure C from seven days before till one day after the dates of inferior conjunctions of

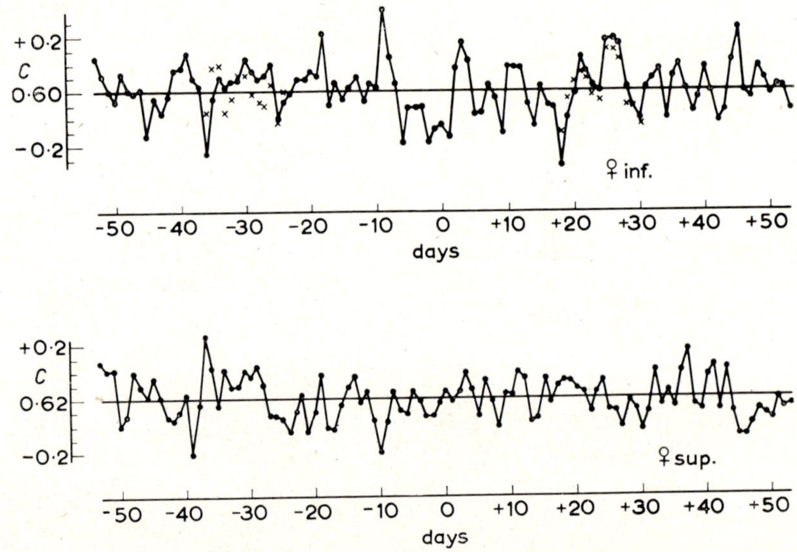

Figs. 1a and 1b.

Venus. It appeared that the dip was relatively more pronounced in the years of lower solar activity, in accordance with the expectation that the dip may be caused by the deflection of solar corpuscules by a magnetic field of Venus. Under very simplifying assumptions, the strength of a dipole field of Venus was deduced, being five times greater than that of the earth.

I have now extended the investigations to a longer interval around conjunction date, for inferior conjunctions of Venus in years with sunspot number $R < 50$, totalling 25 (Fig. 1a), and also for the interval around superior conjunctions for comparison (Fig. 1b). The investigations for inferior conjunctions of Mercury are shown in Fig. 2.

It appears that the curve for superior conjunctions of Venus runs more smoothly than that for inferior conjunctions. The values of C were in both cases corrected for the 27-day period of the sun. This appears to be sound, because in the case of superior conjunctions the value of $|\overline{C - \bar{C}}|$ for uncorrected C's is 0·80 and for corrected 0·64. In the case of inferior conjunctions this correction may not be applied to the values which occur twenty-seven days before

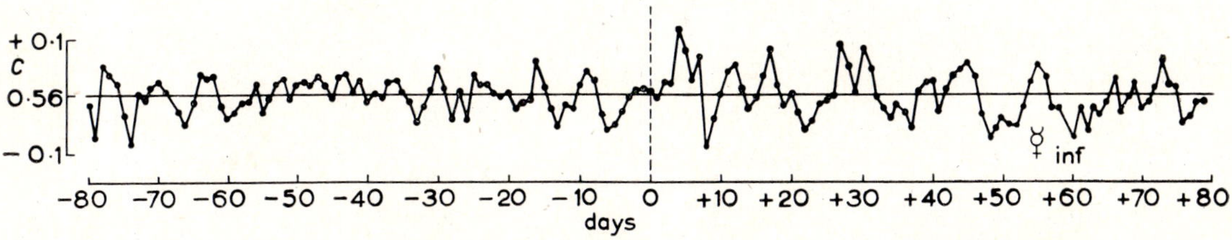

Fig. 2.

and after the dip around conjunction date, if this dip is caused by the deflection of corpuscules by Venus. Indeed, without correction, the curve runs more smoothly in these parts, as is indicated by the crosses in Fig. 1(a). It should be borne in mind that the curve for $R < 50$ does not show the most pronounced decrease in C; if we select smaller angular distances between Venus and the sun at the moment of conjunction, the effect is more clearly shown (HOUTGAST, 1955).

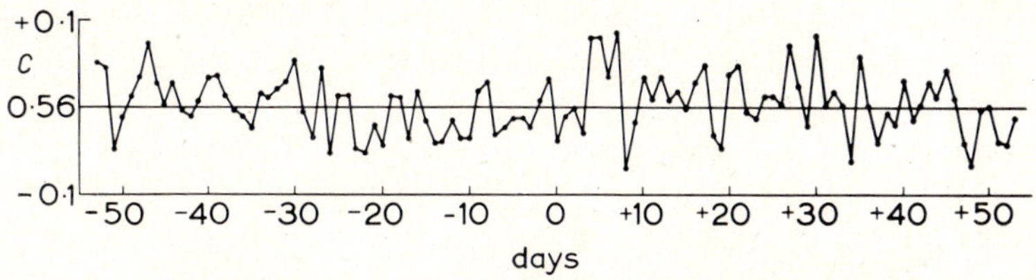

Fig. 3.

The investigation of Mercury applies, as in the case of Venus, to conjunctions during the period 1884–1953, for years in which the sunspot number R is less than 50. The total number of inferior conjunctions amounts then to 117, against 25 in the case of Venus. In this case it seems inappropriate to apply the correction for the 27-day period, since in the superposition of so many series its influence may be obliterated. However, close inspection still shows the effect of the 27-day solar period. The corrected values are shown in Fig. 3, from which at the moment I do not draw any conclusion.*

The occurrence of an influence on the geomagnetic activity of the magnetic field of an inner planet depends on many parameters. If we assume the field to be a dipole field with its axis perpendicular to the orbit, with the same orientation as that of the earth, and the effective corpuscules to be protons, then we can determine the relation between the corpuscular velocity, the dipole strength, and the time at which the decrease in magnetic activity begins relative to the time

* Professor G. KUIPER, in a private communication, did not expect *a priori* a magnetic field of Mercury, because it has presumably a solid core.

of conjunction. With a corpuscular velocity of 500 km/sec, and a magnetic field of the *same* strength as that of the earth, corpuscules deflected over 5° reach the earth not before, but about two days after conjunction in the case of Mercury; for Venus it would then be about one day before conjunction. The general formula for Venus is

$$d = 35{\cdot}4 \sqrt{\frac{p}{v_c \sin \alpha}} - 24{\cdot}6 \sin \alpha - \frac{1166}{v_c}$$

in which d = the number of days before conjunction;

p = the ratio of the magnetic field of Venus to that of the earth;

α = the angle of deflection by Venus of the corpuscules reaching the earth d days before conjunction;

v_c = the corpuscular velocity in km/sec.

The influences of Venus, Mercury, and the moon on geomagnetic activity have previously been announced by SUCKSDORFF (1943). In the case of the planets, he only considered the magnetic indices from four days before till four days after conjunction, which is certainly not enough for a definite statement. SUCKSDORFF found an interesting influence of the moon, and I intend to investigate this point further in the near future.

In view of the fact that a magnetic field of Venus of moderate strength appreciably influences the corpuscular streams from the sun, as is shown by the calculations, a more extensive investigation seems desirable. It may lead to further knowledge about the nature of the solar corpuscular streams, the magnetic fields of the inner planets, and the influence of such a field on the streams. Perhaps a comparison of magnetic indices of observatories at different latitudes may throw more light on the phenomenon. A better index than C may be found by going back to the original records of the magnetic observatories. It should be useful to make calibrated records of magnetic variations at as many observatories as possible during the coming conjunctions of Venus, in periods extending over at least ten days on both sides of conjunction date. The next conjunction dates are: 22 June 1956 and 29 January 1958. It may be expected that during the International Geophysical Year extensive observations for the conjunction of 1958 will become available.

REFERENCES

HOUTGAST, J. 1955 *Nature*, **175**, 678.

SUCKSDORFF, E. 1943 *Die erdmagnetische Aktivität in Sodankylä in den Jahren* 1914–1934, Diss., Helsinki, 1943.

1943 *Geophysica*, **5**, 2.

Solar Eclipses and the Ionosphere (Special Suppl. Vol. 6 to J. Atmosph. Terr. Phys.), Pergamon Press, London, 1956

Chemical Tests and the Study of the Upper Atmosphere

G. PICCARDI
Instituto di Chimica Fisica, University of Florence

(Abstract only)

AN inorganic test which utilizes measurements of the rate of sedimentation of colloidal solutions has been developed which appears to be extremely sensitive to external phenomena. Various correlations have been suggested between the percentage of rapid sedimentations observed and certain geophysical phenomena (magnetic character figure, height of the 96-mb level, cosmic radiation, solar activity, etc.). References to most of these may be found in the following publications.

REFERENCES

BOUTE, C. C.	1955	*Tecnica Italiana*, **10**, 13.
BURKARD, O.	1955	*Archiv. fur Meteorol., Geophys. und Bioklimatol.*, **6**, Ser. B, Vienna, p. 506.
PICCARDI, G.	1951	*Tecnica Italiana*, **6**, 377.
	1952	*ibid.* **7**, 393.
	1954	*ibid.* **9**, 70.
	1953	*Geofis. e Meteorol.*, **1**, 18, 42, 80, 103.
	1954	*ibid.* **2**, 9, 70
	1955	*ibid.* **3**, 31.
	1954	*Naturwissenschaften* **41**, 549.
	1955	*Wetter und Leben* **6**, 217.
PICCARDI, G., and BOUTE, C. C.	1953	*Geofis. e Meteorol.*, **1**, 84.
PICCARDI, G., and DOAT, H. Y.	1953	*Geofis. e Meteorol.*, **1**, 106.
	1955	*ibid.* **3**, 30.
PICCARDI, G., and FRITSCH, V.	1954	*Geofis. e Meteorol.*, **2**, 57.

Solar Eclipses and the Ionosphere (Special Suppl. Vol. 6 to J. Atmosph. Terr. Phys.), Pergamon Press, London, 1956

SECTION IX

CONCLUSION

Concluding Summary

J. A. Ratcliffe
Cavendish Laboratory, Cambridge

In the course of the symposium the author gained the impression that there was fairly general agreement to the conclusions here listed.

The *E* Layer

Sudden changes in the critical frequency observed during eclipses have been associated with known active areas on the sun's disk, and there is little doubt that some of the ionizing radiation comes from them. When, however, attempts are made to account for the shape of the curve near totality, much depends on whether radiation is or is not assumed to come from beyond the visible disk. A fairly self-consistent explanation of most of the results seems to be possible on the supposition that 10 to 15 per cent of the ionizing radiation remains when the visible disk is covered.

The magnitude deduced for the recombination coefficient α depends markedly on whether or not ionizing radiation is present after the visible disk is covered. If it is, α will be larger than has usually been supposed, and may be as great as 10^{-7} cm^3 sec^{-1}. If the *E* layer is supposed to be produced by X-rays, both N_2^+ and O_2^+ ions will be present, and both theory and experiment show that α will be different for recombination to each of these. α has been measured for the process of dissociative recombination, and has been found to be about 4×10^{-8} cm^3 sec^{-1} for O_2^+ and (less certainly) 10^{-7} cm^3 sec^{-1} for N_2^+. During an eclipse, recombination to the N_2^+ would be more rapid, and as the relative amount of N_2^+ decreased, the effective value of α would become smaller.

If, in fact, α in the *E* layer were greater than has usually been supposed (i.e., $>10^{-8}$ cm^3 sec^{-1}), the photon flux required to produce the layer would be greater. This might remove a difficulty encountered by the solar physicists, who usually deduce that the sun would produce more than enough radiation to account for the *E* layer.

Examination of the magnitude of m in the expression $N_p \propto \cos^m \chi$ (see equation 11 of the survey, Section 1) has shown that it varies over the earth as though it were somehow associated with movements of the S_q current system. It is still uncertain, however, what value should be given to m in the interpretation of eclipse results. Nor is it clear whether the eclipse itself modifies the movement of the *E* layer.

The *F1* Layer

All those uncertainties which affect conclusions about the *E* layer also affect consideration of the *F1* layer, and the values deduced from radio measurements for $\alpha F1$ are uncertain. It appears, however, that the ratio $\alpha E/\alpha F1$ can be determined independently, and that it is of the order 1·8.

The *F2* Layer

The behaviour of the *F2* layer during eclipse is largely determined by its movements, and an attempt has been made to measure them.

The $F1\frac{1}{2}$ Layer

It is considered probable that the new critical frequency ($fF1\frac{1}{2}$) occurs during an eclipse at those places where, on a normal day, there is some fairly permanent subdivision of the *F2* layer. At Leopoldville (dip 33·5°) in Africa, $fF1\frac{1}{2}$ was not noticed in the 1952 eclipse, although it was noticed at the other three African stations with dip less than 20°. It is, however, clear that observers have not always described the phenomenon in the same way, and several of them agreed to exchange $h'f$ curves for examination.

The *D* Region

Observations have been made by means of (a) absorption measurements (Piggott), (b) long and very long waves reflected nearly vertically (Weekes), and (c) long and medium waves reflected obliquely (Lauter, Weekes). The observations (b) and (c) showed some phenomena which occurred only very near totality; it will be interesting to examine them on a full-wave theory. Evidence was presented to show that the sources of the *D* and *E* ionizations were located in the same regions on the sun.

Additional Points

Geomagnetic effects

Experiments have shown that, during an eclipse, there is a small change in the geomagnetic field, of the order expected on simple theory. The simple theory has been extended.

Corpuscular eclipse

Experiments made near the auroral zone during an eclipse gave some indication that there might be a corpuscular eclipse effect in the *E* layer.

Eclipse of 1957

The eclipse of 1957 will occur between "layer sunrise" and "ground sunrise," and will present some interesting features.

Literature on Solar Eclipses and the Ionosphere*

CHRONOLOGICAL CATALOGUE OF ECLIPSES

Eclipses of 1912

12–1 Eccles, W. H.; Effect of the eclipse on wireless telegraphic signals. *Nature*, **69** (1912), p. 191.

12–2 Schledermann, H.; Effect of the solar eclipse on wireless telegraph signals. *Nature*, **69** (1912), p. 329.

12–3 Kiebitz, F.; Wireless telegraphic observations during the solar eclipse on 17th April, 1912. *Phys. Z.*, **13** (1912), pp. 890–892.

12–4 Boutaric, A., and Maslin, G.; Influence of the solar eclipse of 17 April 1912 on the propagation of electric oscillations. *Compt. Rend.*, **154** (1912), pp. 1746–1747.

12–5 Flajolet, M.; Reception at the Lyon observatory of radiotelegraphic signals from the Eiffel Tower during the eclipse of the sun on 17th April. *Compt. Rend.*, **154** (1912), 1488–1491.

12–6 Turpain, A.; Influence of the solar eclipse of 17 April 1912 on the propagation of electric waves. *Compt. Rend.*, **154** (1912), pp. 1457–1461.

Eclipses of 1919

19–1 Eccles, W. H.; Radiotelegraphy during the solar eclipse of May 29th, 1919. *Rep. Brit. Assoc.*, (1919), pp. 40–42, and *Nature*, **104** (1919), p. 323–324.

Eclipses of 1921

21–1 Addey, F.; Eclipse of the sun, April 8th, 1921. *Radio Rev.*, **2** (1921), pp. 226–227.

Eclipses of 1925

25–1 Appleton, E. V., and Barnett, M. A. F.; A note on wireless signal strength measurements made during the solar eclipse on January 24th, 1925. *Proc. Camb. Phil. Soc.*, **22** (1925), pp. 672–675.

25–2 Pickard, G. W.; The effect of the solar eclipse of January 24, 1925, on radio reception. *Proc. Inst. Rad. Eng.*, **13** (1925), pp. 539–569.

Eclipses of 1927

27–1 Appleton, E. V. and Barnett, M. A. F.; Wireless observations during the eclipse of the sun, June 29th, 1927. *D.S.I.R.-Radio Research Special Report No. 7* (H.M.S.O.), p. 22.

Eclipses of 1928

28–1 Stchoukin, A.; Observations sur la propagation des ondes électriques courtes pendant l'éclipse solaire du 12 Novembre, 1928. *L'onde élec.*, **8** (1929), pp. 411–419.

Eclipses of 1929

29–1 L'Éclipse du soleil du 9 Mai, 1929. *L'onde élec.*, **8** (1929), pp. 80–84.

29–2 Minohara, T.; Effect of a solar eclipse on long wave propagation. *Naval Tech. Research Lab., Tech. Report No. 391*, June 1929, also *Report on Japanese Research on Radio Wave Propagation*, vol. 1, May 1946.

29–3 Galle, J. B., et Talon, G.; Recherches relatives à la propagation des ondes radio-électriques effectuées à l'occasion de l'éclipse du 9 Mai, 1929. *Compt. Rend.*, **190** (1930), pp. 48–52.

29–4 Stôrmer, C.; Sur l'absence d'échos retardés pendant la totalité de l'éclipse du 9 Mai en Indo-Chine. *Compt. Rend.*, **190** (1930), pp. 106–107.

---

* Compiled, with permission, from Netherlands Postal and Telecommunications Services Document 768a, April 1955. Revised and brought up to date.

29–5 GALLE, J. B.; Observations relatives à la radio-electricité et à la physique du globe: Faites à l'occasion de l'éclipse totale de soleil du 9 Mai 1929 à Poulo Condore-Indo Chine. *L'onde élec.*, **9** (1930), pp. 257–265.

29–6 GALLE, J. B.; Eclipse observations at Poule Condore (Indo-China). *L'onde élec.*, **9** (1930), pp. 312–348.

29–7 KENNELLY, A. E.; Radio during solar eclipses (1929 Poulo Condore results). *Electronics*, **5** (1932), pp. 248–249.

29–8 NAGAOKA, H.; Anomaly in radio transmission during the solar eclipse of May 9 1929. *Proc. Imperial Acad., Tokyo*, **6** (1930), pp. 59–62.

Eclipses of 1930

30–1 BARNETT, M. A. F.; Wireless observations during the total solar eclipse of October, 21st, 22nd, 1930. *N.Z. J. Sci. Technol.*, **12** (1931), pp. 321–331.

Eclipses of 1932

32–1 YOUNG, R. K.; Eclipse of the sun, August 31, 1932, Toronto. *J. Roy. Astron. Soc. Canad.*, **24** (1930), p. 441.

32–2 ROSE, D. C.; Radio observations on the upper ionized layer of the atmosphere at the time of the total solar eclipse of August 31st. *Canad. J. Res.* (1932), pp. 15–28.

32–3 HENDERSON, J. T., and ROSE, D. C.; Fading and signal-strength measurements taken during the solar eclipse of August 31st. *Canad. J. Res.* (1932), pp. 29–36.

32–4 APPLETON, E. V., and CHAPMAN, S.; Suggested wireless observations during the solar eclipse of August 31, 1932. *Nature*, **130** (1932), p. 757–758.

32–5 Electron shadows in "double eclipse" of August 31st. *Electronics* (1932), pp. 249 and 267.

32–6 Eclipse effects on radio (in America). *Electrician*, **109** (1932), p. 304.

32–7 Radio observations during the total solar eclipse of Aug. 31st (and the nature of the ionizing agency). *Nature*, **130** (1932), pp. 385–388.

32–8 The wireless eclipse: corpuscular ionization "not proven"? *Wireless World*, **31** (1932), p. 273.

32–9 Short waves and the solar eclipse. *World Radio*, **15** (1932), p. 541.

32–10 New radio field stations; measurements during eclipse—Bureau of Standards. *Bur. of Stds. Tech. News Bull.* (1932), No. 185, pp. 81–82.

32–11 ALEXANDERSON, E. F. W.; The effect of the sun's eclipse on radio waves. *Radio Eng.*, **12** (1932), pp. 19–20.

32–12 MILLER, J. C. P.; Map of the corpuscular eclipse track of 1932, August. *Monthly Not. Roy. Astron. Soc.*, **92** (1932), pp. 421–422.

32–13 SCHAFER, J. P., and GOODALL, W. M.; An effect of the recent solar eclipse on the ionized layers of the upper atmosphere. *Science*, **76** (1932), pp. 444–446.

32–14 LUGEON, J.; The eclipse of the sun of 31st Aug., 1932 and the "sounding" by atmospheric parasites. *Compt. Rend.*, **195** (1932) pp. 817–819.

32–15 HENDERSON, J.T.; Measurements of ionization in the Kennelly-Heaviside layer during the solar eclipse of 1932. *Canad. J. Res.*, **8** (1933), pp. 1–14.

32–16 WOODWARD, R. W.; Amateur observations during the total eclipse of the sun. *QST*, **17** (1933), pp. 32–38.

32–17 PAUL, H. E.; Beobachtungen an den Kennelly-Heaviside Schichten während der Sonnenfinsternis am 31 August, 1932. *Hochf. tech. El. ak.*, **41** (1933), pp. 81–83.

32–18 HENDERSON, J. T.; Radio test in Canada during the solar eclipse of 31st August, 1932. *Hochf. techn. El. ak.*, **42** (1933), pp. 79–85.

32–19 MIMNO, H. R., and WANG, P. H.; Continuous Kennelly-Heaviside layer records of a solar eclipse (with suggestions of a corpuscular effect on Appleton-layer). *Proc. Inst. Radio Engrs.*, **21** (1933), pp. 529–545.

32–20 KENRICK, G. W., and PICKARD, G. W.; Observations of the effective height of the Kennelly-Heaviside layer and field intensity during the solar eclipse of August 31, 1932. *Proc. Inst. Radio Engrs.*, **21** (1933), pp. 546–566.

32–21 Martin, J. R., and McCuskey, S. W.; Observations in transmission during the solar eclipse of August 31, 1932. *Proc. Inst. Radio Engrs.*, **21** (1933), pp. 567–573.

32–22 Kirby, S. S., Berkner, L. V., Gilliland, T. R., and Norton, K. A.; Radio observations of the Bureau of Standards during the solar eclipse of August 31st, 1932. *Bur. of Stds. J. Res.*, **11** (1933) pp. 829–845. *Proc. Inst. Radio Engrs.*, **22** (1934), pp. 247–264.

32–23 Alexanderson, E. F. W.; The eclipse of August, 1932, observed by radio facsimile. *Proc. Inst. Radio Engrs.*, **23** (1935), p. 454–460.

32–24 Appleton, E. V., and Chapman., S; Report of ionization changes during a solar eclipse. *Proc. Inst. Radio Engrs.*, **23** (1935), pp. 658–669. *U.R.S.I., A.G.* (1934), no. 56 Comm. II.

32–25 Wilson, J. C.; Radio observations during the solar eclipse, Aug. 1932. *Brit. Rad. Annual*, 1934/35, pp. 91–96.

Eclipses of 1933

33–1 Mitra, S. K., Rakshit, H., Syam, P. S. and Ghose, B. N.; Effect of the solar eclipse on the ionosphere. *Nature*, **132** (1933), pp. 442–443.

Eclipses of 1934

34–1 Ito, Y.; Winter state of the ionosphere in the southern seas and observations of the solar eclipse of 14 February, 1934. *Naval Tech. Research Lab. Tech. Report No.* 1268, April 25, 1934. Also *Report on Japanese Research on Radio Wave Propagation*, vol. 1, May 1946.

34–2 Minohara, T., and Ito, Y.; Ionosphere heights measured in the South Sea Islands and radio observations made during the solar eclipse of February 14, 1934. *Rep. Radio Res. Japan*, **4** (1934), pp. 109–120.

34–3 Maeda, K.; Ionospheric measurements at Losap Island during the solar eclipse of February 14, 1934. *Rep. Radio Res. Japan*, **4** (1934), pp. 89–103.

34–4 Nakai, T.; Influence of the solar eclipse upon radio transmission phenomena. *J. Inst. Elec. Engrs. Japan*, **54** (1934), p. 241.

Eclipses of 1935

35–1 Schafer, J. P., and Goodall, W. M.; Eclipse effects in the ionosphere (solar eclipses of February 1935 and August 1932). *Proc. Inst. Radio Engrs.*, **23** (1935), pp. 1356–1360. Also *Bell System Tech. J.*, **15** (1936), pp. 162–166.

35–2 Schafer, J. P., and Goodall, W. M.; Ionosphere measurements during the partial eclipse of the sun of February 3, 1935. *Nature*, **135** (1935), pp. 393–394.

35–3 Kirby, S. S., Gilliland, T. R., and Judson, E. B.; Ionosphere studies during partial solar eclipse. *Phys. Rev., Lancaster, Pa.*, **47** (1935), pp. 632–633.

35–4 Kirby, S. S., Gilliland T. R., and Judson, E. B.; Ionosphere studies during partial solar eclipse of February 3, 1935. *J. Res. Nat. Bur. of Stds.*, **16** (1936), pp. 213–225. Also *Proc. Inst. Radio Engrs.*, **24** (1936), pp. 1027–1040.

35–5 Appleton, E. V., and Naismith, R.; Reports on ionospheric changes during a solar eclipse (Survey, covering eclipses of 1935: discussion of results). *U.R.S.I.; A.G.* 1936, No. 93, Comm. II. Also *U.R.S.I. Proc. of 1936. Gen. Assembly;* Fasc 1, **5** (1938), pp. 314–322.

Eclipses of 1936

36–1 Kirby, S. S., Gilliland, T. R., Smith, N., and Reymer, S. E.; The ionosphere, solar eclipse (of June 19th, 1936) and magnetic storm. *Phys. Rev.*, **50** (1936), pp. 258–259.

36–2 Maeda, K., Isagawa, Y., Minohara, T., Ito, Y., and Nakamura, T.; Ionospheric measurements during the total solar eclipse of June 19, 1936; also ionospheric studies and measurements of the height of the ionosphere (3 articles). *Rep. Radio Res., Japan*, **6** (1936), pp. 91–113, 115–142, 143–150.

36–3 MIYA, K.; On the result of field-intensity measurements of short waves during the solar eclipse of June 19th, 1936. *Nippon. Elec. Comm. Eng.* (1938), p. 26–27.

36–4 Short waves and the eclipse of 19th June. *World Radio,* **22** (1936), pp. 12–13.

36–5 RANZI, I.; Ionospheric observations during the solar eclipse of June 19, 1936. *Nuovo Cimento,* **13** (1936), pp 297–303.

36–6 KESSENIKH, V. N., *et al*; Russian eclipse measurements on the ionosphere. *Nature,* **138** (1936), p. 195.

36–7 ITO, Y., SHINKAWA, H., and YAMAMOTO, M.; Ionospheric observations during the solar eclipse of 19 June, 1936. Part 1. *Naval Tech. Research Lab., Tech. Report No.* 1777, Nov., 1936. Also *Report on Japanese Research on Radio Wave Propagation*, vol. 1, May 1946.

36–8 ITO, Y., SHINKAWA, H., and YAMAMOTO M.; Ionospheric observations during the solar eclipse of 19 June, 1936. *Naval Tech. Research Lab., Tech. Report No.* 1794, Dec. 1936. Also *Report on Japanese Research on Radio Wave Propagation*, vol. 1, May 1946.

36–9 ASCHENBRENNER, H., GOUBEAU, G., PETERSEN, J., and ZENNECK. J.; Influence of the partial solar eclipse of June 19, 1936 on the ionosphere. *Hochfr. Techn. El. Ak.,* **48** (1936), pp. 181–182.

36–10 MAEDA, K., and ISAGAWA, Y.; Ionospheric measurements during the total solar eclipse of 19 June, 1936. *Hiraiso Branch Electrotechn. Lab.*, Dec. 1936. Also *Report on Japanese Research on Radio Wave Propagation*, vol 1, May 1946.

36–11 WHATMAN, A. B., and HAMILTON, R. A.; Radio and magnetic observations at North East Land during total solar eclipse of June 19, 1936. *Nature,* **139** (1937), pp. 69–70.

36–12 MAEDA, K., ISAGAWA, Y., and YUKIO; Ionospheric measurements during the total solar eclipse of June 19, 1936. *Electrotech. J.,* **1** (1937), pp. 12–22.

36–13 KAZIR, T.; On the effects of the eclipse on radio transmission observed in Japan. *Jap. J. Astron. Geophys., Tokyo,* **14** (1937), pp. 219–223.

36–14 MAEDA, K.; Ionospheric measurements during the total eclipse of June 19, 1936. *Jap. J. Astron. Geophys., Tokyo,* **14** (1937), p. 224. (Preliminary report.)

36–15 MINOHARA, T., and ITO, Y.; Ionosphere studies during the total eclipse of June 19th, 1936. *Electrotech. J., Tokyo* **1**, (1937), pp. 159–166. *Jap. J. Astron. Geophys., Tokyo,* **14** (1937), pp. 225–232.

36–16 NAKAMURA, T.; Observations of electric waves in Manchuria during the total solar eclipse of June 19, 1936. *Jap. J. Astron. Geophys., Tokyo,* **14** (1937), pp. 233–254.

36–17 IWATAKE, M., FURUNO, T., TAKEDA, Y., and TERAYE, R.; Measurements of the field intensity and of the ionosphere in the solar eclipse of June 19, 1936. *Jap. J. Astron. Geophys., Tokyo,* **14** (1937), pp. 255–263.

36–18 ARKHANGELSKY, B. F.; Observations of ionosphere during the solar eclipse of June 1936 in Omsk Leningrad. *Glav. Geofiz. Obs., Inf. Sborn. Zem. Mag.* (1937), pp. 67–68. (Russian text.)

36–19 TS'EN, M.-K., CHU, E.-L., and LIANG, P.-H.; Measurements of ionization in the ionospheric layers during the partial solar eclipse of June 19, 1936 at Shanghai. *Chinese J. Phys., Shanghai,* **2** (1937), pp. 169–177.

36–20 KESSENIKH, V. N., BULATOV, N. D., BAERWALD H., and DENISON, V.; Ionospheric observations during the solar eclipse of June, 1936, in Tomsk. *J. Tech. Phys., USSR,* **7** (1937), pp. 466–484, pp. 1237–1252.

36–21 KESSENIKH, V. N.; Study of the ionosphere and solar eclipses. *J. Tech. Phys., USSR* (in Russian), **7** (1937), pp. 1141–1152.

36–22 NAISMITH, R.; Ionospheric conditions during the solar eclipse of 19 June 1936. *Proc. Phys. Soc.,* **49** (1937), pp. 214–224.

36–23 NAKAMURA, T.; Measurements of the height of the ionosphere at Heiho, Manchukuo, during the solar eclipse of June 19, 1936. *Nippon. Elec. Comm. Eng.* (1938), pp. 28–31.

36–24 MIKHAILOV, A. A.; The corpuscular eclipse of 19th June, 1936. *Bull. Komm. po issled. solntsa,* No. XIV (1936).

36–25 PAPALEXI, N. D.; Concerning certain observations during the solar eclipse of 19 June 1936. *Trudy eksped. po nabl. poln. soln. zatm.* 19/6/36, **1** (1938), p. 115–127.

36–26 KESSENIKH, V.N., BULATOV, N. D., and LIKACHEV, A. I.; Ionisation of the upper layers of the atmosphere during a magnetic storm, 18–20 June, 1936, and the solar eclipse, 19 June 1936 at Tomsk. *Trudy eksped. po nabl. poln. soln. zatm.* 19/6/36, **1** (1938), pp. 129–144.

36–27 TETELBAUM, S.; Ionosphere observations during June–July 1936 (at Kiev). *J. Techn. Phys., USSR*, 4, **3** (1937), pp. 200–223. (Abstracted in *Sci. Abstr.* A 4221 (1937).)

Eclipses of 1937

37–1 BERKNER, L. V.; The relations of the Pacific eclipse of June 8, 1937, to investigations of changes of ionization of the ionosphere. *Terrestr. Magn.*, **39** (1934), pp. 317–319.

37–2 WELLS, H. W., STANTON, N. E., and SEATON, S. L.; Ionospheric observations: eclipse of June 8, 1937 (at Huancayo and Watheroo: Data and derived conclusions). *Terrestr. Magn.*, **43** (1938), pp 37–40.

37–3 RANZI, I.; State of the ionosphere during the solar eclipse of 8th June, 1937. *Nuovo Cimento*, **14** (1937), pp. 262–264.

Eclipses of 1940

40–1 HULBURT, E. O.; E region of ionosphere during annular solar eclipse of April 7, 1940. *Terrestr. Magn.*, **44** (1939), pp. 379–381.

40–2 SCHAFER, J. P.; F region critical frequencies at Deal, New Jersey, during partial solar eclipse of April 7, 1940. *Terrestr. Magn.*, **45** (1940), p. 513.

40–3 PIERCE, J. A., HIGGS A. J., and HALLIDAY, E. C.; Ionospheric observations during the solar eclipse of Oct. 1st. 1940 (at 3 points in South Africa). *Nature*, **146** (1940), p. 747. Decrease in ionization of the F2 region during solar eclipse. *Phys. Rev.*, Ser. 2, **58** (1940), p. 1119.

40–4 BERKNER, L. V., and SEATON, S. L.; Ionospheric measurements during the solar eclipse of April 7, 1940. *Trans. Amer. Geophys. U.*, 21st annual meeting, pt. 2, pp. 311–314.

40–5 NATIONAL BUREAU OF STANDARDS; Radio observations during an eclipse (of the sun: period of totality 282 seconds at ground). *J. Franklin Inst.*, **232** (1941), pp. 278–280. See Abstract in *Wireless Eng.* No. 2938 (1943).

40–6 GILLILAND, T. R.; Ionosphere-observations at the 1940 eclipse in Brazil. *Abstract, Trans. Amer. Geophys. U.*, 22nd annual meeting, pt. 2, p. 452.

40–7 HULBURT, E. O.; The E region of the ionosphere during the total solar eclipse of October 1, 1940. *Phys. Rev.*, **55,** pp. 646–647.

40–8 The ionospheric eclipse of 1940, October 1st. *Monthly Not. Roy. Astron. Soc.*, **98** (1938), pp. 664–669.

40–9 The corpuscular eclipse of 1940, October 1st. *Monthly Not. Roy. Astron. Soc.*, **98** (1938), pp. 727–733.

40–10 The ionospheric eclipse of 1st Oct. 1940 (projections on earth of paths of totality at 100–300 kilometres). *Monthly Not. Roy. Astron. Soc.*, **98** (1938). Also *Nature*, **149** (1942) p. 701.

40–11 HIGGS, A. J.; Ionospheric measurements made during the total solar eclipse of 1940, October 1st, South-Africa. *Monthly Not. Roy. Astron. Soc.*, **102** (1942) pp. 24–34. Also *Nature*, **149** (1942) p. 701.

40–12 SMITH, N.; Predictions of normal radio critical frequencies related to solar eclipse in 1940. *J. Res. Bur. Stand.*, **24** (1940), pp. 225–228.

40–13 PIERCE, J. A.; Harvard ionospheric eclipse expedition to South-Africa. *Proc. Inst. Radio Engrs.*, **29** (1941), p. 470.

40–14 GILLILAND, T. R.; Radio observations of the ionosphere—Nat. Geogr. Soc.—N.B.S., Solar eclipse expedition to Brazil. *Nat. Geogr. Soc. Contr. Techn. Paper.*, Solar eclipse ser. no. 2 (1942), pp. 88–95. (Abstracted in *Sci. Abstr.* A 539 (1942).

40–15 PIERCE, J. A.; The ionospheric eclipse of October 1, 1940. *Proc. Inst. Radio Engrs.*, **36** (1948), pp. 8–15.

40–16 BERKNER, L. V.; Infra-red records of the eclipse made in Texas (and report on E and F layer ionization). *Science*, (Supp)., **91** (1940), p. 12.

40–17 HUNTER, A.; The total solar eclipse of October 1, 1940. *Nat. Geog. Soc. Monograph;* also *Nature*, **150** (1942), pp. 512–513.

Eclipses of 1941

41–1 SATO, T.; On the ultraviolet light eclipse at the upper ionized layer, in a solar eclipse (with an example for the coming eclipse of Sept. 21st, 1941). *Sci. Rep. Tohoku Imp. Univ.*, **29** (1941), pp. 649–672.

41–2 OHNO, K.; On the observations of the ionosphere at the total solar eclipse of Sept. 21, 1941. *Radio*, **33** (1941), p. 243.

41–3 SENDA, K., and LI, S.; On the observations of ionosphere and earth's magnetism at the total solar eclipse at Hankow. *Radio*, **34** (1942), p. 232.

41–4 MAEDA, K., KOHNO, T., and SATO, R.; Observations of the ionosphere at the solar eclipse at Taiwan. *Radio*, **34** (1942), p. 236.

41–5 CHAKRAVARTI, S. P.; A note on the field strength of Delhi 3 and Delhi 4 at Calcutta during the solar eclipse of September 21, 1941. *Proc. Inst. Radio Engrs.*, **31** (1943), pp. 269–270. Correction in *Proc. Inst. Rad. Engrs.*, (1943), No. 11, p. 643.

41–6 MAEDA, K., SENDA, K., AONO, Y., NAKANO, T., and AZUMI, C.; Ionospheric observations during the total solar eclipse of 21 September, 1941. *Report on Japanese Research on Radio Wave Propagation*, vol. 1, May 1946.

41–7 MIKHAILOV, A., *et al.*; Eclipse of the sun, 21st Sept. 1941, in the USSR. *Nature*, **152** (1943), p. 658.

41–8 FESSENKOV, B. G.; Preliminary results of observations made during the total eclipse of the sun on 21st September 1941. *J. Phys. USSR*, **6** (1942), pp. 1–5. *Wireless Eng.* Abstract No. 1368 (1943).

Eclipses of 1942

42–1 STOFFREGEN, W.; Ionosphere observations (at Uppsala) during the solar eclipse on September 10, 1942. *Ark. Mat. Astrn. Fys.*, **32** (1946), part 4, section B, No. 9. Abstract in *Wireless Eng.*, **23**, A164 (1946).

Eclipses of 1943

43–1 YOKOYAMA, H.; A new type of ionospheric measuring equipment and some ionospheric observations during the solar eclipse of 5 February, 1943. *Report on Japanese Research on Radio Wave Propagation*, vol. 1, May 1946.

43–2 KOHNO, T.; Ionospheric observations during the solar eclipse of 5 February, 1943. *Report on Japanese Research on Radio Wave Propagation*, vol. 1, May 1946.

43–3 WELLS, H. W., and SHAPLEY, A. H.; Eclipse-effects in *F2*-layer of the ionosphere. *Terrestr. Magn.*, **51** (1946), pp. 401–409.

43–4 YONEZAWA, T.; On the variation of electron and ion densities in the *F2*-layer at the time of the solar eclipse on February, 5, 1943. "IATME"; Oslo meeting, Aug. 1948, Doc. T 59. *IATME Bull.*, **13** (1950), pp. 358–360.

Eclipses of 1944

44–1 LEDIG, P. G., JONES, M. W., GIESECKE, A. A., and CHERNOSKY, E. J.; Effects on the ionosphere at Huancayo, Peru, of the solar eclipse, January 25, 1944. *Terrestr. Magn.*, **51** (1946), pp. 411–418.

44–2 JEN, G. K., CHOW, K. T., LO, Y. T., and KUAN, C. C.; An observation of the ionosphere during the solar eclipse of July 20, 1944. *Phys. Rev.*, **66** (1944), p. 226.

44–3 BARAL, S. S., and MITRA, S. N.; Effect of solar eclipse on the ionosphere. *Science and Culture* (Calcutta), **10** (1944–45), pp. 175–176.

Eclipses of 1945

45–1 Gledhill, J. A. and Szendrei, M.; Behaviour of the F region of the ionosphere over Grahamstown during the partial solar eclipse of January 14th 1945. *Trans. Roy. Soc. South Africa*, **31** (1947), pp. 315–323.

45–2 Grönstrand, H. O.; The total solar eclipse of 1945, July 9. *Stockholms Observatoriums Annaler*, **14** (1944) no. 7.

45–3 Al'pert, Ya. L., and Gorozhankin, B. N.; On the results of radio-observations during the solar eclipse (corpuscular and ultra-violet) of July 9, 1945. *C.R. Acad. Sci. URSS*, **49** (1945), pp. 254–258.

45–4 Harang, L.; Radio-echo observations at Tromsö during the solar eclipse on July 9, 1945. *Terrestr. Magn.* **50**, (1945), pp. 307–310.

45–5 Stoffregen, W.; Records of the ionosphere during the total eclipse in the North of Sweden on July 9, 1945. *Terrestr. Magn.*, **51** (1946), pp. 495–499.

45–6 Waldmeier, M.; Coronal radiation and ionospheric variations during the solar eclipse, July 9, 1945. *Terrestr. Magn.* **52**, (1947), pp. 333–338.

45–7 Gejer, S., and Åkerlind, P.; Some experimental results obtained by ionospheric investigations in Sweden during the total solar eclipse of July 9, 1945. *Terrestr. Magn.*, **52** (1947), pp. 479–491.

45–8 Smith-Rose, R. L.; The solar eclipse of 1945 and radio wave propagation. *Nature*, **157** (1946), pp. 40–42. Also *Alta Frequenza*, **15** (1946), pp. 37–38.

45–9 Papalexi, N. D.; Radio observations during the solar eclipse of 9th July 1945. *Bull. Acad. Sci. URSS, sér. phys.*, **10** (1946), pp. 237–242. In Russian.

45–10 Al'pert, Ya. L., and Gorozhankin, B. N.; On radio observations during the solar eclipse of 9th July 1945. *Bull. Acad. Sci. URSS, sér phys.*, **10** (1946), pp. 245–251. In Russian.

45–11 Grigor'eva, A. A.; Results obtained in observing the propagation of radio waves during the solar eclipse of 9th July 1945. *Bull. Acad. Sci. URSS*, **10** (1946), pp. 253–260. In Russian.

45–12 Kazantseff, A. N.; On the results obtained in the investigation of the ionosphere during the solar eclipse of 9th July 1945. *Bull. Acad. Sci. URSS, sér. phys.*, **10** (1946), pp. 261–267. In Russian.

45–13 Bulatov, N. D.; On the results of the ionosphere measurements made during the solar eclipse of 9th July 1945. *Bull. Acad. Sci. URSS, sér. phys.*, **10** (1946), pp. 269–274. In Russian.

45–14 Kabanov, N. I.; Observations of the ultra-short wave propagation during the solar eclipse of 9th July 1945. *Bull. Acad. Sci. URSS, sér. phys.*, **10** (1946), pp. 275–278. In Russian.

45–15 Osipov, N. V.; Observations of the variations of the ultra-short wave intensity during the solar eclipse of 9th July 1945. *Bull. Acad. Sci. URSS, sér. phys.*, **10** (1946), pp. 281–284. In Russian.

45–16 Rydbeck, O. E. H.; Chalmers solar eclipse ionospheric expedition 1945. *Chalmers tekn. Högsk. Handl.* (1946), no. 53.

45–17 McLeish, C. W.; Solar eclipse observations of the ionosphere. *Canad. J. Res.*, **26** Sec. A (1948), pp. 137–144.

45–18 Jelstrup, H. S.; Observations during the solar eclipse of July 9th 1945. *Avhandlingar Norwegian Scientific Academy*, 1946.

45–19 Solar eclipse observations; effects on the ionization in the E and F layers. *Wireless World*, **51** (1945), p. 240.

45–20 Chakravarti, S. P.; Scientists study of last solar eclipse. *Curr. Sci.*, **14** (1945), p. 283.

45–21 "Mass attack" on secrets of radio waves. *Proc. Inst. Radio Eng., Austr.*, **6** (1945), pp. 9–10.

Eclipses of 1947

47–1 Al'pert, Ya. L., and Einberg, A. A.; Concerning the problem of ionosphere investigations during the total solar eclipse in Brazil on May 20, 1947. *J. Phys. Moscow*, **11** (1947), pp. 94–95.

47–2 ROCARD, Y.; Provisional results obtained by the French mission to Brazil during the total solar eclipse, 20th May 1947. *Rev. sci., Paris*, **85** (1947), p. 618.

47–3 DENISSE, J. F., SELIGMANN, P., and GALLET, R.; Results of ionosphere observations during the total eclipse of the sun, 20th May 1947. *Compt. Rend.*, **225** (1944), pp. 1169–1171.

47–4 AL'PERT, Ya. L., and EINBERG, A. A.; To the question of the coefficient of the ionosphere's recombination and the determination of its quantity at the time of the eclipse on the 20th May, 1947 in Brazil. *Bull. Acad. Sci. URSS, sér. géogr. géophys.*, **11** (1947), pp. 137–140. In Russian with English summary.

47–5 AL'PERT, Ya. L.; On investigations of the *F2*-layer of the ionosphere during the total solar eclipse of 20th May, 1947. *Bull. Acad. Sci. URSS, sér. phys.*, **12** (1948), pp. 44–48. Also *C.R. Acad. Sci. URSS*, **58** (1947), pp. 1919–1922. In Russian.

47–6 SAVITT, J.; Recombination and attachment in the F-region during the eclipse of 20th May, 1947. *J. Geophys. Res.*, **55** (1950), pp. 385–394.

47–7 BOSSON, F., DENISSE, J. F., GALLET, R., and SELIGMANN, P.; Ionospheric observations during the total eclipse of 20th May 1947. *Relations entre les phénomènes solaires et géophysiques*, Lyon, September 1947, pp. 117–127.

Eclipses of 1948

48–1 GROSS, J.; Field-strength observations made during the total eclipse of the sun. *Proc. Inst. Radio Engrs.*, **37** (1949), p. 1447.

48–2 MIYA, K., MURATA, T., and WADA, H.; Study of propagation characteristics of obliquely incident impulse wave in time of the solar eclipse. *Rep. Ionosphere Res. Japan*, **4** (1950), pp. 1–20.

48–3 NAKATA, Y.; Effects on the F-regions of the ionosphere of the solar eclipse of May 9, 1948. *Rep. Ionosphere Res. Japan*, **4** (1950), pp. 21–30.

48–4 OSAWA, K.; A note on the effect of the solar eclipse upon the terrestrial ionosphere. *Rep. Ionosphere Res. Japan*, **4** (1950), p. 53.

48–5 NAKATA, Y.; Ionospheric observation at Wakkanai during the solar eclipse of May 9, 1948. *Rep. Ionosphere Res. Japan*, **4** (1950), p. 54.

48–6 UYEDA, H., KUDO, H., SHIMIZU, T., and SATO, R.; The h′t-measurement of ionosphere at the time of the annular eclipse of May 9, 1948 at Wakkanai, Hokkaido in Japan. *Rep. Ionosphere Res. Japan*, **4** (1950), pp. 75–77.

48–7 LIANG, P. H., *et al.*; Effects on *F2* ionization of the partial eclipse of May 9, 1948 as observed at Wu-chang. URSI VIII-th Gen. Ass. Doc. A.G., 48/no. 127 Comm. II.

48–8 GHERZI, E.; Results of some ionospheric (and solar noise) researches made at Zi-Ka-Wei observatory during the solar annular eclipse of May 9, 1948. *IATME Bull.*, **13** (1950), pp. 360–362.

48–9 MAEDA, K. *et al.*; Combined reports on radio and ionospheric observations during annular solar eclipse 9 May 1948. (Prepared under supervision of GHQ, Supreme Command Allied Powers, Civil Communications System, Research and Development Division).

Eclipses of 1949

49–1 STRANZ, D.; Die ionosphäre während der partiellen Sonnenfinsternis am 28 April 1949. *Z. Naturforsch.* (1950), No. 3, p. 172.

49–2 BRACEWELL, R. N.; Theory of formation of an ionospheric layer below E layer based on eclipse and solar flare effects at 16 kc/s. *J. Atmosph. Terr. Phys.*, **2** (1952), pp. 226–235.

Eclipses of 1950

50–1 NAKATA, Y., and ARAKI, K.; Some experimental results obtained by ionospheric observation during the partial eclipse on September 12, 1950. *Proc. semi-annual meet. CRWO* (1951), No. 2, p. 81. Also *J. Radio Res. Lab.*, **1** (1954), pp. 42–44.

50–2 Kato, Y., Kamiyama, H., and Kikuchi, T.; Report on the result of the observation of the signal strength during the partial solar eclipse of Sept. 12, 1950. *Sci. Rep., Tohoku Univ.*, 5th series, **3** (1951), p. 53.

50–3 Nakata, Y., *et al.*; Provisional reports of observations of the partial solar eclipse of September 12, 1950. Published by Solar Research Committee, Science Council of Japan.

Eclipses of 1951

51–1 Effect of the annular eclipse of March 7, 1951, on radio-wave propagation. *Nature*, **169** (1952), pp. 361–362.

51–2 Wells, H. W.; Ionospheric effects of solar eclipse at sunrise, September 1, 1951. *J. Geophys. Res.*, **57** (1952), pp. 291–304.

51–3 Delobeau, F.; Observation of a partial eclipse of the sun on 1st November 1951. *Ann. Géophys.*, **10** (1954), pp. 148–151.

Eclipses of 1952

52–1 Radio Research Board, The Director of the; The ionosphere and the solar eclipses of 1952. *Radio Research*, 1952. Rep. of the Director of Radio Research for the year 1952. H.M.S.O. London (1952), pp. 22–31. For summary see 52–12 and 52–15.

52–2 Stratton, F. J. M.; Total solar eclipse of February 25. *Nature*, **169** (1952), pp. 259–260.

52–3 Redman, R. O.; The total solar eclipse of February 25, 1952. *Nature*, **169** (1952), pp. 686–688.

52–4 Minnis, C. M.; E region during the solar eclipse of February 25, 1952. *Nature*, **170** (1952), p. 453.

52–5 Klüber, H. von.; Observations of the solar eclipse of 25th February 1952 in the Sudan. *Naturwissenschaften*, **39** (1952), pp. 199–206.

52–6 Banerjee, S. S., Rajan, V. D., and Banerjee, P. C.; Some observations on the effect of solar eclipse on the *F2* region of the ionosphere. *J. Sci. Industr. Res.*, **11B** (1952), pp. 197–198.

52–7 Estrabaud, S.; Effects of the solar eclipse of 25th February 1952 on the ionospheric *F2* region over Equatorial Africa. *Compt. Rend.*, **235** (1952), pp. 1521–1523.

52–8 Estrabaud, S.; Effect of the solar eclipse of 25th February 1952 on the ionosphere E layer in Equatorial Africa. *Compt. Rend.*, **236** (1953), pp. 833–835.

52–9 Ranzi, I.; The ionosphere during the partial solar eclipse observed at Buenos Aires 20th August, 1952. *Alta Frequenza*, **22** (1953), pp. 59–71.

52–10 Delobeau, F.; The solar eclipse of 25–2–1952 at Gao. *Ann. Géophys.*, **9** (1953), pp. 317–324.

52–11 Zanotelli, G.; Anomalies of the ionosphere (*F2*-layer) critical frequencies during a solar eclipse. *Ann. Geofis.*, **6** (1953), pp. 367–372.

52–12 Piggott, W. R.; Ionospheric measurements at Ibadan, Nigeria during the eclipse of the sun, February 25th 1952. *Proc. of the 3rd meeting of the URSI Mixed Comm. on the Ionosphere* (1952), pp. 42–50.

52–13 Minnis, C. M.; Ionospheric measurements at Khartoum during the total solar eclipse on 25th February, 1952. *Proc. of the 3rd meeting of the URSI Mixed Comm. on the Ionosphere* (1952), pp. 38–42. For summary see: 52–4, 52–14, 52–15.

52–14 Minnis, C. M.; Ionospheric behaviour at Khartoum during the eclipse of 25th February 1952. *J. Atmosph. Terr. Phys.*, **6** (1955), pp. 91–112.

52–15 Minnis, C. M.; The ionosphere during the solar eclipse of February 25, 1952. *Nature*, **174** (1954), pp. 1137–1138.

52–16 Estrabaud, S.; Ionospheric measurements at Bangui during the solar eclipse of 25 February 1952. *Proc. of the 3rd meeting of the URSI Mixed Comm. on the Ionosphere* (1952) pp. 34–38. For summary see: 52–7, 52–8, 52–17.

52–17 Estrabaud, S.; The ionosphere at Bangui in February and March 1952 and during the eclipse of February 25. Note Preliminaire du L.N.R., No. 174, Paris (1954).

52–18 MERIAU, P., and RAWER, K.; Ionospheric observations during the eclipse of February 25, 1952 at Jibuti (Br. Somaliland). *Ann. Géophys.*, **10** (1954), pp. 9–18.

52–19 MITRA, S. N.; Partial solar eclipse of 25 February 1952 and its effect on the ionosphere. *J. Sci. Industr. Res., India,* Ser. A, **12** (1953) p. 319.

52–20 PERERS, O. F.; Ionospheric observations made by the Swedish solar eclipse expedition to Italy on February 25, 1952. *Proc. of the 3rd meeting of the URSI Mixed Comm. on the Ionosphere* (1952), pp. 50–62.

52–21 SVENSSON, S. I., HELLGREN, G., and PERERS, O.; The Swedish radioscientific solar eclipse expedition to Italy, 1952. *Trans. Chalmers University of Technology, Gothenburg, Sweden* (1954).

52–22 SEN GUPTA, P. K., and MITRA, S. N.; Corpuscular eclipse in the F2 layer and its association with solar flares and M-regions. *Nature,* **173** (1954), pp. 814–816.

52–23 Catalogue of disturbances in ionosphere, geomagnetic field, field intensity of radio wave, cosmic ray, solar phenomena and other related phenomena. No. 11. 34th Co-operative observation, solar eclipse (Feb. 25th 1952). Published by Ionosphere Research Committee, Science Council of Japan, Ueno Park, Tokyo.

Eclipses of 1953

53–1 SUZUKI, K.; Corpuscular eclipse of the sun on Feb. 14, 1953. *Publ. Astron. Soc., Japan,* **2** (1951), p. 172.

53–2 MIYADI, M.; Travel time of radio time signals during the partial eclipse on February 14, 1953. *Rep. Ionosphere Res., Japan,* **7** (1953), p. 117.

53–3 NAKATA, Y.; Ionospheric observation during the partial solar eclipse of February 14, 1953. *Rep. Ionosphere Res., Japan,* **7** (1953), pp. 157–158. Also *J. Radio Res. Lab.*, **1** (1954), pp. 44–48.

53–4 Catalogue of disturbances in ionosphere, geomagnetic field, field intensity of radio wave, cosmic ray, solar phenomena and other related phenomena. No. 14. 38th Co-operative observation, solar eclipse (Feb. 14th 1953). Published by Ionosphere Research Committee, Science Council of Japan, Ueno Park, Tokyo.

Eclipses of 1954

54–1 GRÖNSTRAND, H. O.; The total eclipse of the sun on June 30, 1954. Prediction for Scandinavia. *Stockholms Observatoriums Annaler,* **16** (1950), No. 2.

54–2 Total solar eclipse of June 30, 1954. *Nature,* **173** (1954), p. 381.

54–3 Lindau observations of the solar eclipse on June 30, 1954. *Ionosphären-Bericht,* **6** (1954), pp. 1–4.

54–4 BIBL, K., and DELOBEAU, F.; Preliminary results of ionospheric observations during the eclipse of 30th June 1954. *Compt. Rend.*, **239** (1954), pp. 1658–1660.

54–5 GHOSE, A. K., ROY, K. M., GUPTA, S. K., DASGUPTA, S. N., SARWATE, V. V., and MITHAL, G. K.; On radio measurements at Jabalpur during the solar eclipse of 30th June 1954. *J. Inst. Telecom. Engrs. (India)*, **1** (1955), pp. 20–26.

54–6 EGEDAL, J., and AMBOLT, N.; The effect on geomagnetism of the eclipse of 30 June 1954. *J. Atmosph. Terr. Phys.*, **7** (1955), p. 40.

MISCELLANEOUS LITERATURE

M 12–1 Eccles, W. H.; Effect of the eclipse on wireless telegraphic signals. *Electrician*, **69** (1912), p. 109.

M 25–1 Addey, E.; Wireless and the eclipse. *Electrician*, **94** (1925), pp. 152–153, p. 208.

M 27–1 Appleton, E. V.; The study of signal fading. *J. Inst. Elec. Eng.*, **66** (1930), p. 872.

M 31–1 Nagaoka, H.; Propagation of wireless waves. *Tokyo Inst. Phys. Chem. Res.*, Sci. Paper No. 297 (1931), pp. 169–188.

M 32–1 Chapman, S.; The influence of a solar eclipse upon upper atmospheric ionization. *Monthly Not. Roy. Astron. Soc.*, **92** (1932), pp. 413–420.

M 33–1 Chapman, S.; Effect of a solar eclipse on the earth's magnetic field. *Terrestr. Magn.*, **38** (1933), pp. 176–183.

M 36–1 Lisman, J. H. C.; Solar eclipse and radio reception. *Radio-centrum*, **2** (1936), No. 31, pp. 425–426.

M 36–2 Leithauser, G., and Beckmann, B.; Ionospheric regions and a solar eclipse. *Z. Tech. Phys.*, **17** (1936), pp. 327–329.

M 36–3 Leithauser, G., and Menzel, W.; The propagation of atmospheric disturbances and the reception of distant stations during the solar eclipse. *Z. Tech. Phys.*, **17** (1936), pp. 330–332.

M 36–4 Maeda, K.; Upper atmospheric ionization and solar radiation. *Rep. Radio Res., Japan*, **7**, No. 3 (1937), pp. 145–151.

M 36–5 Ranzi, I.; Ionizing agents in the upper atmosphere. *Nuovo Cimento*, **14** (1937), pp. 145–170.

M 38–1 The ionospheric eclipse of 1940, October 1. The corpuscular eclipse of 1940, October 1. *Monthly Not. Roy. Astron. Soc.*, **98** (1938), pp. 664–669, pp. 727–733.

M 41–1 Radio observations of the ionosphere during an eclipse. *J. Franklin Inst.*, **232** (1941), pp. 278–280.

M 41–2 Chapman, S.; The sun and the ionosphere. *J. Inst. Electr. Engrs.*, **88** (1941), part 1, pp. 400–413.

M 42–1 Yokoyama, E., and Aono, Y.; Research of the ionosphere at the solar eclipse. *Radio*, **34** (1942), p. 228.

M 44–1 Al'pert, Ya. L., and Gorozhankin, B. N.; Solar eclipses and radio investigations of the ionosphere. *Bull. Acad. Sci., URSS.*, sér. phys., **8** (1944), pp. 85–108.

M 45–1 Marshall, R. K.; Eclipses of the sun. *J. Franklin Inst.*, **239** (1945), pp. 299–324.

M 46–1 Smith-Rose, R. L.; The influence of an eclipse of the sun on the ionosphere. *J. Brit. Instn. Radio Engrs.*, **6** (1946), pp. 82–97.

M 46–2 Appleton, E. V., and Naismith, R.; Report on ionospheric measurements during a solar eclipse. *URSI, A. G.* 1946, Doc. No. 40, Comm. II. *URSI Proc.*, **6**, p. 125.

M 47–1 Seaton, S. L.; Rate of ion production in the ionosphere. *Phys. Rev.*, **72** (1947), pp. 712–714.

M 48–1 British National Committee for Scientific Radio; Report on ionospheric measurements made during recent solar eclipses. *URSI, A.G.* 1948, Doc. No. 113, Comm. II.

M 50–1 Chapman, S.; Corpuscular influences upon the upper atmosphere. *J. Geophys. Res.*, **55** (1950), pp. 361–372.

M 51–1 Baral, S. S.; Title unknown. *J. Sci. Industr. Res.* (*India*), **10A** (1951), pp. 60–64, or *Rep. Ionospheric Res., Univ. of Calcutta*, **1** (1951).

M 51–2 Piddington, J. H.; The modes of formation of the ionospheric layers. *J. Geophys. Res.*, **56** (1951), pp. 409–429.

M 52–1 Bracewell, R. N.; Theory of formation of an ionized layer below E layer based on eclipse and solar flare effects at 16 kc/sec. *J. Atmosph. Terr. Phys.*, **2** (1952), pp. 226–235.

M 54–1 Prabat, K., Sen Gupta, P. K., and Mitra, S. N.; Corpuscular eclipse in the F2 layer and its association with solar flares and M regions. *Nature*, **173** (1954), pp. 814–816.

M 54–2 Nakata, Y.; Short period variations in the ionosphere. Chapter VI, h'f observations during the eclipse. *J. Radio Res. Lab.*, **1** (1954), pp. 34–49.

M 54–3 Rydbeck, O. E. H., and Wilhelmsson, H.; A theoretical investigation of the ionospheric electron density variation during a solar eclipse. *Acta Polytechnica*, **158** El. Eng. Series **5** (1954), No. 12. Reports from the Research Laboratories of Electronics, No. 31, Transactions of Chalmers University of Technology, Gothenburg, Sweden (1954), No 149.

M 55–1 Stoffregen, W. Variation of fEs during solar eclipses. *Nature*, **176,** (1955), p. 610.

M 55–2 Van Wijk, A. M. Magnetic effects during solar eclipses. *J. Geophys. Res.*, **60** (1955), p. 297.

BOOK LITERATURE

B 39–1 WOOLLEY, R. v. d. R., and DYSON, S.; *Eclipses of the Sun and Moon.* Oxford, Clarendon Press, 1939.

B 39–2 FLEMING, J. A.; *Terrestrial Magnetism and Electricity.* Dover Publications, New York, 1939, pp. 454–464.

B 51–1 CHAPMAN, S., and BARTELS, J.; *Geomagnetism.* Clarendon Press, Oxford, 1951, pp. 814–823.

B 51–2 MITCHELL, S. A.; *Eclipses of the Sun.* Columbia University Press, New York, 1951.

B 52–1 MITRA, S. K.; *The Upper Atmosphere.* The Asiatic Society, Calcutta, 1952, 2nd edition, pp. 292–294.

B 54–1 KUIPER, G. P.; *The Sun.* University of Chicago Press, 1954.

AUTHOR INDEX TO BIBLIOGRAPHY

LIST OF DELEGATES ATTENDING THE SYMPOSIUM

Name	Organization
Prof. C. W. ALLEN	University of London Observatory, Gt. Britain
Dr. Y. AONO	Radio Research Laboratories, Tokyo, Japan
Dr. J. BARTELS	University of Göttingen, Germany
Prof. D. R. BATES	Queen's University of Belfast, N. Ireland
Dr. W. BECKER	Institut für Ionosphärenforschung, Lindau, Germany
Mr. J. BELROSE	Cavendish Laboratory, Cambridge, Gt. Britain
Mr. T. W. BENNINGTON	B.B.C. Research Division, Tadworth, Gt. Britain
Mr. P. L. M. VAN BERKEL	Netherlands P.T.T., The Hague, Netherlands
Dr. L. V. BERKNER	Associated Universities Inc., New York, U.S.A.
Dr. W. J. G. BEYNON	University College of Swansea, Gt. Britain
Dr. B. H. BRIGGS	Cavendish Laboratory, Cambridge, Gt. Britain
Mr. G. M. BROWN	University College of Swansea, Gt. Britain
Dr. O. BURKARD	University of Graz, Austria
Dr. B. F. BURKE	Carnegie Institution of Washington, U.S.A.
Prof. S. CHAPMAN	University of Alaska, College, Alaska
Dr. M. CUTOLO	University of Naples, Italy
Dr. H. DAENE	Astrophysikalisches Observatorium, Potsdam, Germany
Mr. M. H. DAVENPORT	Cavendish Laboratory, Cambridge, Gt. Britain
Dr. K. DAVIES	D.R.T.E., Ottawa, Canada
Mr. F. DELOBEAU	S.P.I.M., Freiburg, Germany
Mr. J. EGEDAL	Meteorologiske Institut, Copenhagen, Denmark
Prof. A. EHMERT	Max Planck Institut für Physik der Stratosphäre, Ravensburg, Germany
Dr. M. A. ELLISON	Royal Observatory, Edinburgh, Gt. Britain
Dr. G. ELWERT	University of Tübingen, Germany
Dr. R. FRASER	Liaison Officer ICSU—UNESCO, London, Gt. Britain
Dr. N. FUKUSHIMA	University of Tokyo, Japan
Dr. L. GOLDBERG	University of Michigan Observatory, U.S.A.
Dr. S. D. GOSSNER	U.S. Naval Observatory, Washington, U.S.A.
Mr. F. T. HADDOCK	Naval Research Laboratory, Washington, U.S.A.
Dr. J. P. HAGEN	Naval Research Laboratory, Washington, U.S.A.
Prof. T. HATANAKA	University of Tokyo Observatory, Japan
Dr. J. HOUTGAST	Rijksuniversitat, Utrecht, Netherlands
Mr. H. HVATUM	Chalmers University of Technology, Gothenburg, Sweden
Dr. C. DE JAGER	Rijksuniversitat, Utrecht, Netherlands
Mr. L. JONES	Cavendish Laboratory, Cambridge, Gt. Britain
Prof. J. KAPLAN	University of California, Los Angeles, U.S.A.
Mr. W. W. KELLOGG	The Rand Corporation, Santa Monica, U.S.A.
Mr. J. W. KING	Cavendish Laboratory, Cambridge, Gt. Britain

Dr. M. LAFFINEUR	Institut d'Astrophysique, Paris, France
Mr. B. LANDMARK	Norwegian D.R.E., Kjeller, Norway
Dr. E. A. LAUTER	Meteorologisches Observatorium, Kühlungsborn, Germany
Father P. LEJAY	Bureau Ionosphérique Français, Bagneux, France
Mr. D. LEPECHINSKY	Bureau Ionosphérique Français, Bagneux, France
Mr. A. E. LILLEY	Naval Research Laboratory, Washington, U.S.A.
Dr. R. LINDQUIST	Research Institute of National Defence, Stockholm, Sweden
Mr. A. J. LYON	University of Edinburgh, Gt. Britain
Dr. R. M. MACRAE	David Dunlap Observatory, Toronto, Canada
Dr. P. MANGE	Institut Royal Météorologique, Uccle, Belgium
Mr. J. MAWDSLEY	Cavendish Laboratory, Cambridge, Gt. Britain
Dr. D. H. MENZEL	Harvard College Observatory, Cambridge, U.S.A.
Mr. G. MILLINGTON	Marconi's Wireless Telegraph Company, Chelmsford, Gt. Britain
Mr. C. M. MINNIS	Radio Research Station, Slough, Gt. Britain
Mr. J. C. DE MUNCK	Netherlands P.T.T., The Hague, Netherlands
Prof. T. NAGATA	University of Tokyo, Japan
Mr. S. OLVING	Chalmers University of Technology, Gothenburg, Sweden
Mr. D. M. PARKYN	University College of Swansea, Gt. Britain
Mr. W. R. PIGGOTT	Radio Research Station, Slough, Gt. Britain
Prof. K. R. RAMANATHAN	Physical Research Laboratory, Ahmedabad, India
Mr. J. A. RATCLIFFE	Cavendish Laboratory, Cambridge, Gt. Britain
Dr. K. RAWER	S.P.I.M., Freiburg, Germany
Dr. G. RIGHINI	Osservatorio Astrofisico, Florence, Italy
Dr. R. RIVAULT	Bureau Ionosphérique Français, Poitiers, France
Miss A. R. ROBBINS	Cavendish Laboratory, Cambridge, Gt. Britain
Mr. A. ROHDIN	Board of Telecommunications, Stockholm, Sweden
Mr. M. RYLE	Cavendish Laboratory, Cambridge, Gt. Britain
Prof. J. SAYERS	University of Birmingham, Gt. Britain
Mr. A. H. SHAPLEY	National Bureau of Standards, Boulder, U.S.A.
Mr. N. J. SKINNER	University College, Ibadan, Nigeria
Mr. S. SKRIBELAND	Norwegian D.R.E., Kjeller, Norway
Dr. R. L. SMITH-ROSE	Radio Research Station, Slough, Gt. Britain
Mr. W. STOFFREGEN	Ionosphere Observatory, Upsala, Sweden
Dr. J. O. THOMAS	Cavendish Laboratory, Cambridge, Gt. Britain
Dr. R. TOUSEY	Naval Research Laboratory, Washington, U.S.A.
Prof. L. VEGARD	Universitetets Fysiske Institutt, Oslo, Norway
Mr. A. H. DE VOOGT	Netherlands P.T.T., The Hague, Netherlands
Dr. M. Waldmeier	Swiss Federal Observatory, Zürich, Switzerland
Dr. J. P. WILD	C.S.I.R.O., Sydney, Australia
Mr. A. F. WILKINS	Radio Research Station, Slough, Gt. Britain
Dr. R. W. Wright	University College, Ibadan, Nigeria

NAME INDEX

References to papers are given in Clarendon type